Charge Acceleration and the Spatial Distribution of Radiation Emitted by Antennas and Scatterers

The ACES Series on Computational and Numerical Modelling in Electrical Engineering

Andrew F. Peterson, PhD – Series Editor

The volumes in this series will encompass the development and application of numerical techniques to electrical and electronic systems, including the modeling of electromagnetic phenomena over all frequency ranges and closely related techniques for acoustic and optical analysis. The scope includes the use of computation for engineering design and optimization, as well as the application of commercial modelling tools to practical problems. The series will include titles for senior undergraduate and postgraduate education, research monographs for reference, and practitioner guides and handbooks.

Titles in the Series

K. Warnick, **"Numerical Methods for Engineering,"** 2010

W. Yu, X. Yang and W. Li, **"VALU, AVX and GPU Acceleration Techniques for Parallel FDTD Methods,"** 2014.

A.Z. Elsherbeni, P. Nayeri and C.J. Reddy, **"Antenna Analysis and Design Using FEKO Electromagnetic Simulation Software,"** 2014.

A.Z. Elsherbeni and V. Demir, **"The Finite-Difference Time-Domain Method in Electromagnetics with MATLAB Simulations 2nd Edition,"** 2015.

M. Bakr, A.Z. Elsherbeni and V. Demir, **"Adjoint Sensitivity Analysis of High Frequency Structures with MATLAB,"** 2017.

O. Ergul, **"New Trends in Computational Electromagnetics,"** 2019.

D. Werner, **"Nanoantennas and Plasmonics: Modelling, design and fabrication,"** 2020.

K. Kobayashi and P.D. Smith, **"Advances in Mathematical Methods for Electromagnetics,"** 2020

V. Lancellotti, **"Advanced Theoretical and Numerical Electromagnetics, Volume 1: Static, stationary and time-varying fields,"** 2021

V. Lancellotti, **"Advanced Theoretical and Numerical Electromagnetics, Volume 2: Field representations and the method of moments,"** 2021.

S. Roy, **"Uncertainty Quantification of Electromagnetic Devices, Circuits, and Systems,"** 2021

Charge Acceleration and the Spatial Distribution of Radiation Emitted by Antennas and Scatterers

Edmund K. Miller

The Institution of Engineering and Technology

Published by SciTech Publishing, an imprint of The Institution of Engineering and Technology, London, United Kingdom

The Institution of Engineering and Technology is registered as a Charity in England & Wales (no. 211014) and Scotland (no. SC038698).

First published 2022

The Institution of Engineering and Technology
Futures Place
Kings Way, Stevenage
Hertfordshire SG1 2UA, United Kingdom

www.theiet.org

British Library Cataloguing in Publication Data
A catalogue record for this product is available from the British Library

ISBN 978-1-83953-813-1 (hardback)
ISBN 978-1-83953-814-8 (PDF)

Typeset in India by MPS Limited
Printed in the UK by CPI Group (UK) Ltd, Croydon

Cover Image: Oxygen/Moment via Getty Images

Contents

Journal articles and book chapters that are related to electromagnetic radiation by the author and his colleagues

E. K. Miller and M. I. VanBlaricum (1973), "The Short-Pulse Response of a Straight Wire", *IEEE Transactions on Antennas and Propagation*, AP-**21**, pp. 386–389.

E. K. Miller, A. J. Poggio, and G. J. Burke (1973), "An Integro-Differential Equation Technique for the Time-Domain Analysis of Thin-Wire Structures. Part I: The Numerical Method", *Journal of Computational Physics*, **12**, pp. 24–48.

A. J. Poggio, E. K. Miller and G. J. Burke (1973), "An Integro-Differential Equation Technique for the Time-Domain Analysis of Thin-Wire Structures. Part II: Numerical Results", *Journal of Computational Physics*, **12**, pp. 210–233.

J. A. Landt and E. K. Miller (1974), "Short-Pulse Response of a Circular Loop", *IEEE Transactions on Antennas and Propagation*, AP-**22**, pp. 114–116.

J. A. Landt and E. K. Miller (1976), "Transient Response of the Infinite Cylindrical Antenna and Scatterer", *IEEE Transactions on Antennas and Propagation*, AP-**24**, pp. 246–250.

E. K. Miller and J. A. Landt (1977), "Short-Pulse Characteristics of the Conical Spiral Antenna", *IEEE Transactions on Antennas and Propagation*, AP-**25**, pp. 621–626.

E. K. Miller and D. L. Lager (1978), "Radiation Field Analysis and Synthesis Using Prony's Method", *Electronics Letters*, **14**, pp. 180–182.

E. K. Miller and J. A. Landt (1980), "Direct Time-Domain Techniques for Transient Radiation and Scattering from Wires", Invited Paper in *Proceedings of the IEEE*, **68**, pp. 1396–1423.

E. K. Miller (1980), "Time-Domain Modeling of Wires", in *Applications of the Method of Moments to Electromagnetic Fields*, The SCEEE Press, pp. 213–252.

E. K. Miller (1981), "Direct Time-Domain Techniques for Transient Radiation and Scattering: Wires", in *Theoretical Methods for Determining the Interaction of Electromagnetic Waves with Structures*, Sijthoff and Noordhoff, pp. 213–253.

E. K. Miller, F. J. Deadrick, G. J. Burke and J. A. Landt (1981), "Computer-Graphics Applications in Electromagnetic Computer Modeling, *Electromagnetics*, **1**, pp. 135–153.

E. K. Miller and D. L. Lager (1982), "Inversion of One-Dimensional Scattering Data Using Prony's Method", *Radio Science*, **17**, pp. 211–217.

J. A. Landt, E. K. Miller and F. J. Deadrick (1983), "Time-Domain Modeling of Nonlinear Loads", *IEEE Transactions on Antennas and Propagation*, AP-**31**, pp. 121–126.

E. K. Miller and D. L. Lager (1983), “Imaging of Linear Source Distributions”, *Electromagnetics*, **3**, pp. 21–40.

R. W. Cole, M. A. Soderstrand, and E. K. Miller (1986), “Animated Computer-Generated Instruction: How to Make a Movie”, *ASEE Engineering Education*, 276–281.

E. K. Miller (1987), “An Overview of Time-Domain Integral-Equation Models in Electromagnetics”, *Journal of Electromagnetic Waves and Applications*, VNU Science Press, **1**(3), 269–293.

Rodney W. Cole, Edmund K. Miller, Swapan Chakrabarti and Sivaprasad Gegineni (1990), Learning About Fields and Waves Using Visual Electromagnetics,” *IEEE Transactions on Education*, **33**(1), pp. 81–94.

E. K. Miller, R. Merrill, and R. J. Cole (1988), “Computer Movies for Education”, *IEEE Transactions on Education*, **31**(2), pp. 58–68, Education Transactions Best Paper Award for 1988.

R. W. Cole and E. K. Miller (1988), “Computer Animation of Time-Domain Problems: The Key to Understanding”, in *Proceedings 1988 Frontiers in Education Conference*, Santa Barbara, CA, October 22–25, pp. 157–164.

E. K. Miller (1993), “Computational Electromagnetics,” Chapter 43 in *The Electrical Engineering* Handbook, ed. Richard C. Dorf, CRC Press, Inc, pp. 1028–1049.

E. K. Miller (1994), “Time-Domain Modeling in Electromagnetics,” *Journal of Electromagnetic Waves and Applications*, **8**(9/10), pp. 1125-–172.

E. K. Miller (1997), “An Exploration of Radiation Physics in Electromagnetics,” in *Proceedings of 13th Annual Review of Progress in Applied Computational Electromagnetics*, Naval Postgraduate School, Monterey, CA, pp. 240–247.

E. K. Miller (1997), “Visualization: A Powerful Tool for Understanding Electromagnetics,” in *Proceedings of 13th Annual Review of Progress in Applied Computational Electromagnetics*, Naval Postgraduate School, Monterey, CA, pp. 2–7.

E. K. Miller (1998), “Exploring Electromagnetic Physics Using Thin-Wire Time-Domain (TWTD) Modeling,” *in Proceedings of 14th Annual Review of Progress in Applied Computational Electromagnetics*, Naval Postgraduate School, Monterey, CA, pp. 583–588.

E. K. Miller (1999), “Introduction” to *Time Domain Electromagnetics*, Sadasiva M. Rao, Ed., Academic Press, San Diego, CA, pp. 1–48.

E. K. Miller (1999), Electromagnetics, in *Handbook of Electrical Engineering Calculations*, Edited by Arun G. Phadke, Marcel Dekker, Inc., New York, pp. 61–122.

E. K. Miller (1999), “Modeling Wire Antennas,” in *Wiley Encyclopedia of Electrical and Electronics Engineering*, Vol. 13, John G. Webster, Ed., Wiley Interscience, New York, pp. 417–441.

E. K. Miller (2000), “Some Further Results from FARS: Far-Field Analysis of Radiation Sources,” in *Proceedings of 16th Annual Review of Progress in Applied*

Computational Electromagnetics, Naval Postgraduate School, Monterey, CA, pp. 278–285.

E. K. Miller and F. J. Deadrick (2000), "Visualizing Near-Field Energy Flow and Radiation," *IEEE Antennas and Propagation Society Magazine*, 42(6), pp. 46–54.

E. K. Miller, "PCs for AP and Other EM Reflections," bimonthly column appearing in the *IEEE Antennas and Propagation Society Magazine*, from 1983 to 2000.

E. K. Miller (2001), "Electromagnetics without Equations," *IEEE Potentials*, **20**(2), April/May, pp. 17–20.

E. K. Miller, G. J. Burke, and T. K. Sarkar (2001), "TDFARS: A Procedure to Quantitatively Identify Time-Domain Radiation Sources," International Scientific Radio Union, 2001 Electromagnetic Theory Symposium, May 13–17, Victoria, Canada.

E. K. Miller and G. J. Burke (2001), "A Multi-Perspective Examination of the Physics of Electromagnetic Radiation," *Applied Computational Electromagnetics Society Journal*, 16(3), pp. 190–201.

D. Poljak, E. K. Miller and C. Y. Tham (2002), "Time-Domain Energy Measures for Thin- Wire Antennas and Scatterers," *IEEE Antennas and Propagation Society Magazine*, **44**(1), pp. 87–95.

Edmund K. Miller (2006), "Comparison of the Radiation Properties of a Sinusoidal Current Filament and a PEC Dipole of Near-Zero Radius," *IEEE Antennas and Propagation Society Magazine*, **48**(4), pp. 37–47.

Edmund K. Miller (2006), "Analysis of Radiation from a Traveling-Wave Current and Comparison with the Standing-Wave Sinusoidal Current Filament," *IEEE Antennas and Propagation Society Magazine*, **48**(5), pp. 48–52.

Edmund K. Miller (2006), "The Differentiated On-Surface Poynting Vector as a Measure of Radiation Loss from Wires," *IEEE Antennas and Propagation Society Magazine*, **48**(6), pp. 21–32.

E. K. Miller (2007), "The Incremental Far Field and Degrees of Freedom of the Sinusoidal Current Filament," *IEEE Antennas and Propagation Society Magazine*, **49**(4), pp. 13–21.

Edmund K. Miller (2009), "Development of the Incremental IEMF Method as an Analytical Proof for the Validity of FARS (Far-field Analysis of Radiation Sources)," in *Proceedings of the 25'th Annual Review of Progress in Applied Computational Electromagnetics*, Monterey, CA, pp. 775–780.

E. K. Miller (2010), "Development of the Incremental IEMF Method as an Analytical Proof for the Validity of FARS (Far-field Analysis of Radiation Sources)," *IEEE Antennas and Propagation Society Magazine*, **52**(6), pp. 46–54.

E. K. Miller (2011), "Time-Domain Far-Field Analysis of Radiation Sources," *IEEE Antennas and Propagation Society Magazine*, **53**(5), pp. 81–97.

E. K. Miller (2012), "Time-Domain Far-Field Analysis of Radiation Sources and Point- Source Coherence," *IEEE Antennas and Propagation Society Magazine*, **54**(2), pp. 100–108.

E. K. Miller (2014), "Exploring the Behavior of a Matched Load in the Time Domain," *IEEE Antennas and Propagation Society Magazine*, **56**(3), pp. 103–111.

E. K. Miller (2016), "Electromagnetic Radiation Physics: Demonstrations Using Free Charge and Simple Wire Objects," *Wiley Encyclopedia of Electrical and Electronics Engineering*.

E. K. Miller (2018), "The Proportionality between Charge Acceleration and Radiation from a Generic Wire Object,' *Progress In Electromagnetics Research*, **162**, pp. 15–29.

E. K. Miller (2020), "The Dependence of Time-Domain Radiation Loss on the Circumference and Wire Radius of a Circular Loop Antenna," *Progress In Electromagnetics Research M*, **92**, pp. 1–9.

About the author

Edmund K. Miller earned a PhD in Electrical Engineering from the University of Michigan in 1965 with an emphasis on computational electromagnetics. His working career has been varied including employment at four universities (Michigan Technological University, University of Michigan, Kansas University and Ohio University), three companies (MB Associates, Rockwell International Science Center, and General Research Corporation, all in California) and two national laboratories (Lawrence Livermore and Los Alamos). He was the first president of ACES

He has served two terms on the IEEE Antennas and Propagation Society (AP-S) Administrative Committee and has twice been an AP-S Distinguished Lecturer. He was elected IEEE Fellow in 1984 and received the IEEE Third Millennium Medal in 2000. He was the recipient (with others) of the 1988 Best Paper Award from the IEEE Education Society. Between 1985 and 2002, he served on the editorial board of IEEE Potentials Magazine as Editor or Associate Editor for which he wrote a regular column "On The Job." Dr Miller also wrote a column for the AP-S Magazine "PCs for AP and Other EM Reflections" from 1984 to 2000.

He was an organizer of, served as its first president, and is a Fellow of the Applied Computational Electromagnetics Society (ACES) for which he served two terms on its Board of Directors. His interests include CEM, applied signal processing, visual electromagnetics, and promoting the incorporation of accuracy statements in published CEM numerical results. He has published numerous articles dealing with his major interest, the physics of electromagnetic radiation.

Dedication

This book is dedicated to the memory of Gerald (Jerry) J. Burke, a longtime friend, colleague and collaborator. Jerry was an unexcelled numerical analyst and programmer as well as a gifted electromagneticist. Among the many contributions he made to computational electromagnetics are the computer codes NEC (Numerical Electromagnetics Code) and TWTD (Thin-Wire Time Domain). Jerry implemented FARS (Farfield Analysis of Radiation Sources) in both codes to make possible many of the results presented in this book. It was a privilege to have known and worked with him for more than 50 years.

Presentations

Available at "edmundmiller.academia.edu"

AN EXPLORATION OF RADIATION PHYSICS, APS Distinguished Lecture 2014–2016

WHY IS THERE ELECTROMAGNETIC RADIATION AND WHERE DOES IT COME FROM?, ACES, 2001

Preface

Although this book focuses on radiation, a fundamental property of electromagnetic fields, it does not follow the usual analytical kind of approach to be found in the typical book on electromagnetics. Rather than developing and presenting a formal theoretical foundation of electromagnetic theory, this book instead focuses on various aspects of EM radiation from a variety of perspectives. The goal is to provide the reader with a conceptual basis for thinking about and understanding EM radiation and to introduce some associated computational tools for obtaining relevant quantitative results. This is done as an alternative to a rigorous mathematical formalism that may obscure the simple physical reality of the radiation process. The approach taken here seeks to be relevant about how engineers routinely model and design radiating structures as boundary-value problems.

It has been long established that charge acceleration causes electromagnetic radiation, whether from isolated charges like electrons in free space or from the charge distributions on a dipole antenna or radar target. But finding the fields radiated by such PEC objects does not require knowledge of charge acceleration *per se*. Solving for the current and charge distributions induced on such objects by some specified excitation, typically a tangential electric field, is usually formulated as a boundary-value problem. The fields produced by these induced sources are then obtained by integrating them over the surface of the object being modeled. Whatever charge acceleration may be associated with these induced sources is not specifically involved in determining the radiated field. This is probably why engineering electromagnetics texts rarely mention charge acceleration as the root cause of the radiated field. Knowing that charge acceleration causes electromagnetic radiation may seem to be irrelevant.

Never the less, given that charge acceleration is the cause of all electromagnetic radiation, the question arises about where such acceleration occurs on objects typically modeled and analyzed by electromagnetic engineers. Charge acceleration as the cause of radiation from the typical kinds of objects modeled and analyzed by engineers is examined here on a quantitative basis. This question is considered throughout this book using various analytical and numerical tools primarily developed by the author. The goal is to provide the reader with computational tools for determining quantitatively why and where radiation is emitted by simple wire objects excited as antennas and scatterers. More specifically, this is achieved by demonstrating ways of answering the question “From where, and in what quantity, does EM radiation originate on a per-unit length or per-unit area basis from a perfect electric conductor (PEC) excited as an antenna or scatterer?” Answering this question is shown to be possible using the numerical solutions routinely provided by typical computer models of such problems via straightforward post-processing of their results.

Introduction

The material this book contains is based almost exclusively on material previously developed and written by the author. But much of its content that would not have been generated were it not for collaboration with various colleagues over the years, primarily while at Lawrence Livermore National Laboratory. At the risk of missing some who should be mentioned, among those who contributed to the evolution of some of the concepts presented here are the following:

James Brittingham
Jerry Burke
Fred Deadrick
Daryl Lager
Jeremy Landt
Jeff Lytle
Andrew Poggio
Mike VanBlaricum
Jim Willows

Most instrumental in making possible the results presented here is Jerry Burke. He has been a co-author/collaborator in much of the associated research and development of the two primary CEM models used to develop these results, NEC and TWTD. Jeremy Landt was instrumental in extending the capabilities of TWTD and in developing the time-domain energy measures discussed in Chapter 4. The evolution of RFID tags in research done by Jeremy Landt and his group at Los Alamos National Laboratory was substantially assisted by the TWTD code.

Much of the material included has been described in various articles in the IEEE Antennas and Propagation Society Magazine and other IEEE publications. When that is the case, the material has been reorganized and rewritten with all new graphs, accompanied by additional or updated results. Some results published in PIER publications are used with their permission.

It might seem unusual to produce a book focused entirely on charge acceleration and electromagnetic radiation. This could be considered to be more a physics than an engineering problem. But electromagnetic radiation is where the engineering discipline begins and knowledge about where radiation originates from an antenna or scatterer provides a more complete understanding about the process.

Finally, the various results presented and discussed are for the most part derived from well-validated analytical methods and numerical solutions. FARS, the

IEMF method, the time-domain energy measure, power flow along a conductor represented by the Poynting Vector, are all derived as post-processed quantities from them. One technique, the incremental IEMF (IIEMF) method in Chapter 11, might be viewed with skepticism. But the results derived from it match numerically well the incremental FARS (IFARS) results. Given their generally straightforward implementation to the usual CEM outputs, why not test one or more of these techniques to see what might be learned?

Chapter 1
Overview

Abstract

Radiation is the fundamental physical property of electromagnetic fields from which everything else follows. The goal of the book is to explain why and where radiation occurs from simple perfect-electric conducting objects as exemplified by wire antennas and scatterers. Numerous examples to illustrate how radiation occurs and its relationship with the behavior of the current and charge on such objects are presented. The numerical results are primarily derived using two well-known and well-validated computer models, TWTD (Thin-Wire Time Domain, described in Appendix A) and NEC (Numerical Electromagnetics Code, described in Appendix B) with which the author has been long associated.

Several different methods for determining why and where radiation originates from a perfect electric conductor are examined. While electromagnetic (EM) radiation is caused by acceleration of electric charge, this fact is rarely mentioned in engineering texts. In spite of that, computing the radiation patterns of complex, perfect-electric conductors (PEC) has been long-established and validated using Maxwell's equations (ME). Although charge acceleration is not explicitly included for PEC objects in ME, they obviously contain this effect implicitly in far-field computations based on them. Typical physics texts do discuss charge acceleration as the basis of EM radiation, but in terms of free charge in space, rather than in connection with, for example, how a physical antenna operates.

The goal of this book is to demonstrate the relationship between charge acceleration and how PEC objects behave when excited by localized sources as antennas or by incident waves as scatterers. Various analytical and numerical techniques developed by the author address the question "Where and in what amount is the power (frequency domain) or energy (time domain) emitted by PEC objects for various representative geometries?" A method called FARS (Far field Analysis of Radiation Sources) is demonstrated to provide an answer to this question in both the frequency and time domains. It follows that also answered is where the charge acceleration that produces radiation occurs. FARS also applies to specified currents such as the sinusoidal current filament, long used as an approximation for a dipole antenna.

The techniques described and demonstrated can be incorporated into most computational electromagnetics models with little effort to provide those interested in investigating localized radiation behavior for various situations. Such techniques will be of value in a teaching setting since they will give students a learning tool and unique insight into a fundamental aspect of the electromagnetic discipline. A variety of examples throughout will demonstrate geometrical features that cause charge acceleration or radiation. As a specific example, various published observations that associate current decay along a straight wire with radiation are confirmed to be due to charge reflection, caused by a wave-impedance variation with distance.

Keywords: Differentiated Poynting vector; Distributed radiation resistance; Distribution of radiated power; Electromagnetic radiation; Energy-loss rate; Incremental FARS; Incremental IEMF; Induced electromotive force; Farfield analysis of radiation sources; FARS; NEC; Poynting vector; Radiated energy distribution; Time-domain energy distribution; Time-domain modeling; TWTD

1.1 Introduction

The goal of this book is to describe and demonstrate various ways to determine the quantitative distribution of far-field radiated power or energy from over the surface of a perfect-electric conductor (PEC) excited as an antenna or scatterer. The methods used to do this are straightforward to incorporate in any computer model that yields source distributions over an object and from which the far radiated fields can be obtained. Because electromagnetic radiation is caused by accelerated charge, a common theme throughout is identifying where and why charge acceleration occurs on geometries of interest. As it is desirable to demonstrate the connection between accelerated charge and the physical features that cause it in the clearest way possible, the focus throughout is on fairly simple wire objects. Furthermore, since this connection is easier to show and understand in the time domain, the first several chapters examine radiation from a time-domain perspective. Representative figures from each chapter and appendix are included to give as samples of what is to be covered.

Throughout this book, there will be discussions of electric currents (I) and charges (Q) on wire objects, their propagation speeds and the electromagnetic radiation associated with charge acceleration. This is done from the perspective of assuming these objects are PEC, or using what amounts to a macroscopic, equivalent-source model of the radiation process. While a microscopic picture of an actual physical object having a high, but finite, conductivity might be discussed from the perspective of free electrons as being the radiating sources, this is not necessary. The assumption of perfect conductivity is close enough to physical reality that treating electrons as the actual charge carriers is not necessary. In effect, the PEC assumption means that the mass and charge of an electron are irrelevant to modeling electromagnetic radiation. This is in contrast on the other hand with the situation when designing a particle accelerator.

The fact remains that when modeling in the time domain charge and current pulses are found to move at approximately the speed of light along a uniform-radius, straight wire. However, this phenomenon does not require any physical charge to move at near-light speed. Instead it arises from the fact that the fields launched from the feedpoint of an impulsively excited wire propagate along the wire near or at the speed of light of the medium in which the wire is located. The magnetic and electric fields associated with these pulses must terminate on the wire as required by the boundary conditions of a PEC. This process can be analogized in various ways [16] or as a row of falling dominos. Each domino moves very little but with the overall effect propagates along the entire row.

Since the tangential electric field on the wire surface is zero, there is only a radial field that terminates at the charge on its surface. Similarly, since the normal magnetic field on the wire surface is zero, there is only a tangential, or azimuthal, magnetic field that terminates at the current on its surface. Thus, the moving charge and its accompanying current are attached to the propagating fields. When the field propagation is interrupted by ends, bends, loads, junctions, etc., some or all of the propagating fields and Q/I pulses are reflected with the reflected field lines developing bends as discussed in Chapter 2.

While the magnetic field lines are closed around a wire antenna since there is no magnetic charge, the radiated electric fields from finite objects must eventually form closed lines as well to "break off" from it. This is easily visualized for an impulsively excited object in the time domain. After enough time has passed, the object must return to overall charge neutrality, which means that there are no remaining electric or magnetic fields attached to its surface. Closure of the electric fields occurs when opposite-signed charges can neutralize each other as they pass moving in opposite directions.

It is reasonable to ask whether the equivalent-sources used in boundary-value problems match physical reality, i.e., do their computed results match measurements? The computed radiated fields have been found to match the measured radiation and scattering patterns within their respective numerical and experimental uncertainties since the beginning of electromagnetics as a discipline. Measurements of near-field interactions and input impedance are somewhat more experimentally difficult to determine but they also have confirmed analytical and computational results. One of the more demanding checks on computations is measurement of the source distribution on an object. The time-domain example of Figure 1.1 shows the current response of a Vee-dipole at its connection to a ground plane and illuminated by an impulsive plane wave. Examples of frequency-domain measurements for current distributions can be found in the literature such as [17–22].

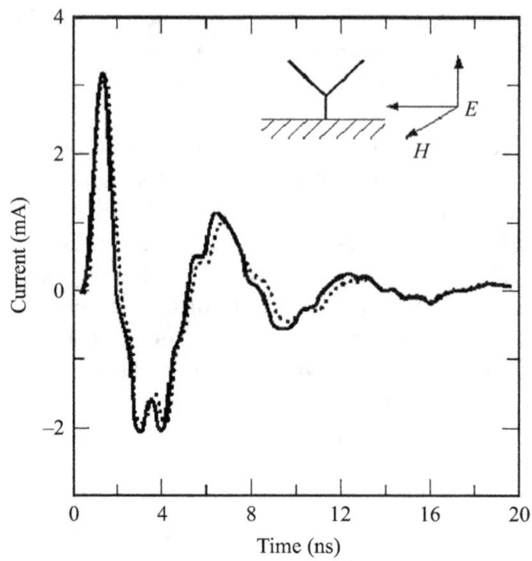

Figure 1.1 The time-domain current measured at the ground-plane connection of a Vee-wire dipole antenna (solid line) (Figure 1.13 of Ref. [2]) and the TWTD result (the dots)

1.2 The electric-field kink model of electromagnetic radiation

The chapter begins by introducing a conceptual and graphical way that shows how time-domain radiation can be explained without complex, mathematical analysis. The E-field "kink" model of radiation exploits two basic facts about electromagnetic fields: (1) that electric-field lines are continuous between their origination and termination on electric charge; and (2) that electromagnetic fields propagate in space at a finite speed, referred to as the speed of light, denoted by c. These properties make it possible to plot the fields of accelerated point charges to thereby illustrate the radiation fields that result.

The fields due to several kinds of charge acceleration are presented in this chapter obtained from a computer program developed at Stanford University [1]. The radiation portions of the electric-field lines develop as "kinks" that move outward at the speed of light for a charge moving in a specified way and maximum speed selected as inputs to the program. A succession of graphs is plotted and displayed as a movie on a computer monitor to demonstrate the radiation the results from the specified motion. Representative plots extracted from such movies illustrate that charge acceleration results in electromagnetic radiation. This is established analytically by an expression for the electric field that has a term whose spatial distance r far from the charge varies as $1/r$. The kink model clearly demonstrates that "wriggling" a charge so as to wriggle its E-field lines essentially produces EM radiation.

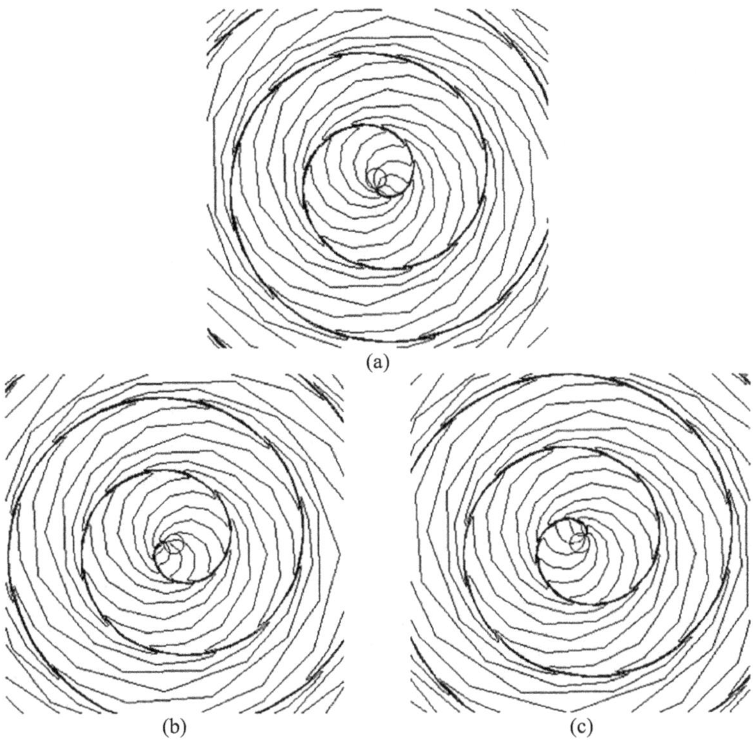

(a) (b) (c)

Figure 1.2 *E-field lines for a charge moving in a circular path at a speed 0.9*c. *The dark bands in this figure, producing an effect called Synchotron radiation that spirals outward with increasing time, indicate high values of electric field*

1.3 The proportionality between charge acceleration and radiation from a generic wire object

The results presented here were developed using the computer model TWTD [9,11]. An examination of how charge acceleration and radiation are investigated for various geometrical features of a wire object is explored in this chapter. The time domain is especially suited for this purpose since the time delay of spatially separated features to the broadside far field for a straight wire results in their time separation too. Consequently, a quantitative relationship between the magnitude of the far field and the accelerated charge that produces it can be established, a quantity denoted here as an acceleration factor (AF).

Values for the AF due to source-induced, end-reflection, propagation, etc. radiated fields and others are derived. Of particular interest is that the AF values for source-induced and end-reflection radiation differ by approximately 2, with end-reflection being the larger. This difference is due to the fact that the excitation

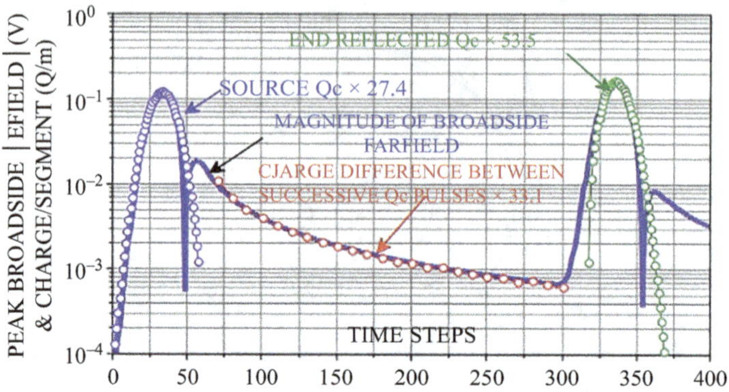

Figure 1.3 Combined plots of the propagation, source and end-reflected broadside E-field magnitudes (solid line) and the respective time varying accelerated Qc *charge pulses (open circles) that produce them on a log plot*

accelerates charge to flow in an outward direction on the antenna while the end reflection causes charge acceleration that reverses the direction of charge flow, or twice as big a velocity change. The Lienard–Wichert potentials predict this phenomenon analytically by a term proportional to charge acceleration. Other features for which the AF is evaluated are wire bends, resistive loads and variable wire radius.

1.4 Time-domain energy measures

Development of an analytical method using the current and charge time-domain solution to obtain a numerical measure of how the electromagnetic energy due to impulsive excitation varies with time leads off this chapter [2]. Differentiation of this energy with respect to time yields an energy loss rate (ELR) that provides a quantitative estimate of how different geometrical features affect the radiated power that an object produces.

As for the frequency-domain Poynting vector discussed in Chapter 6, the time-domain energy is evaluated from the current, $I(t)$, and charge density, $Q(t)$. Being done as a function of time means that separation of various energy changes in time can be correlated with associated geometrical or other features on the object being modeled. One of the more interesting radiation effects that are shown so clearly is a lower-level and decreasing TD energy that occurs between the injected source-excitation energy and end-reflection energy loss. This loss is due to what is referred to as "propagation" radiation caused by a small but continuous reflection of charge as the pulses propagate down the wire. This phenomenon proves conclusively that a straight wire antenna radiates from along its length. Of particular interest is the end-reflection radiation loss that is exhibited as an abrupt decrease in the TD energy. The energy of the propagating Q/I pulses is calculated using $I^2(t) + (Q^2(t)c^2$ for interior segments of a wire. However, using this formula for the last segment of the wire produces an increase in the total

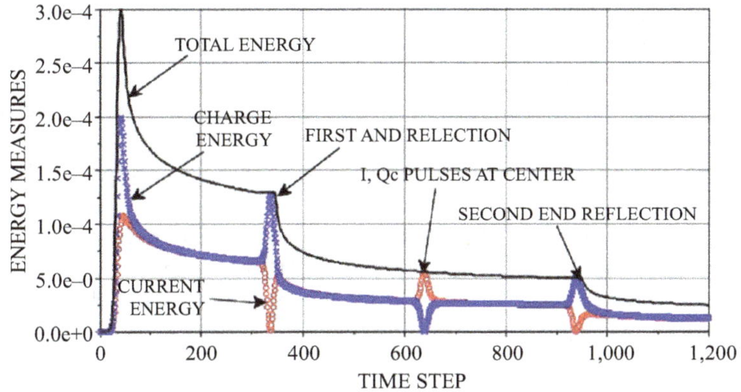

Figure 1.4 Current, charge and total energy measures for impulsively excited, center-fed, 599-segment dipole antenna

energy, a non-physical result. Further analysis [3] shows that the charge contribution upon end reflection is reduced to $Q^2(t)c^2/3$. The 1/3 factor results from modeling the charge speed as reducing smoothly from c to 0 over the last segment.

1.5 The dependence of time-domain radiation loss on the circumference and wire radius of a circular loop

The time-domain behavior of the circular-loop antenna is discussed in this chapter with a particular focus on the loss rate due to radiation caused by the effects of a

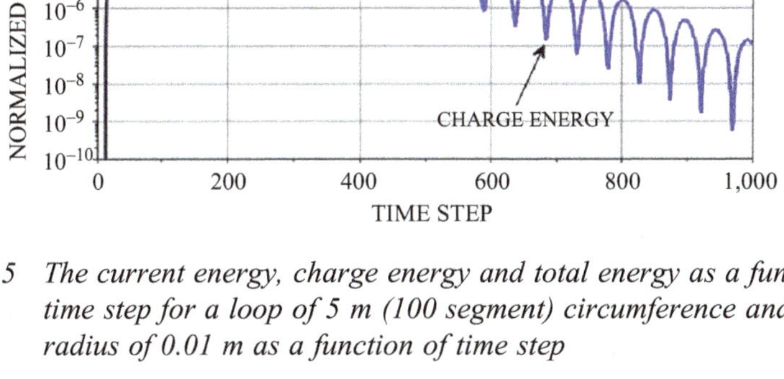

Figure 1.5 The current energy, charge energy and total energy as a function of time step for a loop of 5 m (100 segment) circumference and wire radius of 0.01 m as a function of time step

loop's curvature, or size, and the radius of the wire that comprises it [23]. An added feature of the time-domain loop model is the fact that when a Gaussian-pulse voltage excites a closed loop, its late-time current becomes uniform around the loop. A bonus of this phenomenon is that it provides a way to determine the inductance of loops of relatively arbitrary geometry. And, once the loop inductance and the steady-state, late-time current are known, the loop energy can be numerically calibrated.

Upon determining the ELR with the loop size and wire radius a parameter it is possible to develop analytical expressions to fit these results using model-based parameter estimation. These results for the loop are also compared with those for a straight wire to contrast the effect that the loop curvature has on increasing its ELR.

1.6 The differentiated on-surface Poynting vector as a measure of radiation loss from wires

The Poynting vector has been frequently used as a way to assess and display the electromagnetic power and energy flow for various applications [24]. A particularly useful way of displaying such results is plotting them as streamlines. In this chapter, a specific PV streamline is evaluated on the surface of a wire object, to determine how the spatial variation of power or energy flow might be related to a radiation-caused loss. The on-surface PV does not require any additional computation that would otherwise be necessary to obtain the PV away from a wire. This is because the normal electric field to the wire is determined by the surface charge density Q and the tangential magnetic field similarly is determined by the surface current I.

Several references mentioned in Chapter 6 observe that the current decays with increasing distance away from the feedpoint for a wire antenna, a phenomenon that

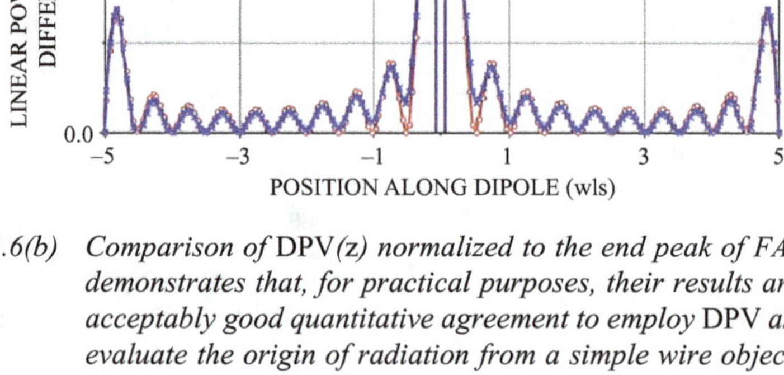

Figure 1.6(b) Comparison of DPV(z) *normalized to the end peak of FARS demonstrates that, for practical purposes, their results are in acceptably good quantitative agreement to employ* DPV *as a tool to evaluate the origin of radiation from a simple wire object*

is attributed to radiation. If that were the case, then the on-surface PV would have to decay as well since there would be a decreasing power flow with distance too. This effect is quantified from the differentiated PV (DPV). The results obtained for straight-wire antennas and scatterers demonstrate that the DPV approach is generally in agreement with those obtained from frequency-domain FARS (FDFARS, introduced in a Chapter 8) in terms of the spatial distribution of radiation from such objects.

1.7 Time-domain far-field analysis of radiation sources (TDFARS) and TWTD

The material in this book was primarily initiated by work originally reported in the Antennas and Propagation Society Magazine column "PCs for AP and Other EM Reflections." While the topic of radiation had been discussed in that column previously, FARS was introduced in a 1999 issue of the magazine [4]. The development of a time-domain version of FARS followed shortly after [14,15].

In both domains, FARS is obtained from the far field of an object whose source distribution is available. The time- and frequency-domain versions are implemented in the same way as their usual far-field results, but with the source integration handled slightly differently. The far-field power flow is normally evaluated at each ϕ,θ observation point on the far-field sphere to obtain the total power radiated by an antenna or a scatterer. The FARS values on the other hand use instead the total electric field with each of the individual segmental electric fields. These computations employ conjugate dot products of the electric fields in the frequency domain while there are only real components and products in the time

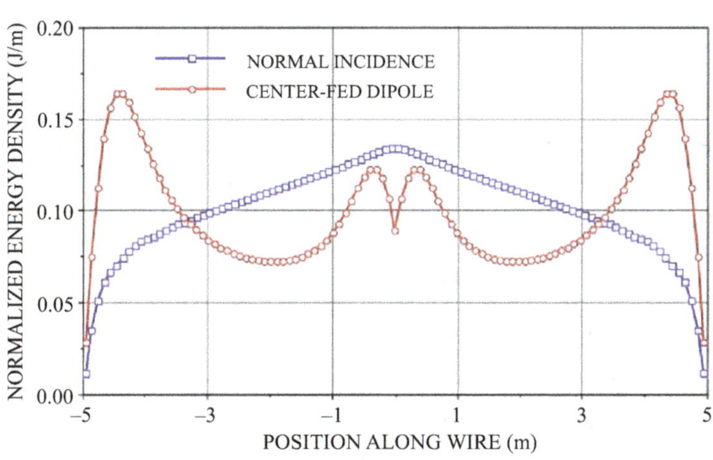

Figure 1.7 The time-domain linear energy densities for a 10-m wire excited at its center as an antenna and by a broadside-incident plane wave as a scatterer for a total energy of 1 J

domain. Integration of these incremental powers over the far-field sphere yields the total power contribution of a given segment. The total radiated power from an object is identical to the standard approach, but it has been decomposed into the power contributed from each individual segment (or patch for a surface).

TDFARS was validated using a result for the normal power flow from a Gaussian current filament [5] and by using a Gaussian-step, time-harmonic excitation for comparison with FDFARS results. Power- and energy-density time plots are included to exhibit the time variation of both quantities for a center-fed, 10-m straight wire. Scattering applications are also shown for both a normally incident and 10-deg off-axis incident Gaussian-pulse plane wave. For the antenna examples, there are two energy peaks near the feedpoint and one at each end joined by a smooth convex-shaped region in between. For normal incidence, there is a center peak in scattered energy that declines smoothly towards each end. In the near-axial incidence case, there are two end peaks joined by a smooth minima with the peak of that on the far end being almost three times the near-end peak. TDFARS energy results for a circular and square loop are found to exhibit larger peaks opposite the feedpoints. How the reflected charge causes these various features is explained in detail.

Observe that the mathematical implementation of FARS is straightforward. References to electromagnetic radiation are voluminous in the electromagnetic literature, but the origin of radiation from even simple objects has remained subject to interpretation. Most work in the literature seems to be in agreement that radiation from a simple dipole antenna occurs at or near its feedpoint and at its ends. However, radiation from along the length of a simple, straight wire as has been shown in this and previous chapters seems to raise varying degrees of skepticism. This is in spite of the fact that there seems to be general agreement that the decay of current along a straight wire provides evidence for radiation. These results are compatible with the interpretation that all radiation from the kinds of wire objects considered here can be concluded to occur because of charge acceleration.

1.8 Frequency-domain far-field analysis of radiation sources (FDFARS) and NEC

This chapter introduces the complementary version of frequency-domain FARS (FDFARS) [25]. The implementation of FDFARS is outlined accompanied by a variety of applications to show how various geometrical and other features affect the distribution of time-domain energy.

Among the specific topics and results included are:

1. Demonstration that radiation from a sinusoidal current filament occurs all along its length as determined by the induced-electromotive-force (IEMF) method.
2. Confirmation of the IEMF results by the initial application of FARS.
3. Showing how the FARS results are affected by the current variation in the center region of the antenna.

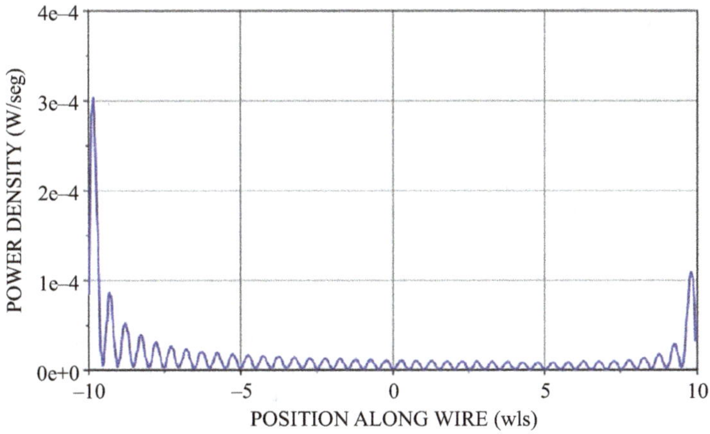

Figure 1.8 FARS power density for scattering of a 1-V/m plane wave incident at 10-deg off axial incidence from the right end of a 20-wavelength straight wire

4. The distribution of FARS power for variable dipole lengths.
5. Comparison of the length-wise radiation distribution of the SCF with a NEC dipole.
6. The radiation distribution for scattering from a long wire at normal and oblique incidence.
7. The effect of a step radius on the distribution of radiation from a straight wire.
8. The close correlation between radial power flow near a straight-wire antenna and the FARS results.
9. The FARS power density of circular-loop antennas decays more rapidly than on a straight wire due to the loop curvature.
10. A square-loop antenna exhibits doublet-shaped FARS radiation at the bends similar to the feedpoint power density on a straight wire having a current-slope discontinuity.
11. Axial-incidence scattering from a circular loop produces a single circular mode in the radiated power while edge-incidence results in multiple lobes.
12. The FARS-radiated powers for several antenna arrays agree within a few percent of the NEC input powers.
13. FARS results provide interesting insights about the radiation distributions of other geometries such as a transmission-line excited dipole, a bent-wire dipole, and dipoles enclosed in wire cages.
14. Equal-area cages are shown to screen the radiated powers of enclosed dipole antennas and the received power for an incident plane wave.
15. Radiation distributions for transverse electric (TE) and transverse magnetic (TM) normally incident on a PEC infinite, circular cylinder reveal the excitation of traveling waves around the circumference.

1.9 A comparison of time-domain and frequency-domain FARS

Comparison of FDFARS and TDFARS results for the wire objects discussed here show that the most distinctive difference between them is that the frequency-domain FARS exhibit one to two lobes per wavelength, a feature absent in the time-domain results as exhibited in Figure 1.9. This is because of the standing-wave current distributions normally found in the frequency domain. These lobes occur at the current maxima caused when counter-propagating waves meet. A similarly caused maximum occurs on a loop opposite the feedpoint in the time domain.

The FARS results for both domains are found in Chapters 7 and 8 to sometimes have negative values over parts of an object. Results presented for a dipole located in a cage of parallel wires demonstrated this outcome as a logical consequence for situations where some parts of an object reduce the radiated field as is the case of the dipole located in a wire cage. Negative FARS values do not represent the radiation of negative power or energy, but simply the reduction of the overall electromagnetic field.

1.10 An examination of the radiation properties of specified currents

This chapter explores the application of FARS and the IEMF method to a variety of specified current filaments. The presented results serve as a mutual validation of these quite-different ways of finding where far-field radiation originates from such sources. These results show as well that radiation generally occurs from over the entire extended source besides more obvious places like current-slope discontinuities.

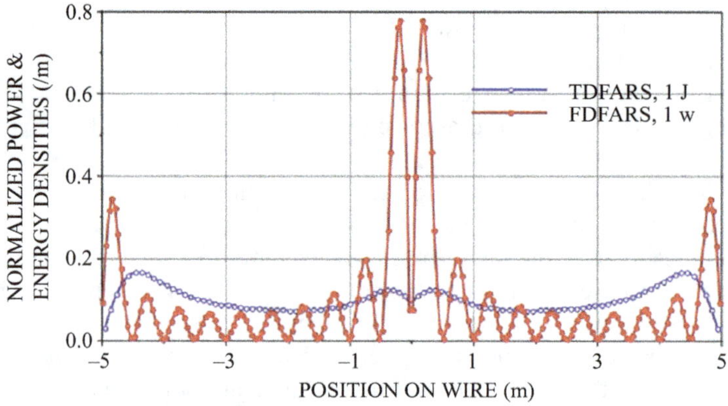

Figure 1.9 Normalized power and energy results for a 10-m dipole radiating 1 W of total power and 1 J of total energy

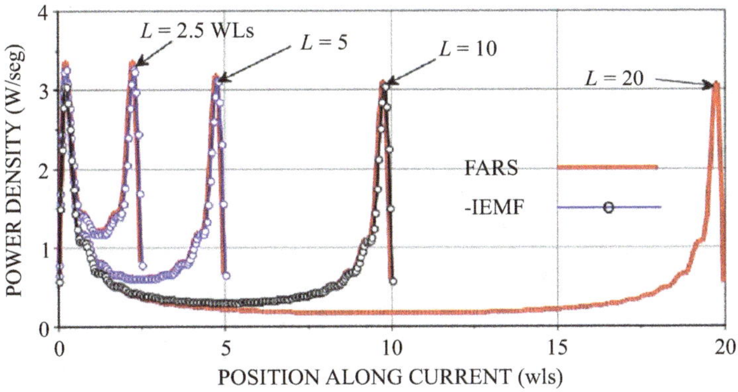

Figure 1.10 The FARS and -IEMF radiation-power densities for traveling-wave currents of variable lengths

The sinusoidal-current filament (SCF) is first examined in detail since it provided a useful approximation of the current distribution on a physical dipole. The length dependence of the power radiated by a unit-amplitude SCF oscillates between two extrema that occur at half-wavelength intervals. For a SCF of length L, maxima occur when $L/2$ is an even number of half wavelengths due to the maxima in the current-slope discontinuity at the center of the SCF. The minima occur when $L/2$ is an odd number of half-wavelengths as this slope discontinuity vanishes. Both extrema increase as $\log(kL)$ due to the radiation that occurs between the center and ends of the SCF. The current, tangential electric field, IEMF and FARS power are plotted to demonstrate their spatial relationships and to show that the input power from the IEMF is balanced by the radiated power determined from FARS. Also demonstrated is the fact that the end radiation from the SCF is independent of the filament length.

Several other variations of the SCF are also investigated to determine how their distributed radiation is affected. It is generally the case that the distributed radiation is spatially similar to the magnitude of the current for the SCF variations as is also found for uniform or triangular current distributions. But a traveling-wave and a linearly varying current had quite different spatial radiation distributions with the traveling-wave current having two end peaks as shown in Figure 1.10.

1.11 Development and comparison of the incremental FARS (IFARS) and incremental IEMF (IIEMF) methods

This chapter introduces the concept of incremental, far-field, paired current segments in a moment-method context as Incremental Far-field Analysis of Radiation Sources (IFARS). This involves exploiting the fact that the conjugate dot product of the total far electric field of a source distribution can be usefully

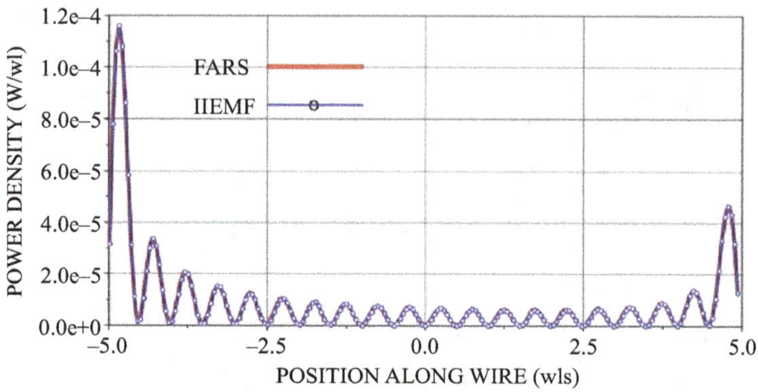

Figure 1.11 The scattered power density for a 10-wavelength straight wire caused by a plane wave incident at an angle of 10 deg from the right end obtained from FDFARS and the IIEMF method

analyzed in terms of the products of fields of incremental source pairs. Also introduced is development of a complementary approach, the incremental IEMF (IIEMF) method. The IIEMF method is shown in Figure 1.11 to offer an independent way of quantitatively determining the distribution of radiated power over a PEC source of interest. This is in contrast with the standard IEMF method that shows zero power on a PEC where there is no excitation and so cannot determine where radiation originates.

1.12 The Schelkunoff–Feldman radiation resistance

In their article "On Radiation from Antennas" [6], Schelkunoff and Feldman introduced the concept of a distributed radiation resistance $R(x)$. They developed the idea analytically for a sinusoidal current filament and conducted some experiments that seemed to confirm the concept. Their analysis was limited to a current an integral number of half-wavelengths long.

Their treatment is extended in this chapter to a current of arbitrary length. More relevant to the FARS approach is showing that when their distributed radiation resistance is converted to power distribution using $I^2(x)R(x)/2$, it numerically matches FDFARS results as illustrated in Figure 1.12.

Also investigated was the feasibility of numerically determining $R_{rad,SF}(x)$ from FARS results using

$$R_{rad,SF}(x) = 2P_{FARS}(x)/I^2(x).$$

This is analytically viable but numerically challenging because it can involve the quotient of two nearly zero quantities along the current and to consequently produce "noisy" results. Averaging or filtering of the noisy results can improve the outcome.

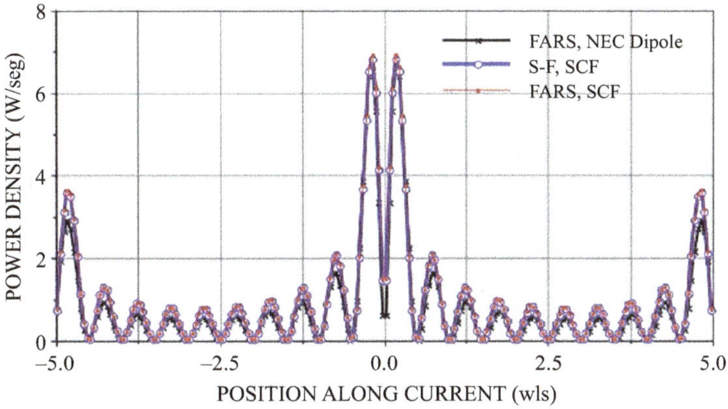

Figure 1.12 The linear power densities for a 1-A 10-wavelength NEC dipole from FARS, from the Schelkunoff–Feldman distributed radiation resistance and FARS for a sinusoidal current

1.13 Comparison of the radiation properties of a sinusoidal current filament and a PEC dipole of near-zero radius

Before effective computer modeling became available for analyzing a wire antenna, the usual practice was to assume that the current on a dipole could be reasonably approximated as a sinusoidal current filament (SCF). The question explored in this chapter is how the accuracy of this approximation varies as the radius of an actual thin-wire antenna modeled using a computer code like NEC is reduced to near-zero radius (NZR).

The radius of a NEC dipole model was varied from 10^{-2} to 10^{-20} wavelengths to compare its properties with those of an SCF [7]. Ways in which the SCF and NZR dipole are found to be similar include their:

- total radiated power as a function of current length as in Figure 1.13;
- radiation patterns;
- length-wise distributions of radiated power, or linear power density [8];
- Schelkunoff–Feldman [6] distributed radiation resistance.

On the other hand, ways in which they differ include their:

- current distributions, which differ slightly, but significantly;
- Poynting vectors parallel to their currents;
- tangential electric fields;
- distributions of induced electromotive force (IEMF) power.

Explanations have been proposed with respect to these observations, specifically in terms of the charge–acceleration, radiation mechanisms of the SCF and dipole. For the SCF, charge–acceleration comes from the applied tangential

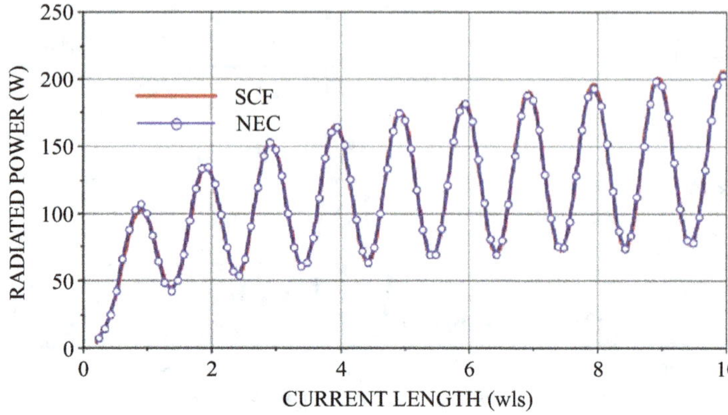

Figure 1.13 The power radiated by a NZR dipole of radius 10^{-20} wavelengths, both having a 1-A maximum current, as a function of length

E-fields needed to maintain the (assumed) sinusoidal spatial variation of the current along the length of the filament as well as at the feedpoint and filament ends. For the PEC dipole, on the other hand, charge acceleration is caused by the reflection of propagating charge along the wire due to its varying wave impedance, as well in the vicinity of the feedpoint and at the dipole ends.

1.14 The incremental FAR field and degrees of freedom of the sinusoidal current filament

The results and observations presented included in this chapter are consistent with various others in supporting the conclusion that radiation originates from along the length of a linear current distribution. This is true whether that distribution satisfies the boundary condition on a PEC wire or one that has no boundary condition, as is the case of the SCF. To summarize [13]:

1. For the most part, the far field at the maximum of the traveling-wave lobe is increased by current all along the SCF.
2. The singular-value spectrum for the SCF in Figure 1.14 extends well beyond the three singular values that apply to the spectrum of three point sources.
3. The total power radiated by a SCF of increasing length grows in proportion to Log (kL) showing a length dependence inconsistent with radiation from three isotropic point sources located at the ends and feedpoint. The power from three point sources as in (4b) instead varies between two fixed limits with increasing length.
4. FARS results exhibit a lobed, distributed radiated power along the entire length of the SCF [26].
5. The IEMF distributed input power is essentially identical to FARS for the SCF, in accord with the fact that the time-average power flow along a standing wave is zero, requiring the input power and radiated power to be equal as a function of position.

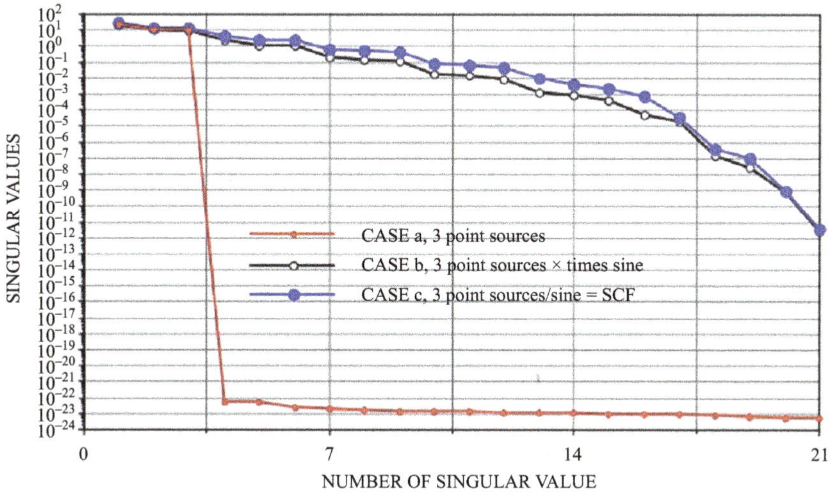

Figure 1.14 The singular-value spectra for three patterns with the open-circle plot being that of three isotropic point sources

6. The Schelkunoff–Feldman (S–F) distributed radiation resistance when multiplied by $[I_z(z_n)]^2/2$ yields results essentially identical to FARS and the IEMF distributed power.
7. The radial power flow a few thousandths of a wavelength from the SCF is within a few percent of the FARS, IEMF, and S–F powers except at the center feedpoint whose effect is included explicitly in the radial power flow whereas it is not in the others.
8. In the time domain, Gaussian-pulse-excited current/charge pulses propagating on a straight PEC wire decrease monotonically in amplitude with increasing distance from the feedpoint, as do their associated energies measured by spatial integrals of I^2 and $(Qc)^2$ [28].

1.15 Appendix A

A brief description of TWTD is presented together with a sampling of some of the results a time-domain model can generate. The code is based on a time-domain version of the electric-field integral equation specialized to thin wires. It originated in the late 1960s inspired by time-domain results reported at the 1968 IEEE AP-S conference. Its initial use was for simple wire antennas and scatterers and subsequently for ElectroMagnetic Pulse (EMP) applications. For the latter problems, a circuit capability was added to evaluate the damage that an EMP might do to electronic systems. This later feature was exploited in later research to design passive electronic tags for Radio-Frequency Identification (RFID) at Los Alamos National Laboratory. A user's manual is also available [10] as well as a Fortran listing of the program [11].

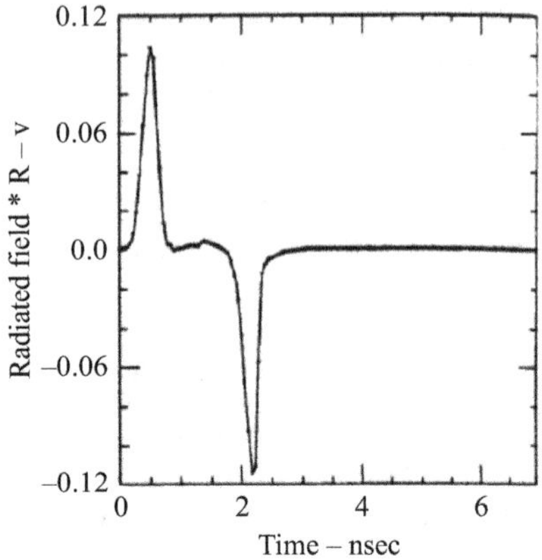

Figure 1.15 *The far radiated field from an impulsively excited dipole loaded with ideal diodes [2]*

1.16 Appendix B

The Numerical Electromagnetics Code is one of the longest-lived computational electromagnetics code, having its origin in the 1960s [26]. As NEC exists now it has little resemblance to its beginnings, benefiting from a changing team of contributors over time. Its primary manager–developer over most of those years was Jerry Burke. Various enhancements and features have been added from the original NEC to NEC-4. Some problems that can arise when validating computer models are demonstrated as part of the process of validating NEC that has continued through various applications as illustrated by some additional examples.

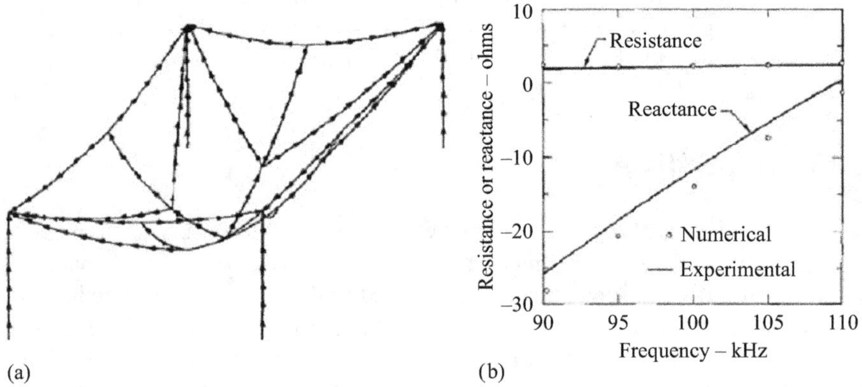

Figure 1.16 *The computer model of the Sectionalized Loran Transmitting antenna (a) and a comparison of numerical and experimental impedance results using a 1/25th scale model from the US Coast Guard and the numerical model (b) [12]*

References

[1] B. Cabrera, "Physics Simulations II: electromagnetism, academic version," Intellimation Library for the Macintosh, Santa Barbara, CA, 1990.

[2] E. K. Miller and J. A. Landt, "Direct time-domain techniques for transient radiation and scattering from wires", Invited Paper in *Proceedings of the IEEE*, 68, 1396–1423, 1980.

[3] E. K. Miller, "The proportionality between charge acceleration and radiation from a generic wire object," *Progress in Electromagnetics Research*, 162, 15–29, 2018.

[4] E. K. Miller, "PCs for AP and other EMReflections," *IEEE AP-S Magazine*, 41(2), 82–86, 1999; Vol. 41(3), 83–88, 1999.

[5] G. S. Smith and T. W. Hertel, "On the transient radiation of energy from simple current distributions and linear antennas," *IEEE AP-S Magazine*, 43(3), 49–62, 2001.

[6] S. A. Schelkunoff and C. B. Feldman, "On radiation from antennas," *Proceedings of the IRE*, 30, 511–516, 1942.

[7] E. K. Miller, "Comparison of the radiation properties of a sinusoidal current filament and a PEC dipole of near-zero radius," *IEEE Antennas and Propagation Society Magazine*, 48(4), 37–47, 2006.

[8] E. K. Miller, "Further investigation using far-field analysis of radiation sources (FARS)," in: *IEEE International Symposium on Antennas and Propagation*, Salt Lake City, UT, Hilton Hotel, July 16–21, 2000, pp. 2058–2061.

[9] E. K. Miller, "Exploring electromagnetic physics using thin-wire time-domain (TWTD) Modeling," in: *Proceedings of 14th Annual Review of Progress in Applied Computational Electromagnetics*, Naval Postgraduate School, Monterey, CA, 1998, pp. 583–588.

[10] J. A. Landt, E. K. Miller, and M. Van Blaricum, "WT-MBA/LLL1B: a computer program for the time-domain electromagnetic response of thin-wire structures," *Lawrence Livermore Laboratory, AFWL Interaction Notes*, No. 210,1974.

[11] FORTRAN version in PDF available as :TWTD FARS: edmundmiller.academia.edu.

[12] E. K. Miller, F. J. Deadrick, and W. O. Henry, "Computer evaluation of large, low-frequency antennas," *IEEE Transactions on Antennas and Propagation*, AP-21, 386–389, 1973.

[13] E. K. Miller, "The incremental far field and degrees of freedom of the sinusoidal current filament," *IEEE Antennas and Propagation Society Magazine*, 49(4), 13–21, 2007.

[14] E. K. Miller, "PCs for AP and other EM reflections," *IEEE Antennas and Propagation Society Magazine*, 42(6), 92–98, 2000.

[15] E. K. Miller, "Time-domain far-field analysis of radiation sources," *IEEE Antennas and Propagation Society Magazine*, 53(5), 81–97, 2011.

[16] C. S. Baird, "What is the speed of electricity," *Science Questions with Surprising Answers*, February 19, 2014.

[17] R. W. P. King and T. T. Wu, "Current measurement-1965-currents, charges, and near fields of cylindrical antennas," *Radio Science*, 69D(3), 429–446, 1965.

[18] W. D. Smith, "Numerical Analysis of Normal Mode Helical Dipoles Antennas," *PhD Dissertation*, Iowa State University, 1971.

[19] T. Yokoyama, T. Hoashi, K. Murata, S. Egashira, K. Egashira, and T. Nakamiya, "Design of multi-band antenna using different radius wires," *PIERS Online*, 5(3), 287–290, 2009.

[20] T. Ishizone, S. Adachi, K. Taira, Y. Mushiake, and K. Mitazaki, "Measurement of antenna current distribution in an anisotropic plasma," *IEEE Transactions on Antennas and Propagation*, 21, 678–679, 1969.

[21] S. Egashira, M. Taguchi, and A. Sakitani, "Consideration on the measurement of current distribution on bent wire antennas," *IEEE Transactions on Antennas and Propagation*, 36(7), 918–926, 1988.

[22] Y. S. H. Khraisat and K. A. Hmood, "Distribution of symmetrical dual and triple feeding full-wave dipole antenna," *Modern Applied Science*, 5(6), 126–132, 2011.

[23] E. K. Miller, "The dependence of time-domain radiation loss on the circumference and wire radius of a circular loop antenna," *Progress in Electromagnetics Research M*, 92, 1–9, 2020.

[24] E. K. Miller, "The differentiated on-surface Poynting vector as a measure of radiation loss from wires," *IEEE Antennas and Propagation Society Magazine*, 48 (6), 21–32, 2006.

[25] E. K. Miller, "Some further results from FARS: far-field analysis of radiation sources," in: *Proceedings of 16th Annual Review of Progress in Applied Computational Electromagnetics*, Naval Postgraduate School, Monterey, CA, 2005, pp. 278–285.

[26] G. J. Burke, E. K. Miller, and A. J. Poggio, "The numerical electromagnetics code (NEC)—a brief history," in: *2004 IEEE Int. Antennas Propagat. Symp. Dig.*, vol. 42, June, 2004, pp. 2871–2874.

Chapter 2

The electric-field kink model of electromagnetic radiation

Abstract

The apparent mathematical complexity of Maxwell's equations may seem to obfuscate the beauty of the physical phenomena that they describe. Yet the defining fundamental property of electromagnetics, radiation, can be illustrated visually and qualitatively independent of its mathematical foundation. The purpose of this chapter is to present a way to do this called the Electric-Field kink model. The E-field kink model relies on two basic properties of electromagnetics: (1) electromagnetic fields propagate at a finite speed, denoted by the symbol "*c*," the speed of light, that is 3×10^8 m/sec in free space; and (2) electric field lines of force originate by convention from positive electric charge and terminate on negative electric charge.

These two facts make possible the development of graphical depictions of the fields radiated by charges whose speed and/or direction of motion change in time, i.e., when they are accelerated. This phenomenon is described in more detail below and illustrated graphically for a variety of charge motions. The goal is to provide a way to visualize how radiation is produced and so to gain an intuitive understanding of the physical process conveyed by Maxwell's equation. Also demonstrated is how the field lines close as they radiate away from a wire antenna in the frequency and time domains.

Keywords: Accelerated charge; Bremsstrahlung radiation; Charge acceleration; Electric-Field Kink Model; Electromagnetic radiation; Oscillatory radiation; Radiation-field plots; Synchotron radiation

2.1 Introduction

As stated in the introduction, the goal of this book is to clarify how antennas radiate and targets scatter electromagnetic fields, i.e., how radiation occurs. Attention throughout this book will be primarily limited to perfect-electric conductors (PEC) excited locally by an electric voltage as antennas, or excited by an incident electromagnetic field as scatterers. Such excitations induce electric current, and charge flows over an object that produce electric, **E**, and magnetic, **H**, fields in response.

The fields near the object comprise both reactive, or non-radiating components as well as radiating components that vary with distance, r, as $1/r$ when far enough away from the object. These radiating fields carry a power density (frequency domain) and energy density (time domain) described by the Poynting vector

$$\boldsymbol{S} = \boldsymbol{E} \times \boldsymbol{H}. \tag{2.1}$$

All electromagnetics (EM) observables arise from the process of radiation. There would be no propagating, scattered or radiated fields were it not for this essential phenomenon. It is true that we are now able to determine, primarily using computer models, the near and radiated fields of rather complex objects subject to arbitrary excitation and can thus perform analysis and design of EM systems. However, it is less clear how to determine quantitatively the spatial distribution of radiation originating from the surface of a given object from such computations.

One way to think about this problem might be to ask, were our eyes sensitive to X-band frequencies and capable of resolving source distributions a half wavelength or so in extent, what image would we see of such simple objects as dipoles, circular loops, conical spirals, log-periodic structures, continuous conducting surfaces, etc. when excited as antennas or scatterers? Many kinds of measurements, analyses, and computations have been made over the years that bear on this question. Various relevant observations concerning radiation physics in both the time and frequency domains for a variety of observables will be summarized or examined in detail throughout this and the following chapters.

We begin by noting that charge acceleration causes electromagnetic radiation. The electric field of a charge q in free space moving at a velocity $\mathbf{u}$ as derived from the Lienard–Wiechert potentials can be expressed analytically as [1]

$$\mathbf{E} = \frac{q}{4\pi\varepsilon_0}\left[\frac{1}{s^3}\left(\mathbf{r} - \frac{r\mathbf{u}}{c}\right)\left(1 - \frac{u^2}{c^2}\right) + \frac{1}{c^2 s^3}\left\{\mathbf{r} \times \left[\left(\mathbf{r} - \frac{r\mathbf{u}}{c}\right)\right] \times \frac{d\mathbf{u}}{dt'}\right\}\right] \tag{2.2}$$

where

$$s = r - (\mathbf{u} \cdot \mathbf{r})/c$$

and the corresponding magnetic field is

$$\mathbf{H} = \frac{\mathbf{r} \times \mathbf{E}}{r\eta_o} \tag{2.3}$$

with $\eta_o = (\varepsilon_o/\mu_o)^{1/2}$ being the impedance and ε_o and μ_o the permittivity and permeability of free space.

The first term of (2.2) decreases with distance as $1/r^2$ while the second term, which is proportional to the time-derivative of $\mathbf{u}$, or the charge acceleration, varies as $1/r$. This results in a far-field power-flow density that falls off as $1/r^2$. Since the surface area of a sphere in which the charge q is located increases as r^2 the total far-field power through that spherical surface is a constant independent of r. This is the definition of radiated power. Thus (2.2) shows that charge must be accelerated for far-field power flow, hence radiation, to occur.

While developing a possible answer to the question of where and how much power (frequency domain) or energy (time domain) is radiated from over the surface of an object might be expected to involve detailed mathematical analysis, surprisingly it is true that a semi-quantitative picture can be derived of how radiation occurs without doing so. The basis for this comes from two fundamental properties associated with electric and electromagnetic fields.

2.1.1 The propagation speed of EM fields is finite

EM fields propagate at the finite speed of approximately 3×10^8 m/s in free space, a quantity conventionally denoted as "c." This means that information about any "disturbance" of the position of the charges that produce electric fields, or of the velocity of charges in motion that produces magnetic fields, cannot be detected by an observer at a distance r away until a time $t = r/c$ has passed. In EM parlance, this delay is called "retarded time" as that distant observer only finds out about changes in charge position and/or velocity at a time retarded by the propagation delay. As a simple example, an eruption on the surface of the sun at time t_o, cannot be observed until later at the time $t_o + 9$ minutes since the earth is about 9 minutes away from the sun at the speed of light.

2.1.2 Electric lines of force are continuous

The other fact is that electric charges produce continuous lines of force that define electric fields. By convention, electric fields originate on positive- and terminate on negative-signed charges. A positive test charge placed in an electric field experiences a force, and thus acceleration, in the direction of the electric fields lines. For charges that are stationary or moving at a constant velocity, the field lines are spatially unchanging and there is no radiation. When the charges are accelerated the field lines change spatially and temporally as radiation is generated.

These two facts make possible the graphical depiction and animation of radiation from an arbitrarily moving charge in a rather straightforward way to provide an intuitively appealing, and physically insightful way to visualize the radiation process. At the simplest level, radiation can be deduced to occur from the "kink" model described below, and from more quantitatively rigorous mathematical analysis, to arise as a result of "wriggling" charges, i.e., subjecting them to acceleration so that their velocities change. A good analogy for EM radiation is to attach a string to a stationary point and wriggle the other end up and down. An outward-propagating wave will be produced comprised of local string motion transverse to the propagation and parallel to the direction of the wriggle. The behavior of the string depicts in a simpler way the EM waves produced by wriggling charge, primarily differing in that the electric-field lines for a charge extend in all directions in contrast with the one-dimensional nature of the string. The results that follow were obtained using a program developed at Stanford University [2].

2.2 The "KINK" model of radiation

The procedure described below was apparently originated by J. J. Thomson [3] who was first to demonstrate the existence of the electron. A detailed account of what is called here "The Kink Model of Radiation" is described by Smith [4] who also lists several additional references where the kink model can be found [5–8]. Further information can be found in [9,10].

That radiation occurs only as a result of charge acceleration can be demonstrated by putting together the two simple facts of a finite propagation speed and continuous electric-field lines. Consider a stationary, isolated charge at the origin, O, of a rectangular coordinate system, whose electric field lines therefore lie along radii terminating at O, as shown in Figure 2.1(a). If the charge is abruptly accelerated to a velocity $v = 0.3c$ along the $+x$ axis at time $t = 0$, then at time $t = t_1$ the field lines will be as shown in Figure 2.1(b). An observer at point $R = \sqrt{(x^2 + y^2)}$ in the x–y plane would not be aware of the charge movement until a time $t = R/c$ has passed when the electric field at the observer's position changes. This change

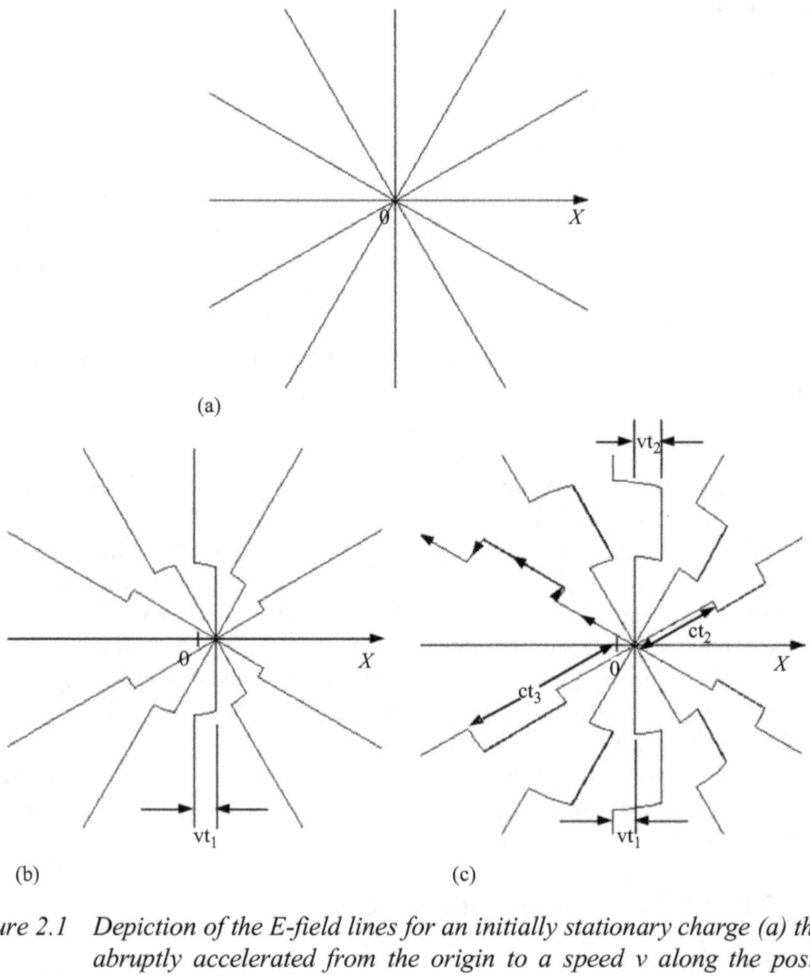

Figure 2.1 Depiction of the E-field lines for an initially stationary charge (a) that's abruptly accelerated from the origin to a speed v along the positive x-axis (b) and after a time t_1 is abruptly stopped at $d = vt_1$ (c)

occurs because the E-field lines that terminated initially on the charge at the origin must "shift" over time with the changing charge position, a process that is not instantaneous because c has a finite value. Also, because of the continuity in the E-field lines, the outer part of a given line will continue to point to the origin while that part near the charge will point to its changing location. Thus, the old and new field lines continue to be joined by a line segment (or kink) that is parallel to neither. The specific shape of the kink depends on the time dependence of the charge acceleration. If the acceleration is instantaneous, the kink lies along the circumference of the sphere (a circle in this two-dimensional plot) defined by $R = ct$ and which moves outward at the speed of light.

These kinks constitute the electric components of an electromagnetic radiation field which are accompanied by magnetic-field components as well, which are not shown here. If the charge is abruptly stopped at time t_1, a second spherical surface of E-field kinks is formed as shown in Figure 2.1(c) at a time t_2 later. This new surface propagates outward from the position $x = vt_1$ about which it has expended to the radius $r = ct_2$ while the original radiation pulse has further expanded to a radius $r = c(t_1 + t_2) = ct$.

There are some other interesting phenomena to observe in Figure 2.1. Note that the E-field lines joining the two kinks continue to move in the $+x$ direction at the speed v to continue pointing at the location where the charge would be had it not been stopped. In addition, observe that the kink portions lengthen as their respective spheres propagate further away from their origins, indicating a decreasing value of their electric fields, which must decay with distance, r, as $1/r$. Further, note that on tracing a field line outward from the charge, the two kinks are oppositely directed because their associated accelerations are of opposite sign. Finally, note that the distance between the acceleration and deceleration kinks in Figure 2.1(c) is shorter in the $+x$ direction than in the $-x$ direction. This demonstrates the well-known Doppler shift exhibited by a moving source of electromagnetic or acoustic energy, i.e., a "squeezing" together of the waveform in the direction of motion and its "stretching" out in the opposite direction.

2.3 Some simple charge motions that produce radiation

The radiation kink model, although not useful for solving boundary-value problems in EM, the solutions of which are needed for determining the scattering properties of radar targets or the current distributions on actual antennas, is never-the-less very helpful for illustrating various kinds of radiation phenomena. The Stanford program [2] develops a series of E-field plots for a point charge undergoing certain kinds of motion selected by the user. These plots can then be viewed as a movie to display the evolution of the field as a function of time in a manner similar to a continuous "film loop." Some sample plots made by that program are presented here to show how simple the radiation process actually is, in contrast with the mathematical complexity needed to describe it rigorously. In the results shown, time advances in proceeding from part (a) to (b) to (c) of each figure. Note that the closer the individual electric-field lines are, the greater the magnitude of the field.

2.3.1 An abrupt start and stop

Information about the changed charge positions of this motion propagates outward from its old and new positions at the speed of light. This is shown in Figure 2.1 by the circular line segments joining the original field lines and those moving with the charge (b) and the moving field lines and those that terminate at the new stationary position (c). The direction of the radiated field caused by charge acceleration is opposite that due to deceleration as can be seen by the reversed directions of the kinks on the field line with the arrows in this plot.

This conceptual demonstration of the "kink" radiation model is qualitatively correct but is not directly produced by Cabrerra's program. The plot in (c) comes from piecing together two different runs to illustrate the reversal in sign of the E-field due to charge acceleration and then deceleration. While not displayed in Figure 2.1, the results that follow do include relativistic effects, although for the cases shown these may not be very dramatic. All results are from the Cabrera program.

2.3.2 A charge abruptly stopped

Abruptly stopping a moving charge is perhaps the simplest kind of radiation-producing acceleration to visualize, for which an example is shown in Figure 2.2.

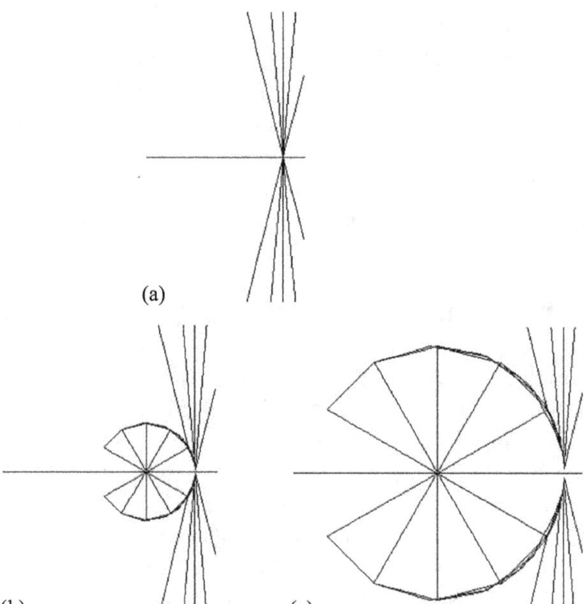

Figure 2.2 E-field lines about a charge undergoing impulse deceleration. The charge starts from a speed of 0.99 c *and abruptly stops to produce the "bubble" of radiation shown expanding from (a) to (c)*

A single burst of radiation is produced in this case for a charge moving near the speed of light, at 0.99*x*. Because of this high speed, a relativistic effect is seen as a "bunching" of the field lines toward a line perpendicular (the *y*-axis) to the direction of motion. The faster the charge speed, the more the field lines bunch around the perpendicular. A practical use of impulsive acceleration occurs when an electron beam impinges on a physical object, producing what is known as Bremsstrahlung (breaking) radiation. High-energy electrons smashing into a dense material such as tungsten are used to produce X-rays through this phenomenon.

2.3.3 A charge given a constant push

Suppose that instead of being impulsively decelerated to zero from some maximum velocity, a charge is continuously accelerated. While relativity theory shows that the charge's final speed cannot exceed that of light, its speed can approach c, increasing in mass as it does so. As seen in Figure 2.3 where the final speed is 0.99*c*, the radiation has the appearance of an EM shock wave with a zone of nearly overlapping field lines (the density of the lines is a measure of the field strength), which is most concentrated toward the forward direction. This kind of radiation is

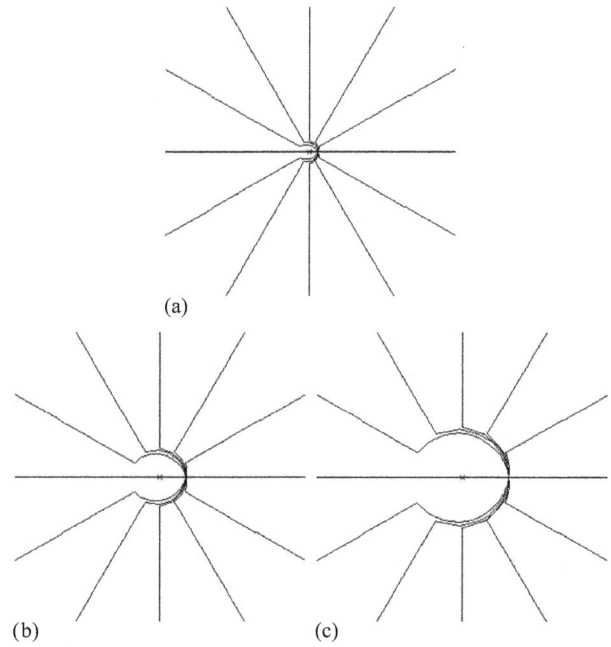

Figure 2.3 E-field lines for a charge accelerated to 0.99c by a constant force. Because its speed is so near c, *the field lines at the charge become concentrated in a small angular sector about the direction of acceleration, creating an EM shock wave*

what would be associated with a linear particle accelerator where charged particles are accelerated to speeds near c over extended distances.

2.3.4 A charge moving at constant speed around a circle

Results are shown in Figure 2.4 for a charge moving around a circle at a constant speed of $0.9c$. There the strongest radiation field can be seen as a spiral of coalescing field lines synchronized with the motion of the charge around the circle. This is the kind of radiation produced in circular particle accelerators and is called "Synchrotron" radiation. It is also sometimes called searchlight radiation, as a burst of radiation is seen at a given observation point with every orbit of the charge, much like the flash of light produced by a continuously rotating searchlight beam. The radiation takes place continuously here since the charge is constantly inwardly accelerated as it moves around the circular path, in contrast with the linear acceleration of the previous examples.

2.3.5 A charge moving at constant speed around a square

Results for a charge moving at a constant speed of $0.5c$ around a square path are shown in Figure 2.5. This kind of charge motion can be especially illuminating

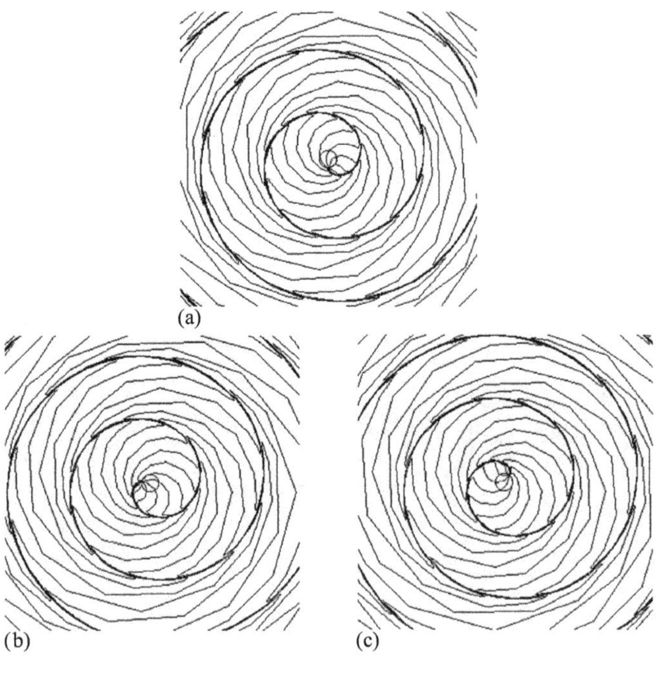

Figure 2.4 E-field lines for a charge moving in a circular path at a speed 0.9c. The dark bands in this figure, producing an effect called Synchotron radiation, indicate high values of electric field. The spiral of maximum radiation spirals outward with increasing time

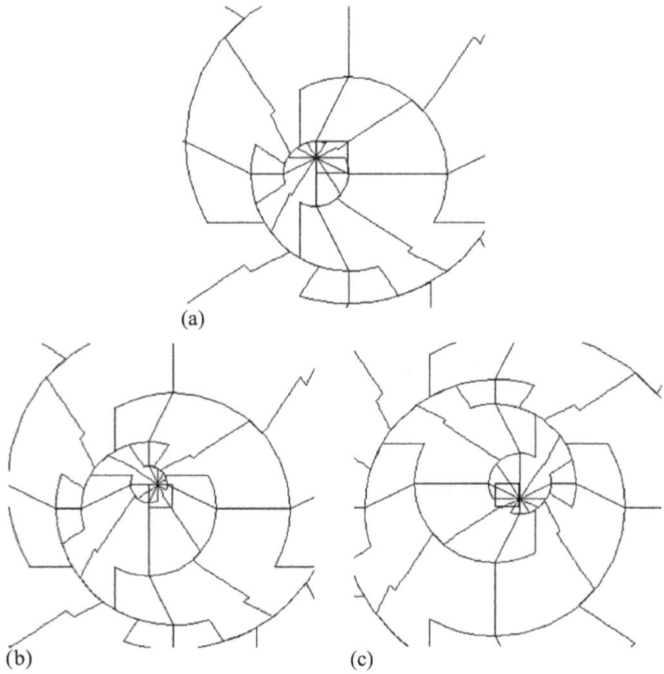

*Figure 2.5 E-field lines for a charge moving around a square path at a constant speed of 0.5*c. *The charge produces an expanding circle of radiation as it goes around each of the right-angle corners of the square*

since the radiation occurs in a series of pulses as the charge moves around each right-angle corner of the square. During the time, the charge moves along the straight side of the square, there is no radiation, but a circular radiation field can be seen centered on each corner as the charge changes direction there. The vectorial nature of this kind of radiation is demonstrated by the fact that these radiation pulses are not closed circles, there being nulls in the radiation along a line bisecting each interior angle of the square.

2.3.6 A charge undergoing oscillatory motion

The kind of motion considered here in Figure 2.6 and one that is most relevant to electromagnetic communications is oscillatory, where a charge moves back-and-forth along a straight line with speed that varies as $V\cos(\omega t)$, with $\omega = 2\pi f$, where f is frequency of the radiation that it produces and $V = 0.5c$. This is not exactly analogous to how the charge motion varies on an actual wire antenna, but it suffices to illustrate the production of a time-harmonic radiation field as shown in Figure 2.6. Not surprisingly, the radiation field is seen to be oscillatory in space, and therefore in time as well at a fixed observation point, as the wave propagates by.

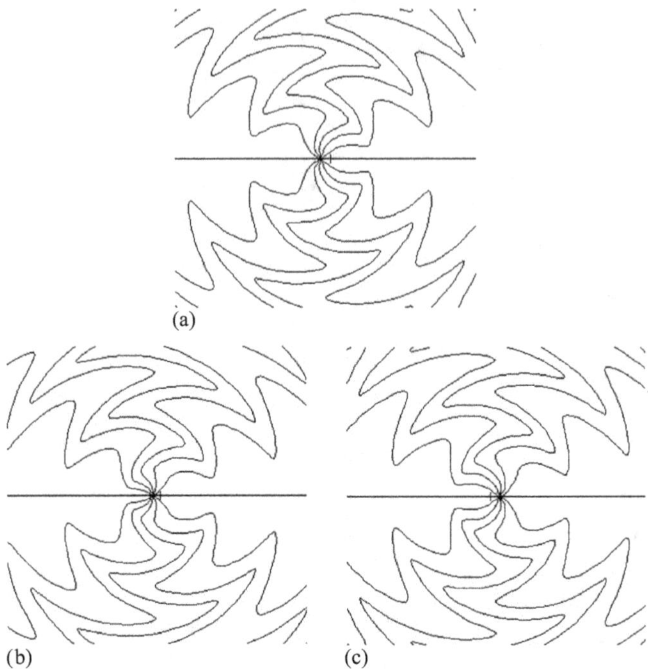

Figure 2.6 *The E-field lines about a point charge undergoing oscillatory motion along a straight line with a maximum speed of 0.5*c. *The generation of outward propagating, time-harmonic waves is clearly discernible, in a fashion somewhat analogous to how an actual wire antenna works*[*]

2.3.7 The effect of increasing speed on an oscillating charge

The E-field lines of the oscillating charge as its speed increases are demonstrated in the following sequence of Figure 2.7 where E-field plots are shown as the maximum charge speed is changed from 0.1x to 0.99c.

[*]The motion of a point charge and the E-field associated with that motion as illustrated here is quite different from the behavior of charge and current on a real conductor such as a straight wire. For example, on a wire assumed to be perfectly conducting and excited at its center by a time-harmonic voltage, a structure known as a dipole antenna, the effective speed of the E-fields along the wire is c, since EM waves propagate at light speed in free space. Since the EM wave E-field lines terminate perpendicularly at charges on the wire, this means the effective charge speed is also c, which is in contrast to the sinusoidal speed assumed here for the charge. But physical charges in the wire cannot move at light speed, and neither does the same physical charge move from one end of the wire to the other. Instead, the charges interact somewhat as a row of falling dominoes to constitute an equivalent current that moves much faster than the individual charge speeds.

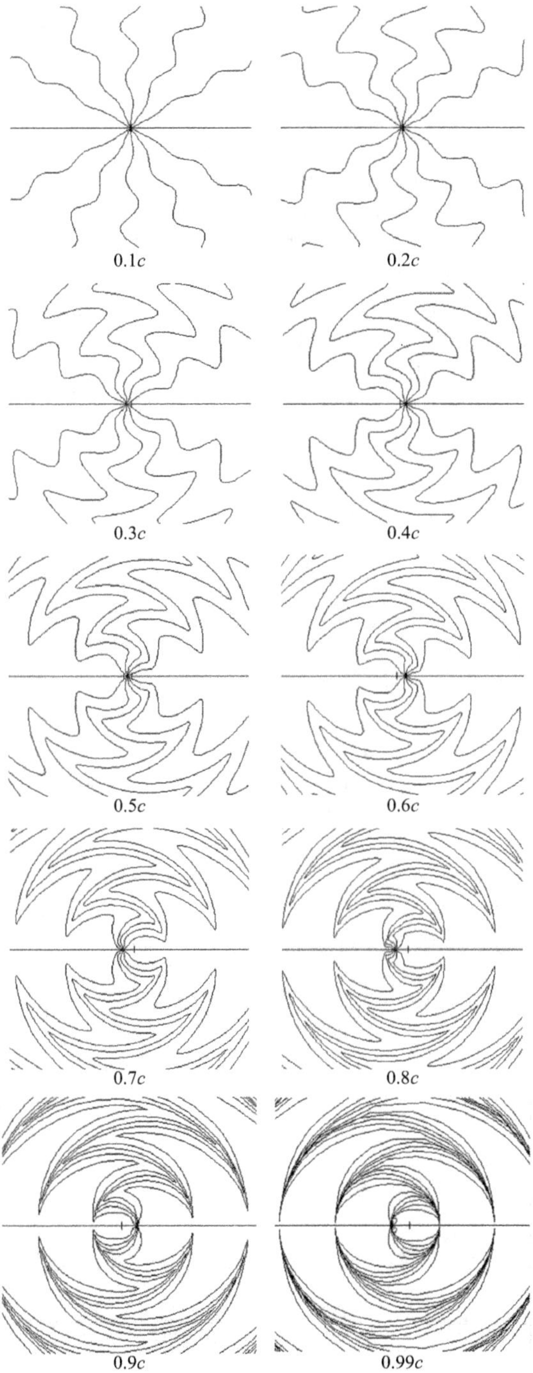

Figure 2.7 *The E-field lines about a point charges undergoing oscillatory motion along a straight line with a maximum speed varying from 0.1*c *to 0.99c to illustrate the effect of increasing charge speeds on the radiated field*

2.4 How field lines close on a physical antenna

The electric-field plots presented above are restricted to a point charge, and consequently are not closed as they propagate away from that charge as would be the case for a physical antenna. As is demonstrated in Figure 3.2 in the following chapter, a dipole excited by an impulsive voltage applied at its center has two opposite-signed charge pulses propagating away from the feedpoint in opposite directions. In this particular case, positive charge pulses are on the right half of the dipole and negative charge pulses are on the left. The expanding electric field that results originates, by convention, from the right-half positive charge pulse and terminates on the left-half negative charge pulse.

This is illustrated by the E-field vectors plotted for the right half of the dipole in Figure 2.8 at several time steps designated by (a) through (d). An expanding

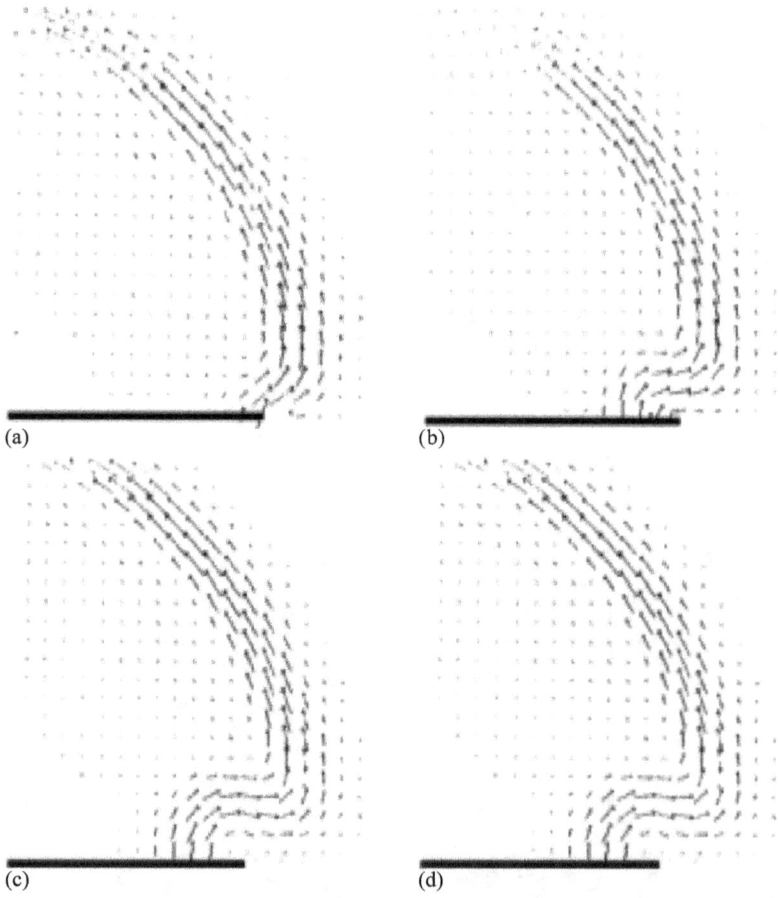

Figure 2.8 Vectorial depiction of electric field samples as the right-half outward-propagating charge pulse reaches the end of the wire and begins to reflect

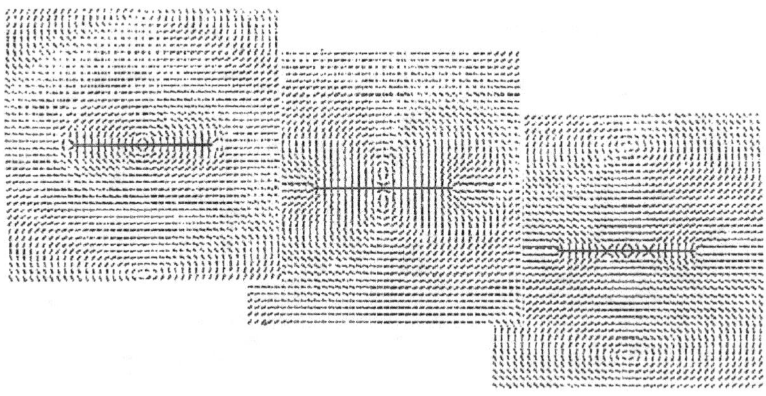

Figure 2.9 Scaled vector plots of the frequency-domain electric field at three times steps about a half-wave, center-fed dipole antenna, after Miller and Deadrick [11]

circular electric field centered on the feedpoint at the center of the dipole has formed while a new circle is developing centered on the end of the dipole. A reversed image of this field structure also forms around the left end of the dipole. The positive and negative pulses continue to propagate back toward the dipole center where they meet and the field lines form closed loops. Note that for these and the next plots, the fields are scaled by multiplying them by their distance from the dipole for display purposes.

The phenomenon of field closure is illustrated for the frequency domain in Figure 2.9 where the E-field vectors are plotted around a half-wave, center-excited dipole [9]. The field vectors are plotted on a regular grid but those closest to the wire are a fraction of a wavelength away from it and so not all of the samples are perpendicular to the wire, possibly having some small tangential component. The closure of the field lines begins at the source and propagates outward as the time advances.

Spatial contours of the near electric field at one time step (a) and the radiated electric field (b) at a latter time step are plotted in Figure 2.10 together with a scaled representation of the dipole. The field originating from the feedpoint due to the impulsive excitation forms the largest circle since it was the first produced. The other two smaller circles, centered at the wire ends, are due to the radiation coming from the first end reflections of the outward-propagating pulses. Note that radiation off the ends of the dipole is zero.

2.5 Summary

The main goal of this chapter is to establish that for there to be electromagnetic radiation, acceleration of electric charge is required. This is proved analytically by an expression for the electric field that has a term whose spatial distance r far from

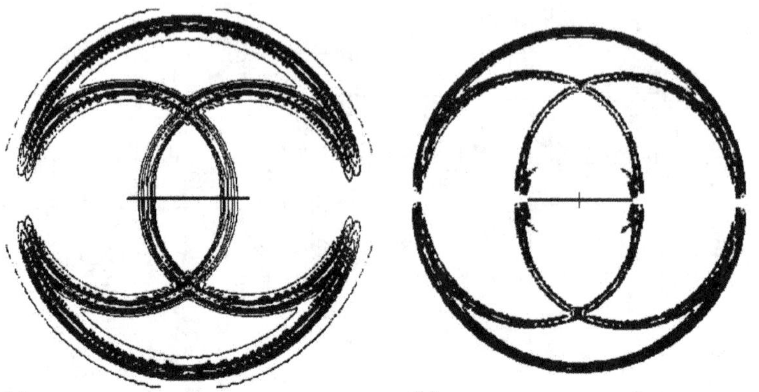

(a) (b)

Figure 2.10 Spatial contours of the scaled near (a) and the far (b) electric fields (with the propagation distance removed) around an impulsively excited, center-fed dipole at one time step of the numerical solution [12]

the charge varies as $1/r$. A way to demonstrate this effect on a graphical basis using a procedure called the E-field kink model was then illustrated for several kinds of charge motion to confirm the analytical expectation that charge acceleration does cause electromagnetic radiation.

At this point, it is appropriate to point out that the kink model clearly demonstrates that "wriggling" a charge so as to wriggle its E-field lines essentially produces EM radiation. As can be seen in Figure 2.6, the outward-propagating waves carried by a line of electric field are tantalizingly similar to the waves caused on a rope tied on the one end by wriggling the other end. This analogy is not entirely superficial. Both waves are transverse, i.e., the displacement of the field line and the rope are perpendicular or transverse, to the direction of wave travel. Also, a field line in the direction of the wriggling charge motion carries no wave and neither does a rope when its end is pushed or pulled on. Thus, any time a charge or current, which is comprised of moving charges, is wriggled, radiation can be expected.

Electromagnetic radiation also involves power flow. The greater the power provided to an antenna, the larger the field at a distant point where its strength varies as the square root of the antenna power. The radiation kink model does not include any consideration of the power needed to make physical charges move in the various arbitrary ways chosen for the above examples. It was mentioned above that power in the EM field is carried by interacting electric and magnetic fields, in an amount and direction given by $\mathbf{E} \times \mathbf{H}$ where $\mathbf{H}$ is the magnetic field and x signifies a vector cross product. The magnetic field has been neglected in this discussion for simplicity.

Finally, EM radiation and scattering problems normally begin without knowledge of what the charge and current distributions may be on objects of interest like a vehicular whip antenna. Before the radiation fields can be obtained, it is necessary that the currents and charges flowing on the whip antenna and the vehicle on which it is mounted be found. This in turn requires solution of what is called a "boundary-value" problem, where use is made of the fact that an electric field tangential to a good conductor such as aluminum or copper is essentially zero. Such materials are usually referred to as perfect electric conductors (PEC). The currents and charges can be obtained when an appropriate "boundary condition" is applied together with specifying how the antenna is "excited" either by an incident EM wave when used in a receiving system, or by a voltage from a generator when used in a transmitting system. These in turn permit evaluation of the receiving or transmitting properties of the antenna. This procedure for solving these kinds of boundary-value problems can be complicated. The kink model, though unsuited for quantitative solution of such problems, can yield insight into electromagnetic radiation that a more rigorous approach may not provide.

Obtaining numerical results of the current and charge for a boundary-value problem is the first step in establishing a quantitative connection between charge acceleration and far-field radiation. This connection will be explored in the following chapters using several different computational and analytical tools. These will include both time- and frequency-domain results. As one example, the variation in time of the energy stored on some radiating object can provide information about from where the object radiates. An example of a method implemented in both domains is Farfield Analysis of Radiation Sources (FARS), a procedure that develops the spatial distribution of radiated power and energy emitted by antennas and scatterers. The derivative of the power flow (Poynting vector) along a wire object is found to provide another way of quantifying the spatial distribution of radiated power in the frequency domain. These and other techniques will be shown to provide ways of relating far-field radiation to charge acceleration on a spatial basis.

References

[1] W. K. R. Panofsky and M. Phillips, *Classical Electricity and Magnetism*, Addison-Wesley Publishing Company, 1955.

[2] B. Cabrera, Physics simulations II: electromagnetism, academic version, Intellimation Library for the Macintosh, Santa Barbara, CA 93116–1530, 1990.

[3] J. J. Thomson, *Electricity and Matter*, Charles Scribner's Sons, New York, NY, 1904. Based on the1903 Stillman Lecture at Yale University.

[4] G. S. Smith, *Classical Electromagnetic Radiation*, Cambridge University Press, 1997.

[5] G. A. Schott, *Electromagnetic Radiation*, Cambridge University Press, Cambridge, 1912.

[6] J. R. Tessman and J. T. Finnell, Jr., Electric fields of an accelerating charge, *Am. J. Phys.*, 35, 523–527, 1967.

[7] R. Y. Tsien, Pictures of dynamic electric fields, *Am. J. Phys.*, 40, 46–56, 1975.

[8] E. M. Purcell, *Electricity and Magnetism*, Berkeley Physics Course, Vol. 2, 2nd ed., Chapter 5, McGraw-Hill, New York, NY, 1985.

[9] E. K. Miller, R. Merrill, and R. J. Cole, Computer movies for education, *IEEE Trans. Educ., May*, 58–68, 1988.

[10] B. Cabrera and E. K. Miller, Macintosh movies for teaching undergraduate electricity and magnetism, in: 1986 *International IEEE AP-S Symposium*, Philadelphia, PA, June 9–13.

[11] E. K. Miller and F. J. Deadrick, Visualizing near-field energy flow and radiation, *IEEE Antennas Propag. Soc. Mag.*, 42(6), 46–54, 2000.

[12] E. K. Miller and J. A. Landt, Direct time-domain techniques for transient radiation and scattering from wires, Invited Paper in *Proceedings of the IEEE*, 68, 1396–1423, 1980.

Chapter 3

Charge-acceleration and radiation from a generic wire object*

Abstract

The Lienard–Wichert potentials show that radiation is caused by charge acceleration. The question arises about where charge acceleration occurs on the most basic of antennas, a center-fed, perfectly conducting dipole for which there are two obvious causes. One is the feedpoint exciting voltage that sets into motion an outward-propagating charge and current wave at light speed c in the medium. A second is at the dipole ends where the outgoing wave is totally reflected producing a change in charge speed of $2c$. A third, possibly less-obvious cause is the deceasing amplitude of the propagating wave with distance due to its partial reflection along the wire. That reflected charge also undergoes a speed change of $2c$. This is the reason why the decay of current flowing along a straight wire antenna has been attributed as being due to radiation. Radiation caused by these and other kinds of charge acceleration due to resistive loads, right-angle bends, and radius steps are investigated.

These phenomena are examined primarily in the time-domain where they are more observably separable in time and space than in the frequency domain. The current and charge induced on an impulsively excited wire antenna and its broad-side radiated E-field are computed using a time-domain, integral-equation model, thin-wire time domain (TWTD). The computed results are used to derive a numerical relationship between the amount of accelerated charge and its radiated field. This relationship is denoted as an acceleration factor (*AF*) and is derived for various charge-accelerating features of a generic wire object. Their values are normalized to those of the exciting sources to standardize their respective values.

Keywords: Acceleration factor; Charge acceleration; Electromagnetic radiation; Field-charge proportionality; Impulsive excitation; Lienard–Wichert potentials; Propagation radiation; Reflection radiation; Time-domain modeling; TWTD

*This chapter is largely based on "The Proportionality between Charge Acceleration and Radiation from a Generic Wire Object,' *Progress In Electromagnetics Research*, Vol. 162, 2018, 15–29, by E. K. Miller, with some additions and deletions including correction of acceleration factors for resistance-loaded and bent wires that were too low by a factor of 2.

3.1 Introduction

The preceding chapter demonstrated that charge acceleration is the fundamental cause of electromagnetic radiation. This was done using the E-field kink model and graphical results obtained from it for several kinds of accelerated free charge. The relationship between charge acceleration and radiation for an actual antenna, a straight-wire dipole and variations thereof, is considered in this chapter. This will be done using a time-domain computer code, thin-wire time-domain (TWTD) [1]. In effect, the results of computer experiments are presented and discussed to confirm how the effects of various kinds of charge acceleration are exhibited in the radiation fields that result. As discussed in Chapter 2, the Lienard–Wichert potentials and the field expressions derived from them show analytically that charge acceleration is required to produce electromagnetic radiation.

Thus, if we desire to determine where electromagnetic radiation originates from some object, one way to accomplish this is to find where charge acceleration occurs on it. This goal is made somewhat simpler when the object under consideration is a perfect-electric conductor (PEC), as is assumed here. It is clear that charge acceleration on a dipole occurs at its feedpoint where charge is set into motion and at its ends where charge is reflected, so it follows that these are places where radiation originates. But there has been some disagreement about whether it radiates elsewhere [2,3].

Numerous authors [4–13] associate the current decay that is found in the frequency domain to occur with increasing distance down a long-wire antenna from its feedpoint as being due to radiation. Similar observations have been made with respect to time-domain current behavior [14–17]. This seems plausible since the current decay implies that the decreasing current and its associated charge carry less power down the wire the farther they propagate away from the feedpoint. Since the charge does not stick to the wire or somehow leak off it, then it follows that some charge is being is being reflected back toward the feedpoint point. The corresponding tangential, on-surface Poynting's vector consequently carries less power or energy with increasing distance [13]. This reflected charge is accelerated to produce radiation along the wire similar to what happens at the wire ends, an observation made by only one of the references [16]. Feedpoint and various reflection-caused radiation mechanisms are explored quantitatively in the following.

3.2 Charge reflection and radiation

Detailed attention in this chapter is primarily focused on a PEC straight wire for simplicity. However, it is illuminating to consider conceptually a more general wire object as sketched in Figure 3.1. The radiation due to charge acceleration associated with each highlighted feature denoted by alphabetical letter is briefly listed below.

(A) Any electric field applied to the object in Figure 3.1, whether from some external excitation such as an incident plane wave or from a localized source such as the feedpoint of an antenna that will cause a current to flow in

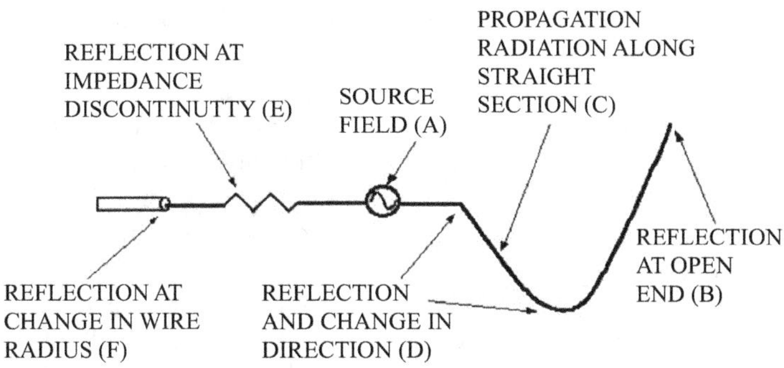

Figure 3.1 A generic wire object exhibiting various features that cause radiation

response. Setting charge into motion produces accompanying acceleration that causes radiation.

(B) Charge reflection at an open end of a wire is another cause of radiation. It is worth noting that reflection produces a charge speed change of $2c$ whereas a source field as at A causes a change of c.

(C) A spatially varying wave impedance that produces charge reflection along a straight section of wire.

(D) A directional change in charge velocity at a sharp bend or smoother curvature produces charge reflection and acceleration that causes radiation (velocity is used here when a directional effect occurs, but when only straight-line motion is involved the term used is speed).

(E) Charge reflection from an impedance discontinuity.

(F) Charge reflection at a change in wire radius.

That charge acceleration that causes radiation was demonstrated graphically in Chapter 2 using the E-field kink model [18]. This is compatible with the Lienard–Wiechert potentials [18] from which the electric field due to a charge q can be written as

$$\mathbf{E} = \frac{Q}{4\pi\varepsilon_0}\left[\frac{1}{s^3}\left(\mathbf{r} - \frac{r\mathbf{u}}{c}\right)\left(1 - \frac{u^2}{c^2}\right) + \frac{1}{c^2 s^3}\left\{\mathbf{r} \times \left[\left(\mathbf{r} - \frac{r\mathbf{u}}{c}\right)\right] \times \frac{d\mathbf{u}}{dt'}\right\}\right] \tag{3.1a}$$

so that the far, radiated field is given by

$$\mathbf{E} = \frac{Q}{4\pi\varepsilon_0 c^2 s^3}\left[\mathbf{r} \times \left[\left(\mathbf{r} - \frac{r\mathbf{u}}{c}\right)\right] \times \frac{d\mathbf{u}}{dt'}\right] = qK\left[\mathbf{r} \times \left(\mathbf{r} - \frac{r\mathbf{u}}{c}\right) \times \mathbf{a}\right] \tag{3.1b}$$

and with

$$s = r - \frac{\mathbf{u}\bullet\mathbf{r}}{c}. \tag{3.1c}$$

Equation (3.1) yields the electric field $\mathbf{E}$ of a charge Q moving with velocity $\mathbf{u}$ and acceleration $\mathbf{a}$ along a prescribed path in space. Its applications include accelerator design and analysis of electron lasers. Note that charge acceleration is not explicitly included in a computer model such as thin-wire time domain (TWTD) [1] that was developed for modeling a PEC object in the time domain. But, as shown here, the effect of acceleration can be derived from the results that TWTD provides. Our primary goal in the following is to develop a quantitative relationship between $\mathbf{E}$ and Q for the various kinds radiation that can occur on the type of object in Figure 3.1. The well-validated computer code TWTD is used to study these effects.

Radiation due to charge acceleration caused by a source field, and several kinds of reflection phenomena are now discussed. Computed TWTD results are used to develop a quantitative estimate of the relationship between the accelerated charge Q, and $E_{||}$, the broadside, electric field parallel to a straight section of a wire that radiates it. This relationship, developed as the proportionality between the numerically computed charge and radiated electric field values is expressed here as an AF. Thus, the absolute value of $|E_{||}|$ can be written as

$$|E_{||}| = AF \times |Qc|, \tag{3.2a}$$

$$\text{so that } AF = \frac{|E_{||}|}{|Qc|}, \tag{3.2b}$$

where a subscript will be used on AF to denote the particular kind of acceleration. In (3.2b) Q is the total accelerated charge density on the wire associated with a particular $|E_{||}|$. The charge is multiplied by light speed c to normalize its value to amperes for comparison with the current. Since charge pulses of opposite signs simultaneously accelerated in opposite directions can produce additive fields of either sign, absolute values are used here for Qc and $E_{||}$. For the case of source, end-reflection and propagation radiation (A, B, and C of Figure 3.1) the pulse shapes from TWTD are used from (3.2b) to illustrate the temporal correlation between $E_{||}$ and Qc. Otherwise, the numerical AF values are obtained using the corresponding peak values of these quantities to show the effect of some parameter variation such as length, radius or resistance loads. Note that the absolute AF values obtained for the various cases investigated are not themselves of primary interest, but instead are derived to establish the extent to which $E_{||}$ and Qc are proportional in time or to a particular parameter. Also of interest are their values relative to that of an exciting source. Observe that the source-caused acceleration involves a change of c in the charge speed as it leaves the region of the source while any reflection-caused acceleration results in a $2c$ speed change, a phenomenon that is illustrated further below.

A wire radius of 10^{-3} m is used throughout unless specified otherwise, with a segment length, Δx, in the frequency domain of 0.01 wavelengths and in the time domain of 0.01 m where a time step of $\Delta t = \Delta x/c$ is used. Some other material related to the discussion that follows can be found in Refs. [19,20]. The excitation is a 1-V peak Gaussian pulse for all time-domain results presented in the following.

3.3 A preview of radiation and current decay for a straight wire

Before developing *AF* estimates for the various cases to be considered, it is instructive to first examine some preliminary results for long straight wires. The current I and charge-density Q times light speed c are shown at several time steps in Figure 3.2 on a dipole 599 segments long excited at its center by a 1-V Gaussian pulse. The positive I and Qc pulses are numerically equal on the right-hand side of the dipole (i.e. $Qc = I$) but of opposite signs on the left-half side. This is because a positive charge moving to the right produces a positive current, as also does a negative charge moving in the opposite direction to the left. Their numerical equality implies that the current and charge carry the same amount of energy, an effect that is demonstrated explicitly in Section 3.11 and explored further in Chapter 4. It is also clear that the magnitudes of the current and charge decay monotonically with increasing distance from the feedpoint. This decay exhibits the effect of partial reflection of the propagating Q pulses. This reflection produces the smaller negative current components trailing the positive current pulses in Figure 3.2.

Observe that on the impulsively excited, center-fed dipole used to obtain the results of Figure 3.2, the partial reflection of the rightward-moving positive-charge pulse and the partial reflection of the leftward-moving negative-charge pulse cause acceleration of both reflected pulses. This acceleration produces kinks in their electric field lines in accordance with the E-field kink model [18], to produce $1/r$ outward-propagating, radiated electric fields. Since the charges are of opposite sign and accelerated in opposite directions their radiation fields have the same polarity and are thus additive, a result demonstrated in Figure 3.4.

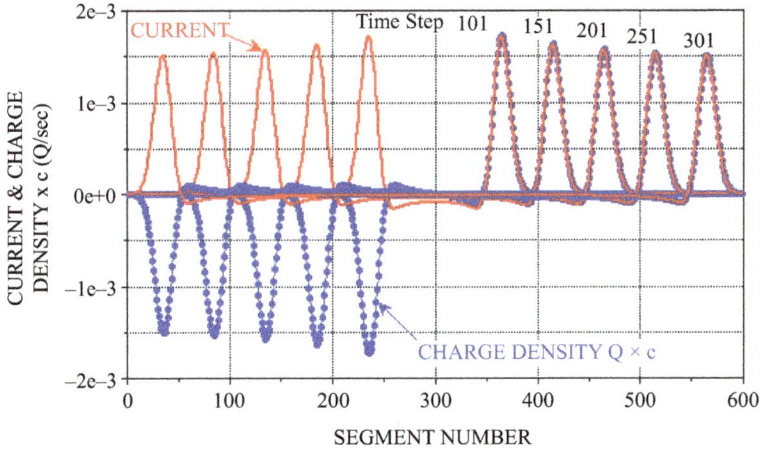

Figure 3.2 Charge density Q *times light-speed* c *and the current* I *for a 599-segment wire excited at its center by a Gaussian voltage pulse at several time steps*

The *Qc* and *I* pulses in Figure 3.2 show somewhat indirectly that they propagate at the speed of light. This is demonstrated more explicitly in the plot of Figure 3.3 where the position in space of the rightward-propagating *Qc* pulse peak is plotted as a function of time for the a 1,200-segment wire of variable radii (a) and the 599-segment wire of Figure 3.2(b). A propagation speed of 2.9885×10^8 m/sec is deduced by developing a best-fit straight line of Figure 3.3(a), a value that is 99.62% of the light speed of 3×10^8 specified in TWTD. Considering the approximations inherent in a discretized model, this slight difference seems reasonable. Part (b), where the arrow indicates the time at which the *Qc* pulse reaches the end of the 599-segment wire, is included to

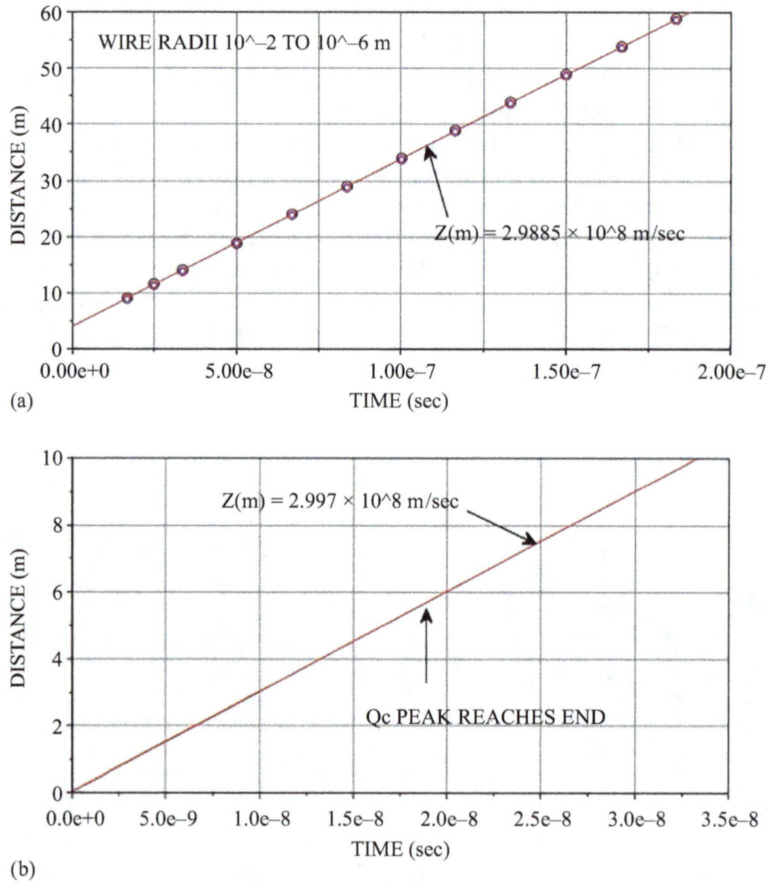

Figure 3.3 A distance-time plot for charge and current pulses on a 1,200-segment wires of variable radii (a) and a 599-segment straight wire (b) with both excited near there left ends at segment number 49

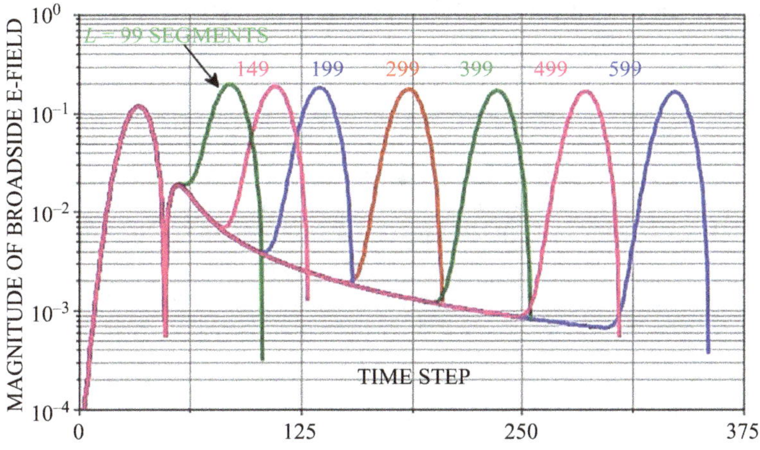

Figure 3.4 Magnitudes of the broadside radiated E-fields versus time step for impulsively excited, center-fed straight wires of various lengths up the time of the first end reflection

show that there is no discernable evidence of the effect of end reflection on the speed.

In Figure 3.4, are plotted the radiated, broadside E-fields for the time between turning on of a center, impulsive excitation and the first end reflection of the outgoing current and charge pulses for various wire lengths. The first E-field pulse occurs as the exciting source initially sets current and charge into motion. The second pulse is produced by reflection of the outward-propagating current and charge pulses from the ends of the dipole. Connecting these two radiation pulses is a continuous, decreasing radiation field due to the partial reflection of the outward-propagating charge pulses, denoted here as propagation radiation.

The reflection radiation from a partially reflected charge pulse propagating along the wire is qualitatively the same as the end-reflection radiation that occurs when the entire pulse is reflected from the open ends of the wire. In both cases, the total change in speed of the reflected charge is $2c$. However, observe that the propagation radiation that results is less than end-reflection radiation because less charge is reflected in the propagating phase. Also, while the total speed changes are the same in both cases, their associated accelerations are not necessarily the same. This is possible if the time over which their speed changes differ because their respective reflection processes differ. This is discussed further in connection with source radiation in Section 3.5 and end refection in Section 3.11 of this chapter. Since the source-accelerated charge undergoes a speed change of only c in contrast with the end-reflection change of $2c$, the end-reflected radiated peak field in Figure 3.4 is almost two times that of the source field.

3.4 Comparing current decay on a long wire and circular loop

Current decay over a greater distance and time duration is shown in Figure 3.5a. A sequence of current pulses at 100 time-step intervals is plotted for a straight wire 1,199 segments in length excited near its left end at segment 49 by a 1-V Gaussian voltage pulse. Also plotted in Figure 3.5a for comparison are equivalent results for a circular loop 1,199-segments in circumference. For clarity, both plots include only the positive current pulses propagating away from the feed-point to the right for the dipole and clockwise for the loop. The speed of the

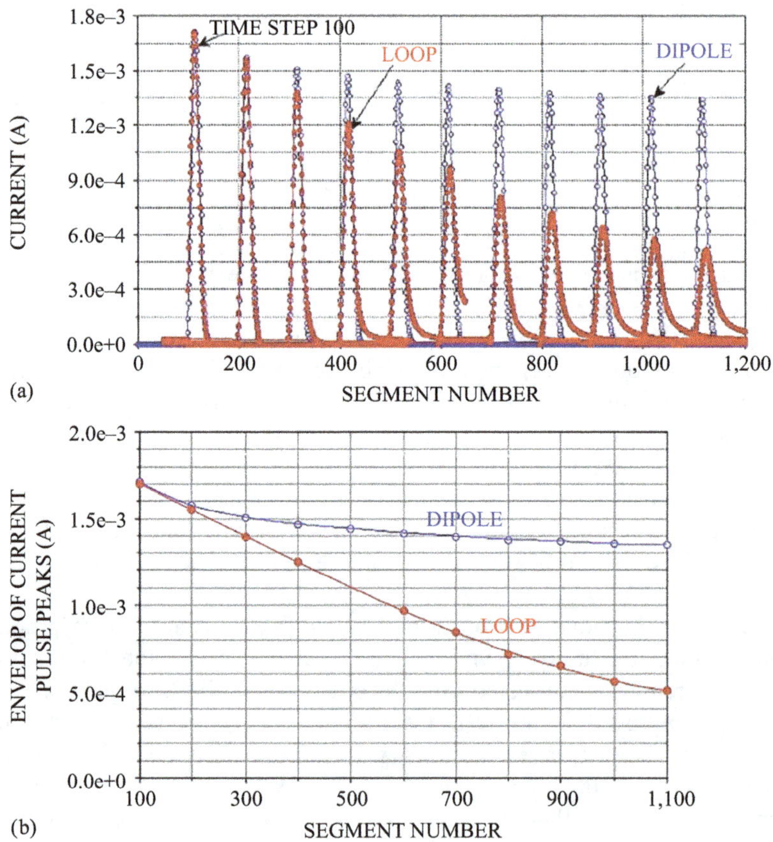

Figure 3.5 The rightward-propagating positive current pulses on a 1,199-segment straight wire (open circles) and clockwise on a circular loop (solid circles) at intervals of 100 time steps excited by a Gaussian voltage pulse (a) whose peaks are plotted as a function of distance from segment 100 (b)

loop-current peaks is ~3.0168×10^8, or somewhat faster than that on the straight wire, as can be seen in Figure 3.5a. This is apparently due to the fact that the straight-line distance between two points on an arc is less than that on a straight wire of the same length so that the interaction fields have a shorter distance to travel. This effect also accounts for the gradual spreading out of the leading edge of the loop current.

The normalized peaks of the current-pulse amplitudes obtained from the results of Figure 3.5a are presented in Figure 3.5b. Current-pulse decay on the circular loop exceeds that on the straight wire because the loop curvature produces increased charge reflection. Subsequent attention is limited hereafter to the straight wire because its broadside far field is more easily related to its propagating current and charge due to the more straightforward time-delay differences from various parts of the straight wire.

3.5 Source-region radiation

The broadside E-field magnitude caused by the feedpoint *Qc* pulse is plotted in Figure 3.6 where an AF_S of 29.4 is obtained, a result that obviously is independent of wire length as demonstrated in Figure 3.4. This independence of wire length occurs so long as the time duration of the excitation is less than the propagation time along the wire. On the other hand the exciting-source radiation pulse is dependent on wire radius as demonstrated in Figure 3.7. The magnitude of the peak radiation E-field pulse for various wire radii is plotted as a function of the charge-pulse peak produced at the exciting source. The average source–AF for the variable-radius case, AF_{SR}, is 29.2. This result confirms the linear relationship

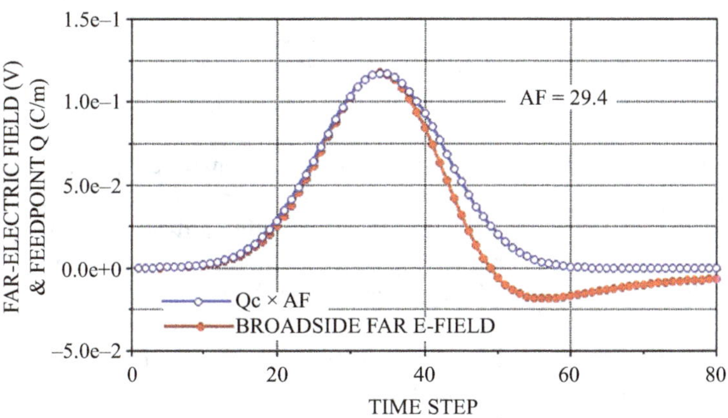

Figure 3.6 Time plots of the magnitudes of the broadside E-field and the source-induced charge of the 599-segment dipole as a function of time to exhibit their proportionality with the a source AF_S *of 29.4*

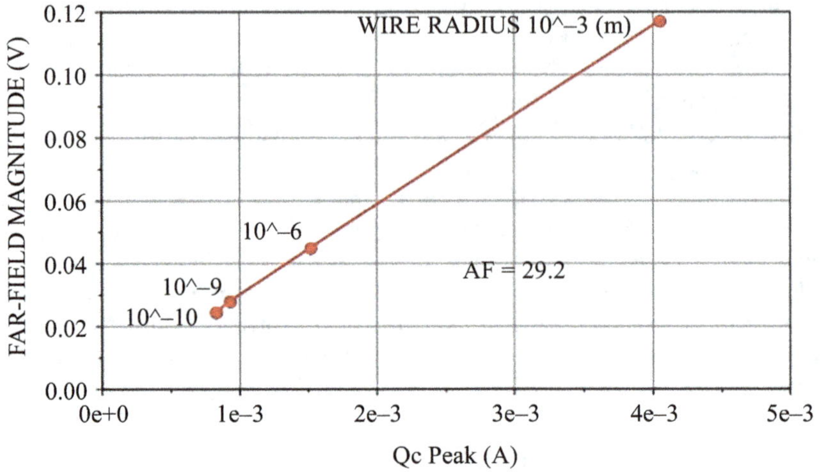

Figure 3.7 The magnitude of the peak broadside electric field due to the exciting-source charge acceleration as the dipole wire radius is varied over seven orders of magnitude to exhibit their proportionality with an average $\mathrm{AF}_{\mathrm{SR}}$ *of 29.2*

expected from the Lienard–Wiechert potentials that the radiated field is proportional to the charge.

3.6 Propagation radiation

In the time-domain results of Figures 3.2, 3.4, and 3.5 reflection of a propagating charge pulse is exhibited directly as a decay of the outward propagating current and charge pulse amplitudes with distance. This is also indicated indirectly as a negative trailing component of the outward-propagating positive current pulses in Figure 3.4.

The reflection of the current and charge demonstrates that the wave impedance presented by the wire does not match that of the propagating pulses, as discussed in [9]. The integrated charge in the rightward-propagating Qc pulse is plotted as a function of time in Figure 3.8a for time steps T and $T + 1$. The difference between the two plots of Figure 3.8a is shown as the amount of charge reflected per time step in Figure 3.8b. A comparison of the broadside propagation-radiation E-field with the reflected Qc charge of a 599-segment wire multiplied by an AF_P of 33.1 is plotted in Figure 3.9. Observe that this AF_P value is not twice the 29.4 for AF_S of Figure 3.6 as might be expected, given that the speed change for reflection is $2c$ compared with c for that caused by the source. This implies that their respective accelerations processes must be different, one possibility being that end-reflection is more abrupt than for propagating-wave reflection.

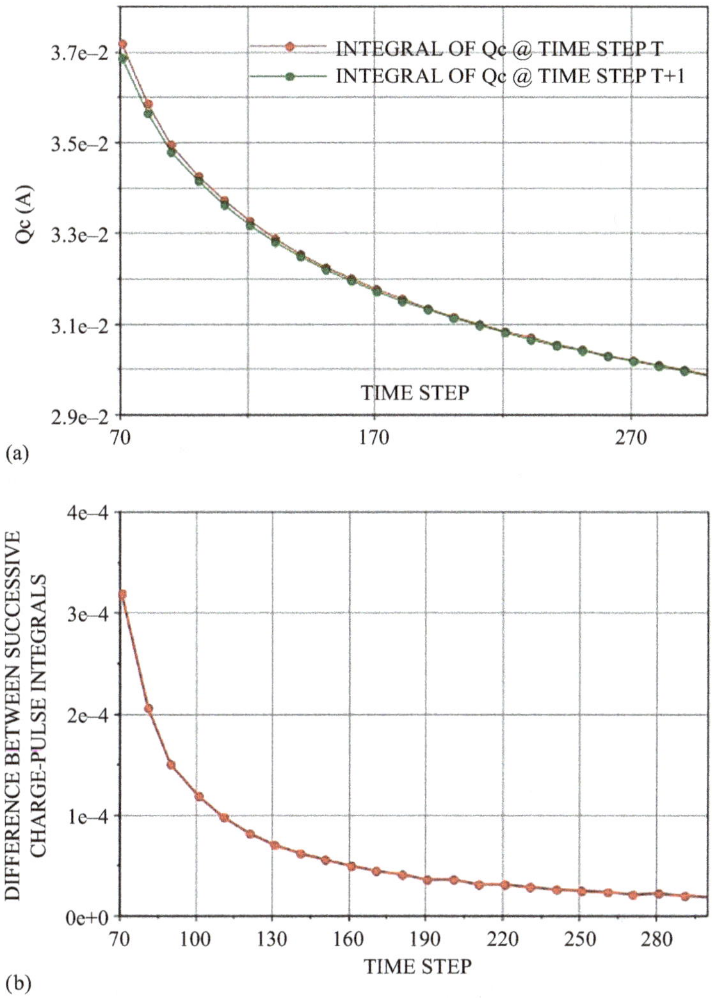

Figure 3.8 The positive space integrals of Qc *at successive time steps on the right-hand half of the 599-segment dipole (a) and (b) the difference between successive positive* Qc *pulses on the right-hand half of the dipole in (a) as a function of time*

3.7 Reflection radiation from the end of a wire

The reflection radiation from a wire end is plotted for a 599-segment wire as a function of time step in Figure 3.10 where the AF_E is 53.5. This is 1.83 times that of the average exciting-source radiation, AF_S of Figure 3.9. This ratio reflects the fact that the source-accelerated charge undergoes a speed change of c while the end-reflection charge experiences a speed change of $2c$. The fact that

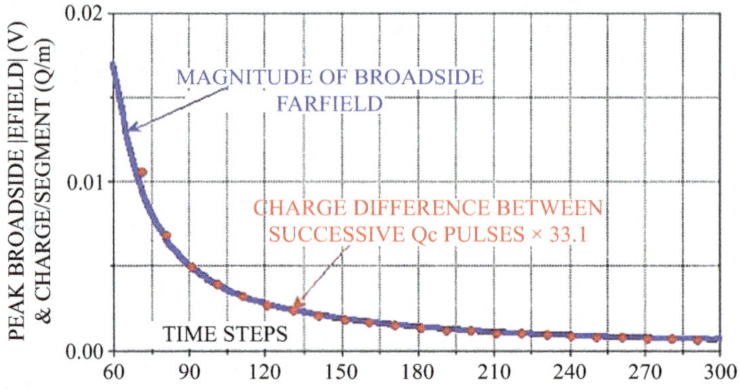

Figure 3.9 Time plots of the magnitude of the broadside E-field (continuous line) and the reflected charge (the solid dots) on a 599-segment dipole as a function of time to exhibit their proportionality with an $\mathrm{AF_P}$ *of 33.1*

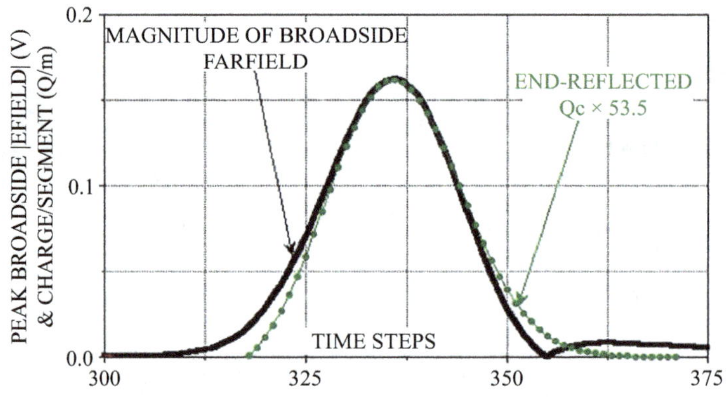

Figure 3.10 Time plots of the magnitude of the end-reflected E-field and the end-incident charge on a 599-segment dipole as a function of time to exhibit their proportionality with an $\mathrm{AF_E}$ *of 53.5*

this ratio is not 2 apparently demonstrates that end-reflection acceleration is not quite as abrupt as the acceleration due to the exciting source. This point is further discussed in Section 3.12 in connection with the energy carried by the current and charge. The end-reflection radiation-field peak was shown to decrease with increasing dipole length in Figure 3.6, due to the decreasing amount of charge that reaches its end due to charge reflection along the wire. This is demonstrated quantitatively in Figure 3.11. The results for the propagation, source and end-reflection radiation are illustrated together in the combined plots of Figure 3.12.

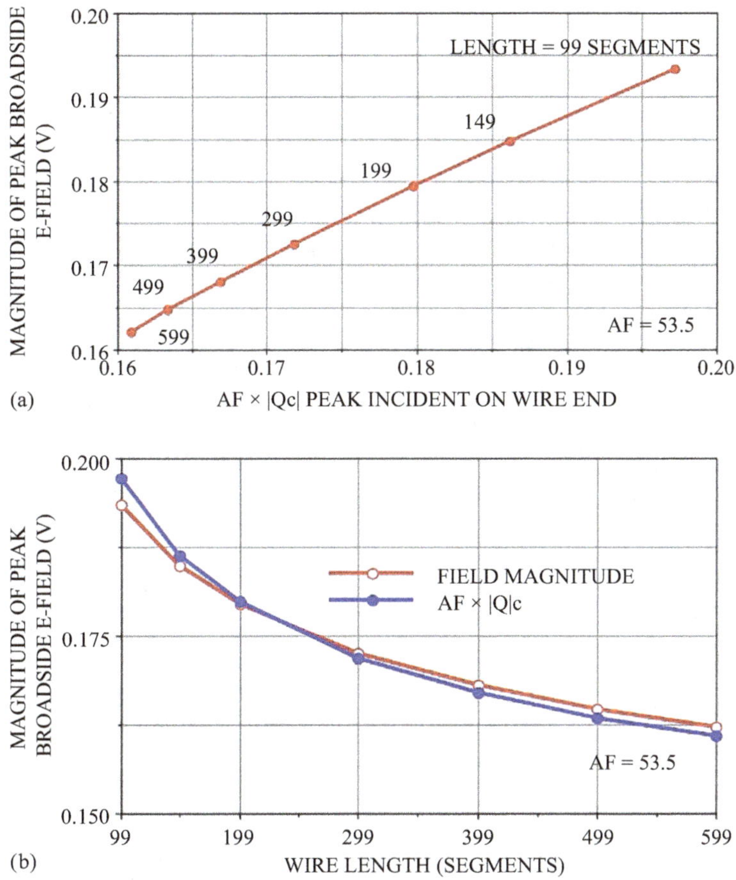

Figure 3.11 The decrease in the peak broadside electric field caused by a decreasing end-reflected Qc *pulse with the wire length a parameter (a) and a comparison of the broadside E-field and* $AF_E \times /Qc/$ *variation with wire length (b)*

3.8 Reflection radiation from resistive loads

The magnitude of the peak broadside radiated E-field caused by charge reflection from a lumped, resistive load is presented in Figure 3.13 for a wire 199 segments long. A Gaussian voltage pulse is applied at segment 75 with the load located at segment 100. As for the various similar plots above, the radiation field is proportional to the reflected charge with an average AF_R for this case of 59.4.

One particular interesting value of a resistance load is that for which half the charge in an incident Qc pulse is reflected from it and half is transmitted through it. On a transmission line, this would be considered a matched condition. A similar

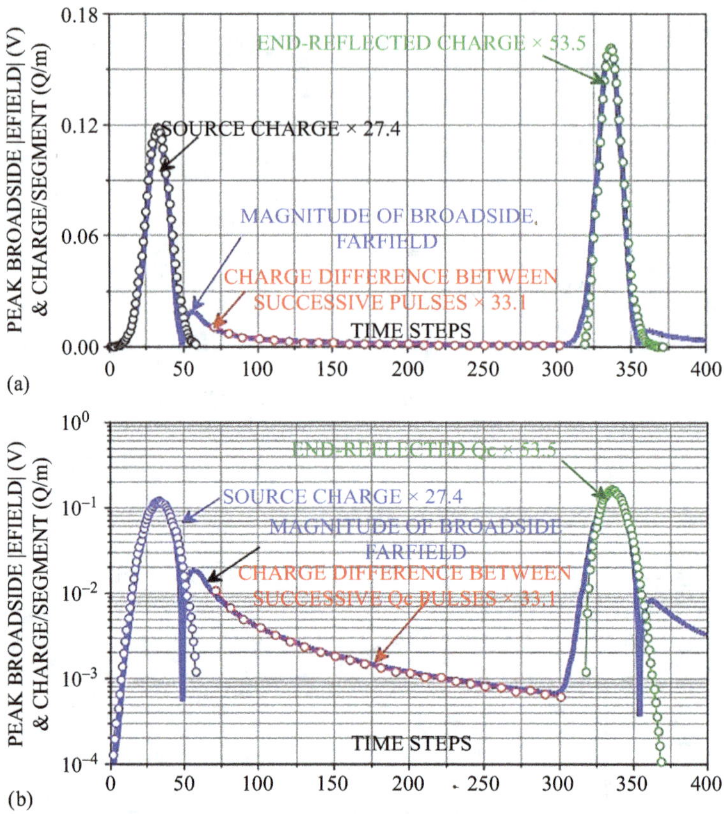

Figure 3.12 Combined plots of the propagation, source and end-reflected broadside E-field magnitudes (solid line) and the respective time varying accelerated Qc *charge pulses (open circles) that produce them on a linear plot (a) and log plot (b)*

situation can be implemented on a straight-wire antenna as demonstrated in Figures 3.14 and 3.15. A 199-segment dipole with a variable resistance load at its center is excited by an impulsive 1-V source at segment 75. The reflected and transmitted *Qc* pulses are plotted in Figure 3.14 as a function of the center-resistance value. A resistance load of 1,500-ohm can be seen to produce a matched condition, i.e., one where the reflected and transmitted *Qc* pulses are equal. When a second 1-V source is introduced at segment 125, its negative *Qc* pulse (see Figure 3.2) and the positive *Qc* pulse from the source at segment 75 are absorbed in the center resistance, as demonstrated in Figure 3.15.

The broadside far fields for the one- and two-source situations are plotted in Figure 3.16. The far-field for the single source is about ½ that of the two-source case, and is due to a 2*c* speed change of a single reflected pulse. The two-source far field is twice that of the one-source case to represent a factor of 4 increase in the

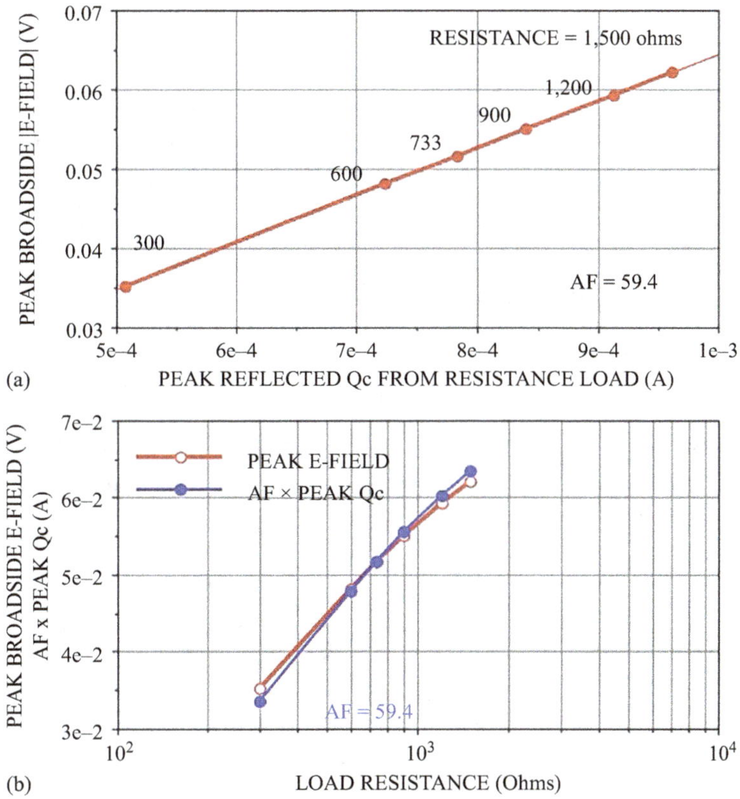

Figure 3.13 The peak broadside E-field as a function of resistance load (a) and a comparison of the broadside E-field and $AF_R \times /Qc/$ *variation with resistance load (b)*

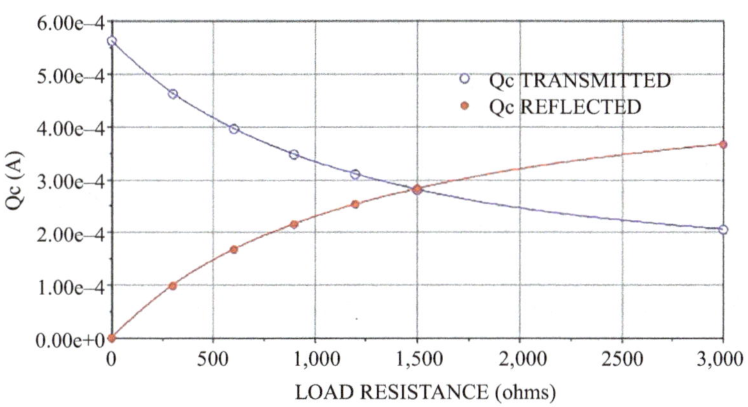

Figure 3.14 The reflected and transmitted Qc *pulse amplitudes as a function of load resistance for a straight wire of radius* 10^{-3} *m*

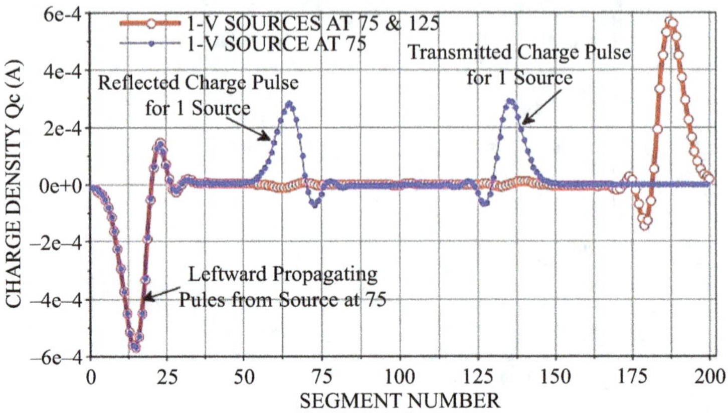

Figure 3.15 A 199-segment straight wire impulsively excited a segment 75 and at segments 75 and 125 with a matched load of 1,500 ohms at segment 100 at tune step 75

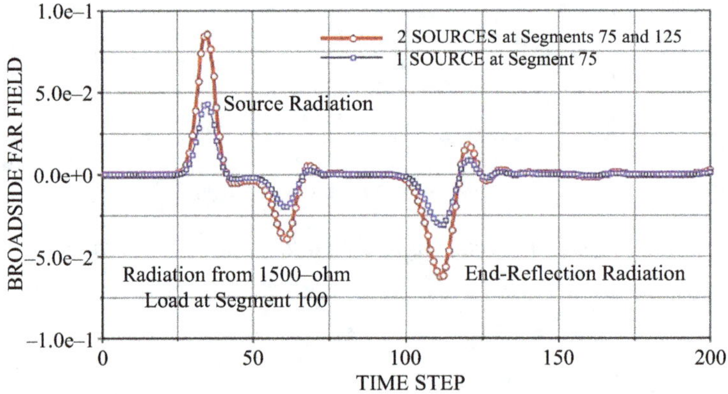

Figure 3.16 The broadside far field due to a matched load on a straight wire

radiated power, even though the respective *Qc* pulses undergo only a 1*c* speed change. This occurs because the two pulses are absorbed at the same location, the matched load, and produce spatially coherent far fields.

3.9 Radiation from directional reflection at a 90-deg bend

Reflection radiation caused by a 90-deg bend is plotted with the wire radius a parameter in Figure 3.17 for a 199-segment wire excited impulsively at segment 75 with a bend at segment 100. The acceleration factor, AF_B, is found to be 59.4. Only

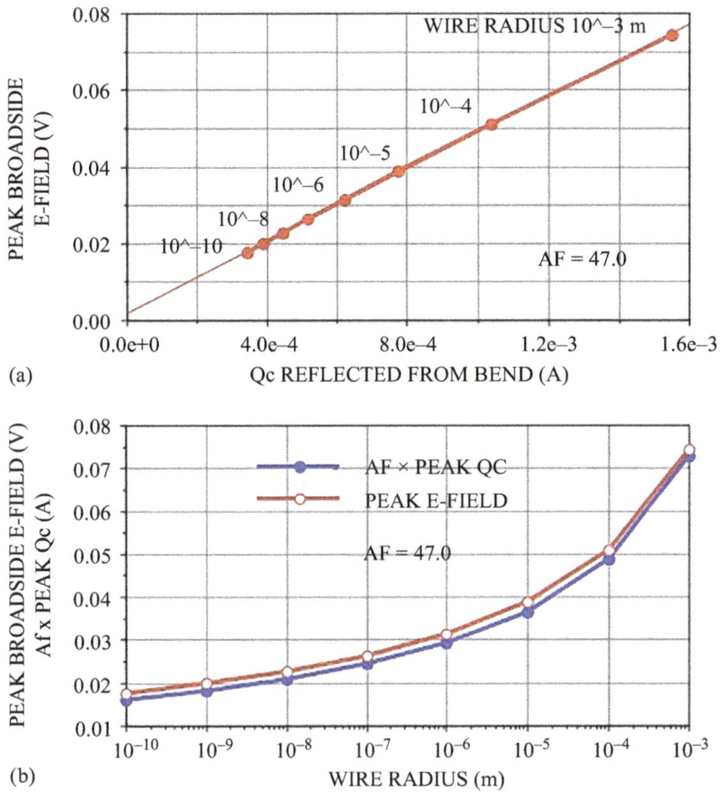

Figure 3.17 The peak magnitudes of the broadside E-field and the reflected Qc *pulse from a right-angle bend for an impulsively excited wire 199 segments long with wire radius a parameter*

a 90-deg bend is examined here as this results in radiation from the bent portion of the wire being orthogonal to that from the other wire half. If the bend were different from 90-deg, identifying what part of the radiated E-field comes from the two halves of the wire would be less clear. The plot of Figure 3.17 continues the relationship established in the previous results that show an essentially linear proportionality between the peaks of the reflected *Qc* pulses and broadside radiated E-field pulses.

3.10 Reflection radiation from a step in the wire radius

The remaining example for a radiation feature from the generic object of Figure 3.1 is a step change in wire radius. But in contrast with previous results where the reflected *Qc* pulses are clearly defined, this was not found to be the case for the

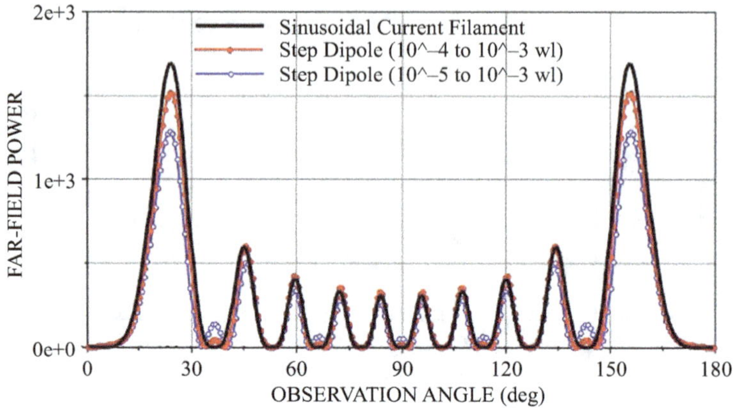

Figure 3.18 The radiation patterns for a 10-wavelength sinusoidal current filament and two NEC dipoles having radius steps at ¼ and ¾ of their lengths with each having a 1-A maximum current

step-radius wire. Instead, the step effect is more clearly demonstrated in a frequency-domain radiation pattern such as shown in Figure 3.18. The radiation pattern of a sinusoidal current filament (SCF) is plotted for comparison together with the patterns of two center-fed, 200-segment dipoles modeled using NEC (Numerical Electromagnetics Code) (Appendix B). Each dipole is 10 wavelengths long with a center section of 100 segments and a radius of 10^{-3} wavelengths. The end sections are 50 segments long and have radii of 10^{-4} and 10^{-5} wavelengths respectively. The current magnitudes for each are 1-A.

The SCF and the step-dipole patterns in Figure 3.18 are similar with respect to their main lobes. The effect of the radius steps is to cause 5 smaller lobes to appear at two-lobe intervals between the larger lobes beginning at broadside, an effect that increases with an increase in step radius. These results provide indirect evidence for the radiation caused by a step in the wire radius.

3.11 An energy measure that demonstrates radiation loss

One way to evaluate the energy radiated by an impulsively excited object is to integrate the far fields over an enclosing sphere as a function of time, something that can be computationally expensive. An alternative introduced here is to instead integrate the boundary sources over the object at each time step of the numerical solution [25], a topic that is discussed in more detail in the next chapter.

This basic idea involves spatial integrals of the square of the time-dependent current and charge

$$W_I(t) = \int_0^L I^2(x.t)dx \tag{3.5a}$$

and

$$W_{Qc}(t) = c^2 \int_{-\frac{L}{2}+\Delta}^{\frac{L}{2}-\Delta} Q^2(x,t)dx + \frac{c^2}{3}\left[\int_{-\frac{L}{2}}^{-\frac{L}{2}+\Delta} Q^2(x,t)dx + \int_{\frac{L}{2}-\Delta}^{\frac{L}{2}} Q^2(x,t)dx\right] \tag{3.5b}$$

where $W_I(t)$ and $W_{Qc}(t)$ are the current (or magnetic) and charge (or electric) energy measures at time t and Δ is the segment length in the numerical model. The two terms in the bracketed portion of (3.5b) account for the fact that the propagation speed of the charge pulse going to zero at the ends of the wire cannot happen instantaneously. This fact is approximated for by assuming that the squared speed falls linearly to zero over the last segment of the numerical model. The $c^2/3$ is an average that comes from

$$c_{ave}^2 = \frac{1}{\Delta}\int_0^{\Delta}\left[\frac{cx}{\Delta}\right]^2 dx = \frac{c^2x^3}{3\Delta^3}\bigg|_0^{\Delta} = \frac{c^2}{3} \tag{3.6}$$

Observe that if the 1/3 is not used in (3.5) the total energy no longer decreases monotonically, but rises at end reflection, a non-physical result. This result (3.6) also offers some indication about how the acceleration term in the Lienard–Wiechert potentials might be influenced by the particular details of the radiation features discussed in connection with Figure 3.1.

Results for these two quantities and their sum are presented for a center-fed dipole 199 segments long in Figure 3.19. After the startup transient, $W_I(t)$ and $W_{Qc}(t)$ are equal until the I and Qc pulses reach the end of the wire where $W_{Qc}(t)$ becomes predominant as the charge accumulates there and the current goes to zero. Upon reflection, their equality is reestablished until the oppositely propagating pulses meet at the wire center where $W_I(t)$ becomes predominant since the current is additive while the charge pulses cancel. After each end reflection, the total energy exhibits a marked decline due to the

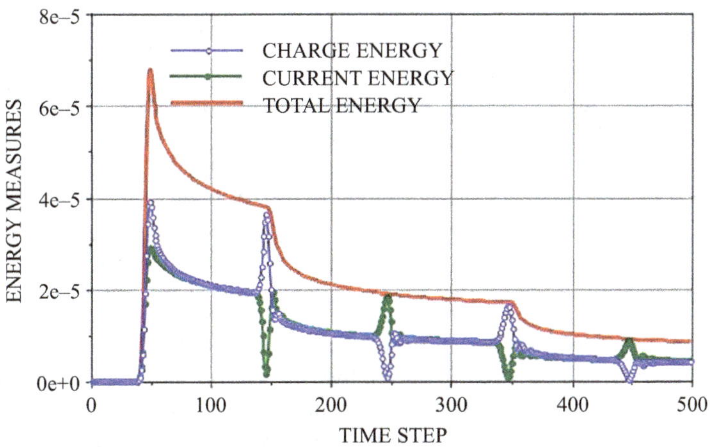

Figure 3.19 The current, charge, and total energy measures as a function of time for an impulsively excited straight wire 199 segments long

Table 3.1 Acceleration factors developed from the above results

Radiation type	AF	Normalized AF
Source: AF_S	29.4	1.0
Source: variable wire radius: AF_{SR}	29.2	0.99
Right-angle bend: AF_B	47.0	1,60
Propagation: AF_P	33.1	1.13
Resistance load: AF_{Rs}	59.4	2.02
End reflection: AF_E	53.5	1.83

end-reflection radiation. This behavior continues periodically until the wire returns to charge neutrality. As a result, the total propagating energy decreases monotonically due to the effects of end-reflection and propagation radiation.

3.12 Summary of the derived AFs

The various AFs developed from the above results are summarized for reference in Table 3.1.

3.13 Summary

This chapter has summarized a number of issues concerning the relationship between charge acceleration and the radiation that results in various situations. The question arises about where charge acceleration occurs on the most basic of antennas, a center-fed, perfectly conducting dipole, for which there are two obvious causes and one more subtle but important factor:

1. The feedpoint exciting voltage that sets into motion an outward-propagating charge and current wave at light speed c in the medium.
2. A second is the dipole ends where the outgoing Q/I wave is totally reflected producing a change in charge speed of $2c$.
3. A third is the deceasing amplitude of the propagating wave with distance due to its partial reflection along the wire, a situation where the charge also undergoes a speed change of $2c$. This is the reason why the decay of current flowing along a straight wire antenna has been attributed as being due to radiation.

Radiation caused by these and other kinds of charge acceleration due to resistive loads, right-angle bends, and radius steps are compared. These phenomena are examined primarily in the time-domain where they are more observably separable in time and space than in the frequency domain. The current and charge induced on an impulsively excited wire antenna and its broadside radiated E-field are computed using a time-domain, integral-equation model. The computed results are used to derive a numerical relationship between the amount of accelerated charge and its radiated field. This relationship, denoted as an *AF*, is obtained for various charge-accelerating features of a generic wire object and normalized to that of the exciting source for comparison with their respect speed changes.

The computer model TWTD [1] was used to develop an estimate of the quantitative relationship between an accelerated charge pulse on a PEC wire and the radiation that results. The estimates developed for what is denoted here as an *AF* cannot be considered numerically precise. This is because there is no clear-cut way for choosing at what time a reflected *Qc* pulse should be "measured" for comparison with the time variation of its radiated E-field. This is true whether the pulse shapes themselves are used or when it is only their peak amplitudes are needed to estimate an *AF*. Thus, the *AF* values given throughout the text and summarized in Table I, being derived from a computational "experiment," must be considered as approximations. Also an issue in this respect is the fact that the actual acceleration depends on the specific nature of the physical feature that determines it. This is demonstrated, for example, in the values of AF_E and AF_S that might be expected to be in a ratio of 2:1 given their respective speed-change ratio of $2c/c$. The numerical value actually obtained is somewhat less at 1.83. In spite of the limitations involved, the numerical results obtained do provide a useful insight into the relative effects of various radiating features on a generic wire object and do demonstrate a consistent proportionality between the radiated E-field and the charge that produces it.

References

[1] J. A. Landt, E. K. Miller, and M. Van Blaricum, *WT-MBA/LLL1B (TWTD): A Computer Program for the Time-Domain Electromagnetic Response of Thin-Wire Structures*, Lawrence Livermore Laboratory Report No. UCRL-51585, 1974.

[2] R. Gómez Martin, A. Rubio Bretones, and S. Gonzalez Garzia, Some thoughts about transient radiation by straight thin wires, *IEEE AP-S Mag.*, 41(3), pp. 24–33, 1999.

[3] C. C. Bantin, Radiation from a pulse-excited thin wire monopole, *IEEE AP-S Mag.*, 43, pp. 64–69, 2001.

[4] S. A. Schelkunoff, *Advanced Antenna Theory*, John Wiley & Sons, New York, NY, p. 190, 1952.

[5] C. Manneback, Radiation from transmission lines, *AIEE J.*, 42, pp. 95–105, 1923.

[6] J. D. Jackson, *Classical Electrodynamics*, John Wiley & Sons, New York, NY, p. 401, 1962.

[7] Y. S. Yeh and K . K. Mei, Theory of conical equiangular-spiral antennas. Part II—Current distributions and input impedances, *IEEE Trans. AP-S*, 16(1), pp. 14–21, 1968.

[8] L-C. Shen, T. T. Wu, and R. W. P. King, A simple formula of current in dipole antennas, *IEEE Trans. AP-S*, 16(5), pp. 542–547, 1968.

[9] J. Bach Anderson, Admittance of infinite and finite cylindrical metallic antenna, *Radio Sci.*, 3 (6), pp. 607–621, 1968.

[10] D. S. Jones, *Methods in Electromagnetic Wave Propagation*, Clarendon Press, Oxford, p. 295, 1994.

[11] A. V. Kudrin, E. Y. Petrov, G. A. Kyriacou, and T. M. Zaboronkova, Insulated cylindrical antenna in a cold magnetoplasma, *Prog. Electromag. Res.*, 53, pp. 135–166, 2005.

[12] J. D. Jackson, How an antenna launches its input power into radiation: the pattern of the Poynting vector at and near an antenna, *Am. J. Phys.* 74, pp. 280–288, 2006.

[13] E. K. Miller, The differentiated on-surface Poynting vector as a measure of radiation loss from wires, *IEEE Antennas Propag. Soc. Mag.*, 48(6), pp. 21–32, 2006.

[14] B. Kordi, R. Moini, and V. A. Rakov, Comment on 'Return stroke transmission line model for stroke speed near and equal that of light' by R. Thottappillil, J. Schoene, and M. A. Uman, *Geophys. Res. Lett.*, 29(10), p. 1369, 2002.

[15] B. Kordi, R. Moini, and V. A. Rakov, Comparison of lightning return stroke electric fields predicted by the transmission line and antenna theory models, in *Proc. 15th Int. Zurich Symp. Electromagn. Compat.*, 102P2, Zurich, 2003, pp. 551–556.

[16] Y. Baba and M. Ishii, Characteristics of electromagnetic return-stroke models, *IEEE Trans. Electromagn. Compat.*, 45(1), pp. 129–135, 2003.

[17] Y. Baba and V. A. Rakov, On the mechanism of attenuation of current waves propagating along a vertical perfectly conducting wire above ground: application to lightning, *IEEE Trans. Electromag. Compat.*, 47(3), 2005.

[18] G. S. Smith, *An Introduction to Classical Electromagnetic Radiation*, Cambridge University Press, Cambridge, 1997.

[19] E. K. Miller and G. J. Burke, A multi-perspective examination of the physics of electromagnetic radiation, *Appl. Comput. Electromag. Soc. J.*, 16(3), pp. 190–201, 2001.

[20] E. K. Miller and J. A. Landt, Direct time-domain techniques for transient radiation and scattering from wires, *Proc. IEEE*, 68, pp. 1396–1423, 1980.

Chapter 4
Time-domain electromagnetic-field energy measures

Abstract

Among the options offered as outputs of the thin-wire time-domain (TWTD) computer model are the current and charge distributions as a function of time over the wire object for which a numerical solution has been obtained. These quantities are useful in various ways, ranging from providing the frequency dependence of an antenna's input admittance to providing the near and far fields that the object produces as a function of time or frequency. Another benefit provided by the far fields is that they can reveal from where on the object that they originate, using Time-Domain Far-field Analysis of Radiation Sources (TDFARS) as will be discussed in Chapter 7. One disadvantage of FARS in either the time or frequency domain, however, is that it requires integrating the radiated fields over the far-field sphere, an exercise that can be computationally expensive for a large or complex object.

A simpler alternative to TDFARS is presented in this chapter, time-domain energy measures (TDEM). These involve using only the solved-for current and charge space-time variations, a concept that was introduced in [1] and expanded upon further in [2,3]. The total time-dependent energy decays due to radiation loss over time, and its differentiation with respect to time yields the rate of energy loss. Furthermore, where the loss occurs is discernible from where the current and charge pulses are located at a particular time. The following discussion first summarizes the basic idea and then demonstrates this approach with a variety of applications. The advantage of TDEM is its computational efficiency compared with TDFARS, but it does have some limitations that are also addressed.

Keywords: Charge distribution; Current distribution; Electromagnetic energy measures; Energy-loss rate; Energy measures; Impulsive excitation; Radiation-loss rate; Time-domain electromagnetics; Thin-wire time domain; Time-domain energy measures; TWTD

4.1 Introduction

The previous two chapters introduced the fundamental fact that electromagnetic radiation is caused by charge acceleration. This was illustrated in Chapter 2 using the E-field kink model to show how acceleration of free charge produces radiating electric fields that vary with distance as $1/r$. This was further demonstrated in Chapter 3 using the time-domain computer model, TWTD, for an impulsively excited wire antenna to examine the quantitative correlation between charge acceleration and the radiated field that it produces and to introduce charge and current energy measures. In this chapter, the relationship between the decay of current and charge energies on an impulsively excited wire that is caused by radiation is discussed. One way to evaluate the energy radiated by an impulsively excited object is to integrate the far fields over an enclosing sphere as a function of time, something that can be computationally expensive.

An alternative demonstrated here is to instead integrate the boundary sources over the object at each time step of the numerical solution [1]. This is done by developing expressions for the energies stored in the current and charge distributions as functions of time and evaluating their numerical results for a variety of situations, extending the initial discussion in Chapter 3.

4.2 Stored-energy measures

The derivation of quantitative measures for the current and charge energies follows readily from the development presented by Miller and Landt [1] beginning with the Maxwell curl equations

$$\nabla \times \mathbf{E} = -\mu_o \frac{\partial}{\partial t}\mathbf{H} \tag{4.1}$$

$$\nabla \times \mathbf{H} = \varepsilon_o \frac{\partial}{\partial t}\mathbf{E}. \tag{4.2}$$

From the **H** dot product of (4.1) and the **E** dot product of (4.2), their subsequent difference yields

$$\nabla \cdot \mathbf{E} \times \mathbf{H} = -\mu_o \mathbf{H} \cdot \frac{\partial}{\partial t}\mathbf{H} + \varepsilon_o \mathbf{E} \cdot \frac{\partial}{\partial t}\mathbf{E} \tag{4.3}$$

on using the identity

$$\mathbf{A} \cdot \nabla \times \mathbf{B} - \mathbf{B} \cdot \nabla \times \mathbf{A} = \nabla \cdot \mathbf{A} \times \mathbf{B}.$$

Integrating (4.3) over a volume V that contains a conducting body of interest, there is obtained

$$\int_V \nabla \cdot \mathbf{E} \times \mathbf{H} dv' = -\frac{1}{2}\int_V \frac{\partial}{\partial t}[\mu_o \mathbf{H} \cdot \mathbf{H} + \varepsilon_o \mathbf{E} \cdot \mathbf{E}]. \tag{4.4}$$

Using the Gauss divergence theorem and integrating with respect to time, we obtain

$$\int_{-\infty}^{t} dt' \oint_A \mathbf{E} \times \mathbf{H} \cdot d\mathbf{a}' = -\frac{1}{2}\int_{\infty}^{t} dt' \int_V \left[\mu_o \frac{\partial H^2}{\partial t^2} + \frac{\partial E^2}{\partial t^2}\right] dv'$$

$$= -\frac{1}{2}\int_V \mu_o [H^2 + \varepsilon_o E^2] dv' \tag{4.5}$$

Now let the volume V be made congruent with the volume (V_c) of the conducting body, taken here to be a wire and extending a distance d beyond its surface. Then on the right side of (4.5), with $\mathbf{n}$ and $\mathbf{t}$ unit normal and tangent vectors to the surface, $\mathbf{H}$ and $\mathbf{E}$ can be approximated by

$$\mathbf{H} \approx \mathbf{I} \times \mathbf{n}/2\pi r = \mathbf{t} \times \mathbf{n}I/2\pi r \tag{4.6a}$$

$$\mathbf{E} \approx \mathbf{n}Q/2\pi r \varepsilon_o \tag{4.6b}$$

since $\boldsymbol{n}\cdot\boldsymbol{H} = 0$ and $\boldsymbol{t}\cdot\boldsymbol{E} = 0$ in the space between V and V_c. The right side of (4.5) then takes the form

$$RHS \approx -\frac{1}{2}\int_C \int_0^{d+a} \int_0^{2\pi} \left[\mu_o \frac{I^2}{(2\pi r)^2} + \varepsilon_o \frac{Q^2}{(2\pi r \varepsilon_o)^2}\right] dw' dr' r' d\phi' \tag{4.7}$$

with C being the wire contour and contributions to the volume integral of its ends (if the wire is open ended) are neglected. On forming the r' and ϕ' integrals of (4.7) it follows that

$$RHS \approx -\frac{1}{4\pi} \log\left(\frac{d+a}{a}\right) \int_C (\mu_o I^2 + Q^2/\varepsilon_o) dw'. \tag{4.8}$$

A measure of the total energy due to $I(w,t)$ and $Q(w,t)$ at space–time at time t stored in the fields near the wire, is thus proportional to

$$W_T = W(t)|_{Wire} \propto \frac{1}{4\pi}\int_{C(r)} \left[\mu_o I^2(w',t) + \frac{Q^2(w',t)}{\varepsilon_o}\right] dw'$$

$$= 10^{-7}\int_{C(r)} [I^2(w',t) + Q^2(w',t')c^2] = 10^{-7}[W_I + W_Q]. \tag{4.9}$$

The current and charge energies are denoted by W_I and W_Q in the following, with their sum W_T being the total energy due to $I(w,t)$ and $Q(w,t)$, the current and charge space-time distributions. Observe that these quantities are proportional and not actual values. Never the less they are useful in showing how the global energy stored in the time-varying current and charge on an object decrease with time due to varying energy-loss processes.

The derivation is completed by similarly simplifying the left side of (4.5) to obtain

$$LHS \approx \int_{-\infty}^{t} dt' \int_C \boldsymbol{E}\cdot\boldsymbol{I} dv' \tag{4.10}$$

Because

$$\boldsymbol{E}\cdot\boldsymbol{I} = E_{\tan}I = (E_{exc} + E_{scat})_{\tan}I = 0$$

on a perfect conductor, the expression (4.11) for collected energy (4.11) immediately follows:

$$W_C = W(t)|_{Collected} = \int_{-\infty}^{t}\int_{C(r)} \boldsymbol{E}(w',t')\cdot\boldsymbol{t}I(w',t')dw'dt' \tag{4.11}$$

If the conductor is lossy, then

$$E_{\tan}I = E_{loss}I = (IR)I$$

from which (4.12) also follows:

$$W_D = W(t)_{Dissipated} = \int_{-\infty}^{t}\int_{C(r)} I^2(w',t')R(w',t')dw'dt' \tag{4.12}$$

These measures, as discussed earlier by Miller and Landt (1980)* and introduced previously in Chapter 3, are given by the quantities

$$W_I(t) = \iint_{C(\vec{r})} I^2(s,t)ds \tag{4.13}$$

and

$$W_{Qc}(t) = \int_{C(\vec{r})} [Q(s,t)c]^2 ds, \tag{4.14}$$

where c is the speed of light in the medium in which the object is located and the integrals are over the object's surface.

4.3 Implementation of time-domain energy measures

One distinct advantage of using a time-domain (TD) model is the possibility that it offers for obtaining a better understanding of radiation phenomena, as illustrated elsewhere in this book (Chapters 3, 5, 7, and 9). Discussed here is how TD results can provide measures of the energy stored in an object's near fields as represented by the time-dependent electric current and charge flowing on it and how such

*Miller and Landt computed W_I and $W_Q' = W_{Qc}/c^2$, which results in the average value of $W_Q' \approx W_I/c^2$. The c^2 proportionality factor not only puts Qc and I in the same units but then yields $W_I \approx W_Q$ which seems reasonable if the energy is approximately shared between the electric field of moving charge and the magnetic field due to the charge motion. Bevensee [6] derived a result containing I^2 and $(Qc)^2$ as an energy measure, but using their difference rather than their sum. Consideration of E and H energy integrals would also involve the summed result.

measures may yield insight into where and how much the object radiates as a function of time. Of particular interest is the sum

$$W_T(t) = W_I(t) + W_{Qc}(t). \tag{4.15}$$

of the spatially integrated energy densities. While W_I and W_{Qc} individually can be useful, it is their sum that is needed if the rate of total energy loss, and therefore, of radiation, is to be determined. One approach, but a computationally expensive one to obtain the energy radiated by an impulsively excited object, is to integrate the far fields over an enclosing sphere as a function of time. The alternative of integrating the boundary sources over the object at each time step of the numerical solution provides the equivalent information but with less computation.

It is assumed here that the total energy stored in the near fields of an object is proportional to W_T, with W_I representing the magnetic-field energy and W_{Qc} the electric-field energy. Of course, there may be some uncertainty about whether the energy stored in the near fields can be determined from the source distribution (i.e., of current and charge) itself, rather than from a volumetric integral of E^2 and H^2. But it should be obvious that the near fields account for the storage of energy/power near the object, with any decrease of their overall value occurring because of radiation or other losses. Furthermore, were there no boundary sources, i.e. current and charge, there would be no near field. Thus it appears reasonable, since the sources terminate the fields through the appropriate boundary conditions, to proceed as though the stored, or near-field, energy is proportional to the quantities given in (4.9). Note that if $W_T(t)$ does provide a measure of the total energy stored in an object's near fields, then after some initial excitation has decayed back to zero and is no longer providing energy to the object, then $W_T(t)$ should exhibit a monotonic-decaying behavior. As illustrated below, this turns out not always to be the case, the implications of which discussed shortly.

Results obtained using TWTD are shown for a representative problem, a dipole antenna center-excited by a Gaussian voltage pulse. The antenna numerical model, as are other wire objects modeled using TWTD unless otherwise indicated, has a wire radius of 10^{-3} m, and space-segment lengths, dx (or Δ), of 0.05 m, The associated time step, dt, is given by dx/c, with c the speed of light, or 3×10^8 m/sec. The current and charge pulses that result are shown at several time steps in Figure 4.1 with the energy measures for this problem, W_T, W_I, and W_{Qc} plotted in Figure 4.2.

However, contrary to the physical requirement that W_T should decay monotonically if it actually represents the energy stored by the dipole, the result in Figure 4.3 does not do so after the driving voltage goes to zero. Instead it increases during each end reflection of the propagating charge pulses. As expected, during the time intervals an end reflection occurs, there is a near zero in W_I, since the current goes to zero at an end of the wire resulting in W_T being dominated by W_{Qc}. Recall that W_{Qc} involves a proportionality constant, c^2, which results in equality between W_T and W_{Qc} away from the wire ends and away from the source location where the pulses meet, as can be seen in Figure 4.2, i.e., the propagating current and

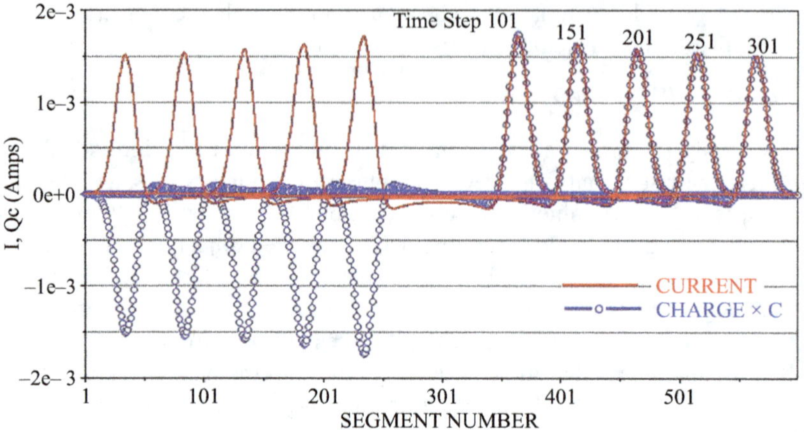

Figure 4.1 Current and charge pulses at several time steps on a 599-segment, center-excited dipole

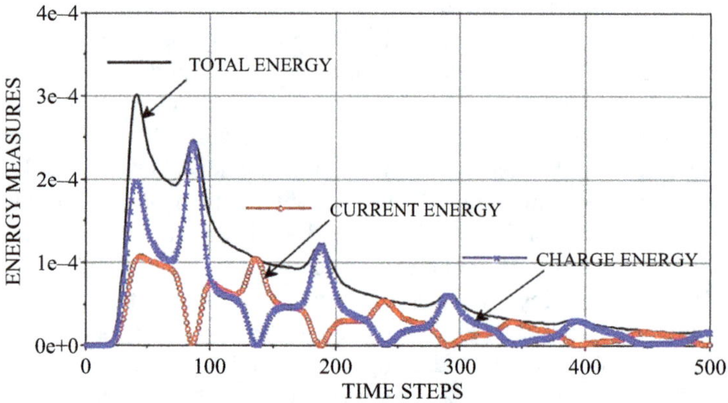

Figure 4.2 The current, charge, and total energy measures on an impulsively excited 99-segment dipole antenna

charge pulses have equal power contents. This equality cannot be expected at end reflection where the charge accumulates and reverses its direction of motion. It appears that a smaller value than c must apply to the computation of W_{Qc} in the vicinity of the end to avoid a nonphysical increase of the total energy.

Referring to the discussion of Chapter 3, current energy measure is simply obtained as

$$W_I(t) = \int_0^L I^2(x,t)dx \tag{4.16a}$$

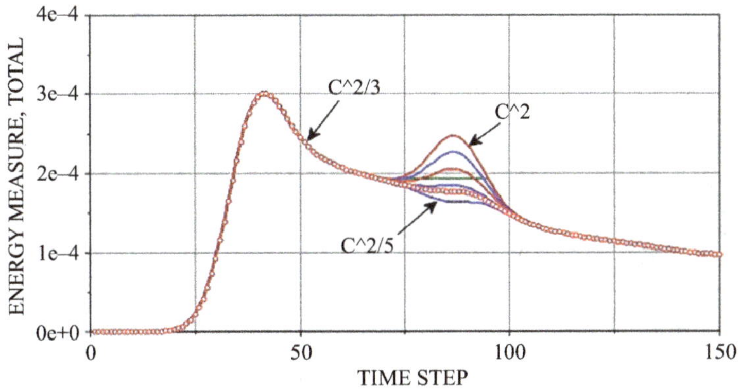

Figure 4.3 An expanded view of the first end-reflection, total-energy results for the end-segment c^2 *factor varying from 1, 0.8, and 0.6, monotonic, 0.4, 1/3, and 0.2 top to bottom*

while the charge energy measure at open ends of a wire object is modified as shown in (4.16b)

$$W_{Qc}(t) = c^2 \int_{-\frac{L}{2}+\Delta}^{\frac{L}{2}-\Delta} Q^2(x,t)dx + \frac{c^2}{3}\left[\int_{-\frac{L}{2}}^{-\frac{L}{2}+\Delta} Q^2(x,t)dx + \int_{\frac{L}{2}-\Delta}^{\frac{L}{2}} Q^2(x,t)dx\right]. \tag{4.16b}$$

The two terms in the bracketed portion of (4.16b) account for the fact that the propagation speed of the charge pulse at the end of the wire cannot go to zero instantaneously. This fact is approximated for by assuming that the squared speed falls linearly to zero over the last segment of the numerical model to have an average value given by

$$c_{ave}^2 = \frac{1}{\Delta}\int_0^{\Delta}\left[\frac{cx}{\Delta}\right]^2 dx = \frac{c^2x^3}{3\Delta^3}\Big|_0^{\Delta} = \frac{c^2}{3}. \tag{4.17}$$

Observe that if the 1/3 is not used in (4.16b) the total energy rises at end reflection, a non-physical result. This result (4.17) also offers some indication about how the acceleration term in the Lienard–Wiechert potentials might be influenced by the particular details of the radiation features discussed in connection with Figure 3.1.

A numerical experiment was conducted to determine the effect on the total charge energy of varying the c^2 factor for the end-segment, charge-energy computation from 1 to 0.2 with the result shown in Figure 4.3. The total energy curve using the 1/3 factor appears to provide the most physically realistic of the various factors used in Figure 4.3 to support (4.17). The monotonic curve of Figures 4.3 and 4.4 comes from limiting the total-energy result to being non-increasing, which avoids the non-physical increase in total energy, but yields a less satisfactory result

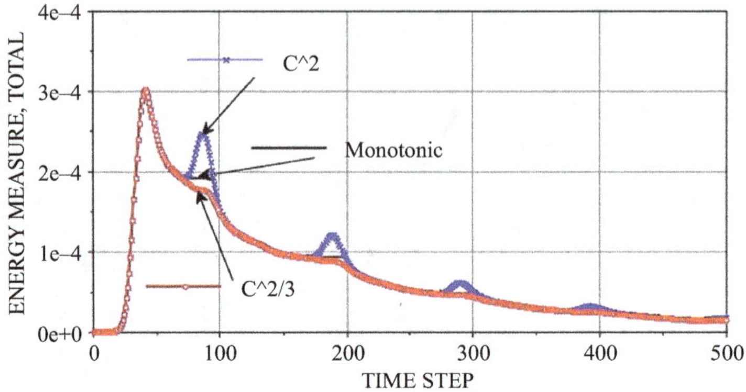

Figure 4.4 The total energy measure for the 99-segment dipole with a variable end-segment treatment

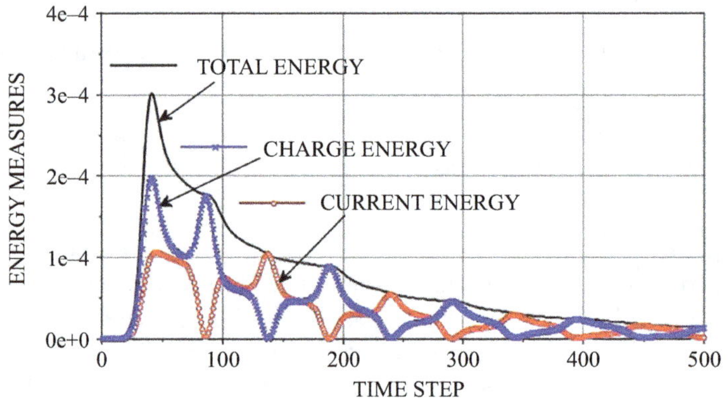

Figure 4.5 A corrected version of Figure 4.2 using an end-segment factor of $c^2/3$

that will be discussed in connection with Figure 4.7 and beyond concerning differentiating the total-energy curve to estimate the loss rate due to radiation.

Figure 4.4 includes plots of the entire total-energy curve where c^2 factors varying from 1, to monotonic to 1/3 are used. This result emphasizes the validity of using a $c^2/3$ end-reflection factor for producing a more physically realistic total-energy time curve. This point is further illustrated in Figure 4.5 where the corrected current, charge and total-energy curves are plotted for comparison with Figure 4.2.

Observe that the TDEMs are global in the sense that their time plots come from integrating over the entire object. It might seem that they could provide little information spatially about details of where radiation originates. But consider that when local, Gaussian-pulse excitation is used as for antenna modeling, the local spatial extent of the exciting pulse as a function of time does provide some spatial resolution concerning where radiation loss is occurring. Figure 4.6 is for a

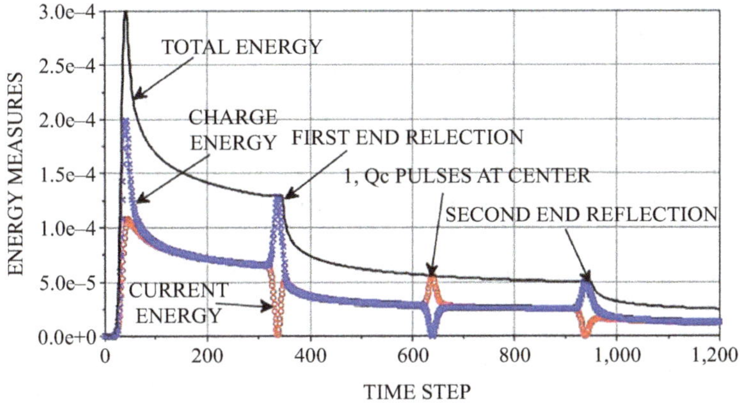

Figure 4.6 Current, charge and total energy measures for impulsively excited, center-fed, 599-segment dipole antenna

599-segment, center-fed dipole for which the spatial width of the exciting pulse is about 20 segments. The loss rate increases markedly after the peak in the total energy that occurs about the time that the Gaussian excitation becomes effectively zero. It then smoothly decreases as the *Qc/I* pulse propagates down the wire due to what can be defined as propagation radiation. Each end reflection results in another sharper decrease in the total energy, an effect caused by the radiation that occurs due to charge acceleration when the Qc/I pulse reflects and reverses direction. This behavior continues until the dipole becomes charge neutral as all of the energy provided by the Gaussian excitation is radiated away.

The monotonic declining behavior exhibited by the total energy measure suggests differentiating this quantity to determine the rate at which energy is being radiated by the antenna, denoted here as ELR, the energy-loss rate. This is straightforward to do numerically by computing the difference between successive samples of the total energy as a function of time, i.e.,

$$ELR(t) = \frac{dW_T(t)}{dt}. \tag{4.18a}$$

This could be conveniently done numerically using the finite difference approximation

$$ELR(t_i) = \frac{dW_T(t)}{dt} \approx \frac{W_T(t_{i+1}) - W_T(t_i)}{t_{i+1} - t_i}. \tag{4.18b}$$

But since the quantity of interest is the difference in successive energy values, the results for $ELR(t_i)$ that follow do not include the time-interval divisor, so that what is displayed is simply

$$ELR_{diff}(t_i) \approx |W_T(t_{i+1}) - W_T(t_i)| \tag{4.18c}$$

or the differentiated total energy. Although the *ELR* of (4.18b) would be negative for monotonic-declining energy measures, the results as given (4.18c) are plotted throughout as positive for comparison with related quantities while also being able to use a log scale to display a wider dynamic range.

An example of using (4.18c) is displayed in Figure 4.7 for a shorter, 199-segment dipole along with the associated energy measures. The *ELR* as indicated by the differentiated total energy is closely correlated with changes that occur in the total-energy curve as expected. The larger loss is associated with end reflection of the *Qc*/*I* pulses where the current energy is minimized. A similar effect occurs in the charge energy when the *Q*/*I* pulses meet at the center of the wire, which is associated with an approximate doubling of the *ELR*.[1]

Referring to Figure 4.6 where the same alternating behavior of the *Qc*/*I* pulses is seen, observe that the charge pulses cancel a little after time step 600 when the reflected charge pulses overlap at the feedpoint. There the oppositely signed and oppositely propagating charge pulses momentarily cancel while the current pulses, being same-signed are therefore doubled. But it is not obvious what also happens until the loss rate per time step normalized by the total energy of Figure 4.6 is examined as plotted in Figure 4.8,† where a small increase in the loss rate coincides with the overlapping of the current pulses as in Figure 4.7. This prompts the question: Is this a numerical artifact, or is this a physically real effect?

Observe that as the pairs of counter-propagating *Qc*/*I* pulses move along the wire their spatial fields periodically add in some directions but cancel in others due to the differences in the propagation times of these separated sources. But when they overlap, their far-fields are additive everywhere in space, a coherence that

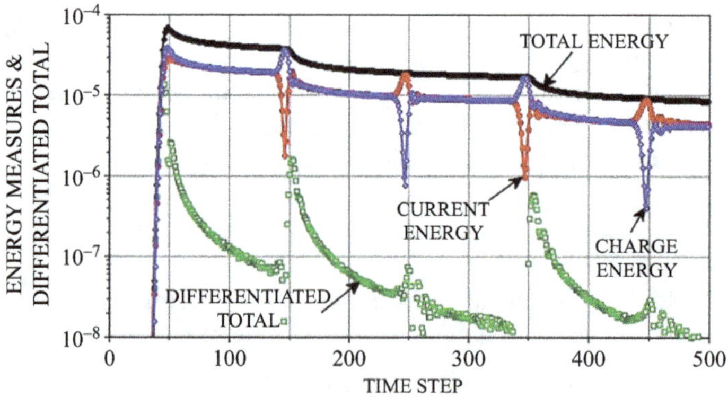

Figure 4.7 The total, current, and charge energy measures and the differentiated total energy as a function of time step for a 199-segment dipole

†These plots can become noisy as their levels decrease because their computation involves taking the difference of more nearly equal numbers.

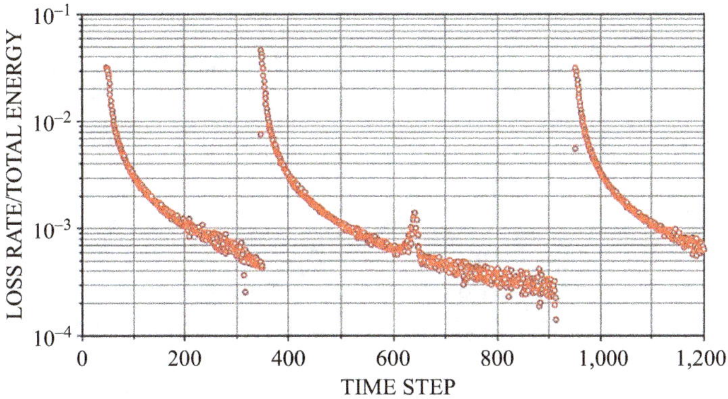

Figure 4.8 The normalized, differentiated energy-loss rate per time step for the total-energy measure of Figure 4.6 divided by the total energy

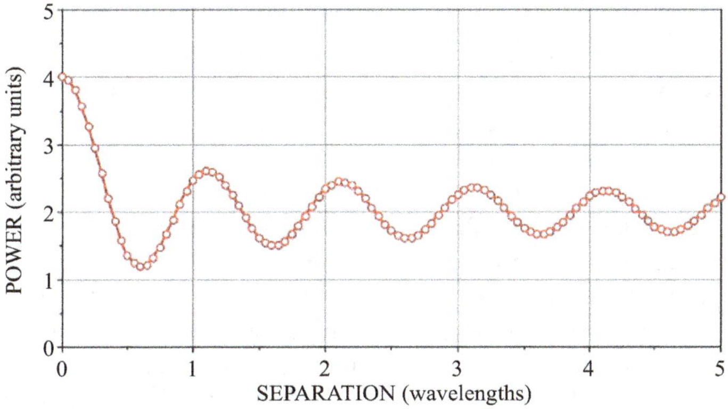

Figure 4.9 Total power radiated by two, unit-amplitude, isotropic point sources as a function of their separation

effectively doubles their otherwise spatially incoherent contributions. This doubling is confirmed as a doubling in value of the loss rate of the blip at about time step 640 in Figure 4.8, an effect not obvious in the total energy plot of Figure 4.6. This phenomenon of increased radiation is similar to that exhibited in Figure 4.9 where the total power radiated by two isotropic point sources is shown as a function of their separation. When they become co-located, their radiated power is found to similarly double.

The energy-loss rate of Figure 4.8 is repeated in Figure 4.10 together with plots of the total energy and squared, broadside far field. For convenience in comparing these quantities, their values normalized to a maximum of unity, rather than actual quantitative amounts, and shown on a log scale to cover the wide range of values of these results. That the peaks in the radiated field are correlated with both of the

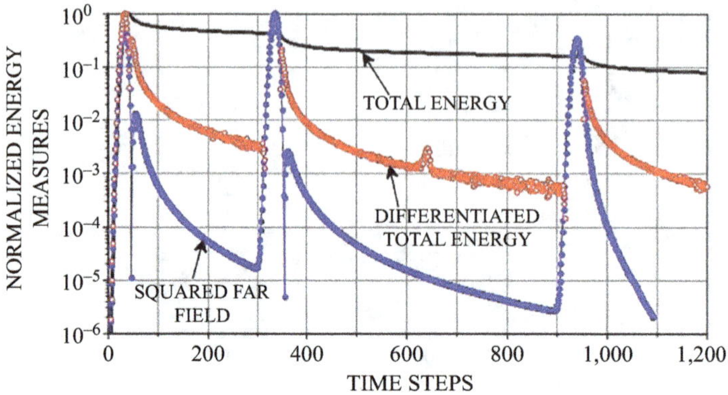

Figure 4.10 Normalized total and differentiated total energy measures and squared broadside, radiated far field for a 599-segment, center-excited dipole antenna

energy curves demonstrates their causative connection. As is done throughout the time delay of the far field is removed to demonstrate more clearly the connection of energy loss from the antenna and the resulting radiation.

There are some interesting observations that can be made. During the time that the antenna total energy is increasing due to the Gaussian-pulse excitation, the far field reaches a value of about 0.5 on this normalized scale. This is followed by a larger field peak of unity about 300 time steps later due to the first end reflection. This first end-reflected is about twice that of the initial source-generated pulse, due to experiencing a 2*c* speed change as compared with a speed change of *c* for the initial pulse. The smaller pulse that immediately follows the source pulse, of about 0.01 in magnitude, is due to the reflection of charge that occurs from the outgoing *Qc/I* pulse, as previously discussed in Chapter 3. The radiated energy between the far-field peaks is correlated with the differentiated-energy curves.

Note that numerically differentiating a numerically generated result can be somewhat uncertain. This is because any errors in that result are amplified when they are successively subtracted point-by-point. This effect can be observed in the differentiated curves of Figures 4.7, 4.8 and 4.10.

In order to show in more detail, the small peak in energy loss that occurs when the counter-propagating *Qc/I* pulses overlap as seen in the previous figures, this area of Figure 4.6 is presented on an expanded scale in Figure 4.11.

A further example of the kind of result shown above is next presented for a more complicated situation. This is again for a 599-segment, impulsively excited wire but where the exciting voltage is applied at segment 50, near the left end of the wire. Shortly after this asymmetric excitation is turned, two *Qc/I* pulses propagate toward the right separated by about 100 time steps, producing the energy-loss result of Figure 4.12. The first end reflection occurs when the leftward-propagating *Qc/I* pulse reaches the left end. A propagation-radiation sequence then occurs, followed

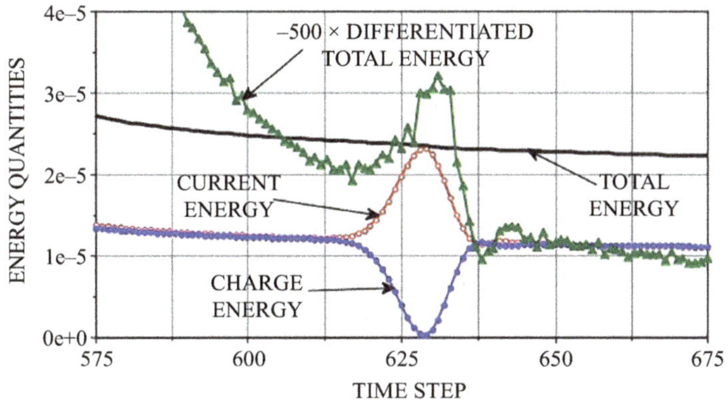

Figure 4.11 Energy measures and differentiated total energy for the 599-segment dipole during the time where the current and charge pulses cross at the feedpoint

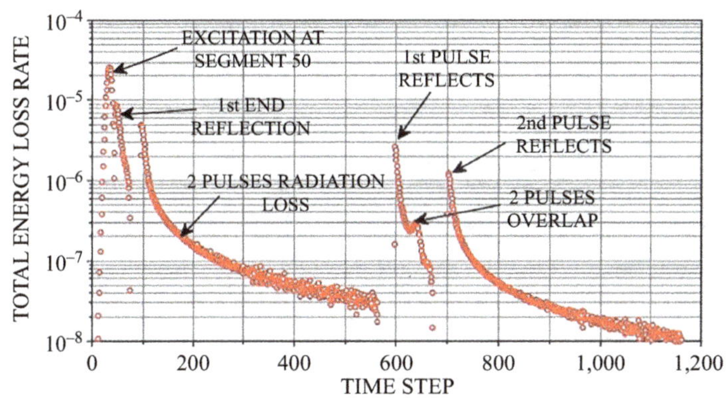

Figure 4.12 The time-dependent loss rates for asymmetric excitation of a 599-segment wire

by the two separated pulses reflecting at the right end at time steps 600 and 700, with another overlapping energy loss taking place near the right end between those two events.

A graph showing the normalized total and differentiated energies and broadside, radiated far field for this case is shown in Figure 4.13, which exhibits a correlation similar to that of Figure 4.10 for the case of symmetric excitation.

Results for a horizontal, center-fed dipole located above a PEC ground are presented in Figures 4.14 and 4.15. The 3.82-m dipole is modeled using 149 segments, a wire radius of 2×10^{-3} m and is 0.125 m above the ground. A comparison of Figure 4.14 with the dipole antenna in free space of Figure 4.6 demonstrates the much-reduced rate of radiation loss for the latter case due to proximity to the PEC

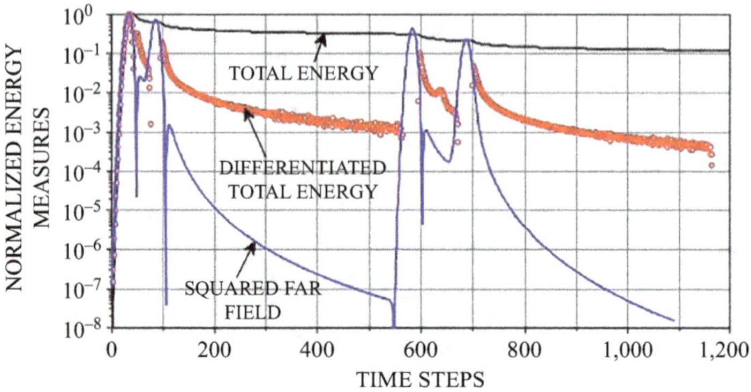

Figure 4.13 Normalized total and differentiated total energy and squared, broadside radiated far field for the asymmetrically excited 599-segment antenna of Figure 4.12

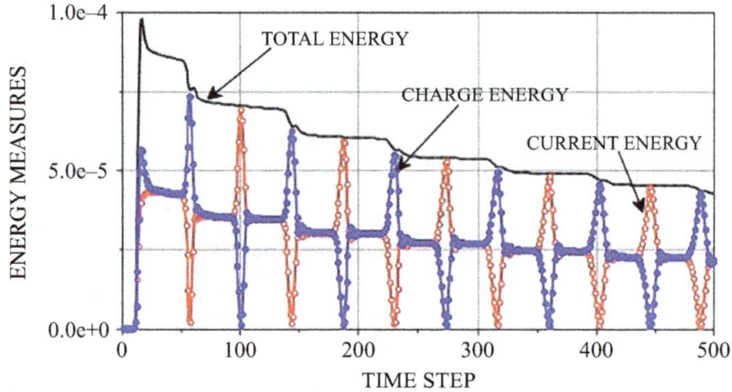

Figure 4.14 The total, current and charge energy measures for a 149-segment, 3.82 m long, center-excited dipole antenna of 2×10^{-3} m radius and 0.125 m above a perfect ground

ground. The end-reflection energy loss and propagation loss are both about ¼ that of the free-space dipole.

Only the total and differentiated energies are shown in Figure 4.15 as the radiated field is substantially reduced by the image radiation. Differentiation of the total-energy end-reflection radiation produces doublets in the loss rate, evidently as a result of the ground interaction. The doublet peaks are spaced about 10 time steps apart, equivalent to the time separation between the dipole and its image.

A last TDEM example for a straight wire is for a normally incident plane wave scattered from a straight wire in Figures 4.16–4.19. A 199-segment wire is used to exhibit more of the effects of multiple end reflections. The current and charge density time light speed c induced on the wire are shown for 9 time steps in

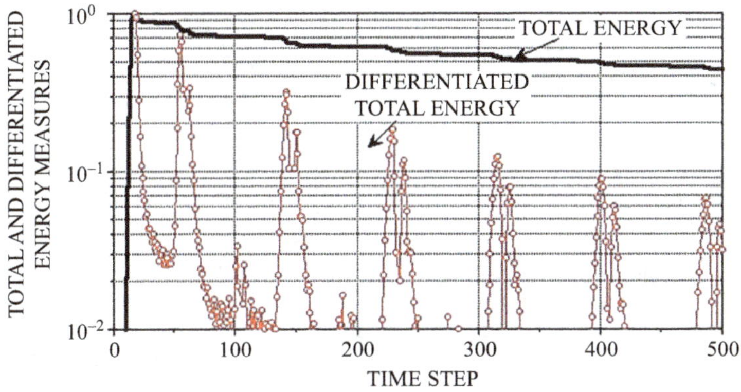

Figure 4.15 The total energy measure and differentiated total energy for the 149-segment wire over a perfect ground

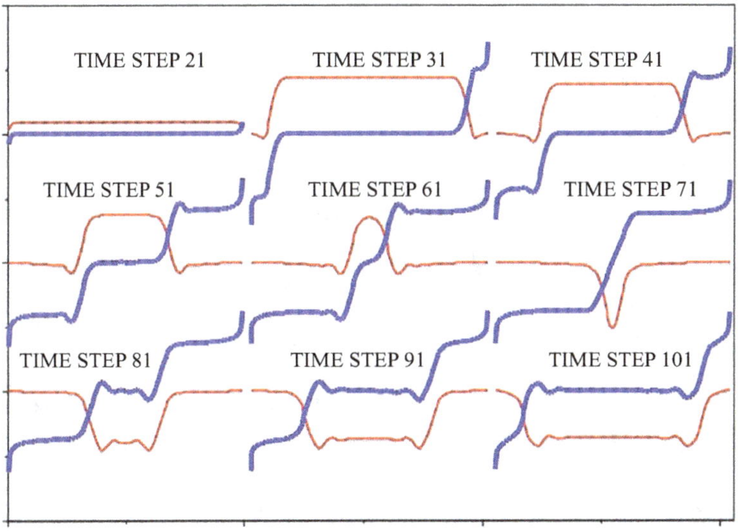

Figure 4.16 Snapshots of a series of current (lighter curves) and charge distributions on a 199-segment wire induced by a normally incident plane wave

Figure 4.16. The charge (the heavy lines) and current (lighter lines) are initially uniformly distributed along the wire for time step 21. The current continues to increase in time step 31, where the effects of end reflection are becoming more pronounced and the current and charge distributions are symmetric and anti-symmetric respectively about the center. For time steps 41 through 71, the current collapses through zero and reverses sign with the charge becoming the dominant

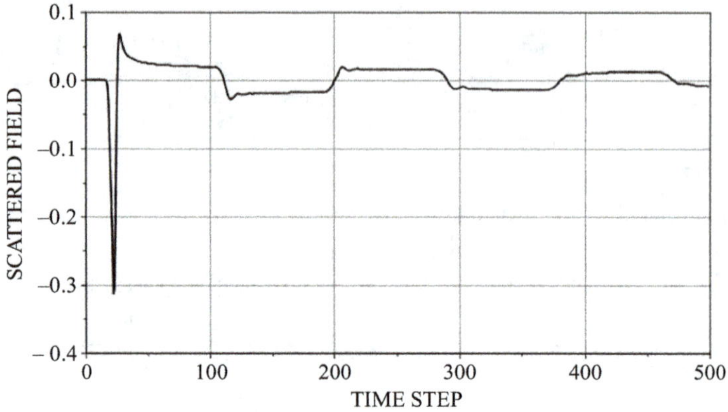

Figure 4.17 The broadside, backscattered field for the 199-segment wire as a function of time

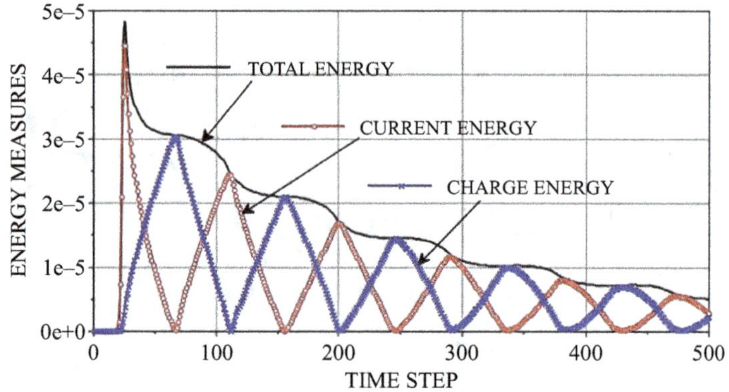

Figure 4.18 The energy measures of the 199-segment wire illuminated by a normally incident plane wave

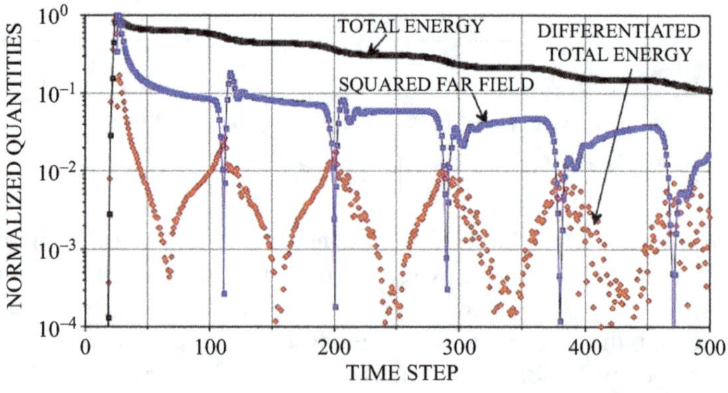

Figure 4.19 The normalized total and differentiated energies and squared far field for the 199-segment wire illuminated from broadside by a plane wave

component. As time advances, the current increases to a maximum negative value while the charge becomes concentrated at the wire ends.

The broadside, backscattered field from the wire is plotted in Figure 4.17 where a negative-signed "specular" flash is first followed a decreasing nearly rectangular-shaped alternating scattered field.

The energy measures for the scattering problem of Figure 4.18 are insensitive to the computation of the charge energy on the wire-end segments because the charge is not as concentrated there as for the antenna case. The alternating current and charge distributions exhibited by the plots in Figure 4.18 result in an alternating pattern of their associated energies. The normalized plots of the total and differentiated energies and the squared far field are shown in Figure 4.19. The decreased slope in the total energy curve results in differentiated-energy plots that differ substantially from the previous examples.

4.4 Some additional examples of TDEMs

In the preceding section, only straight-wire antennas and scatterers were investigated in connection with testing the use of time-domain energy measures as a way of determining where energy loss and its accompanying radiation originates on such objects. In this section, the application of TDEM to more general geometries and other straight-wire configurations are explored beginning with the circular-loop antenna. It is appropriate to mention that straight wires and circular loops might be considered the canonical geometries of wire-like problems. One example of a circular loop is included here, with the next chapter devoted entirely to loop geometries. Some other wire problems conclude this chapter, including another example where the TDEM may not be suitable for some geometries, at least not as clearly useful as has been generally found so far.

An energy-measure plot for a 100-segment circular loop having a Gaussian-pulse voltage applied to a single segment is shown in Figure 4.20. The current and

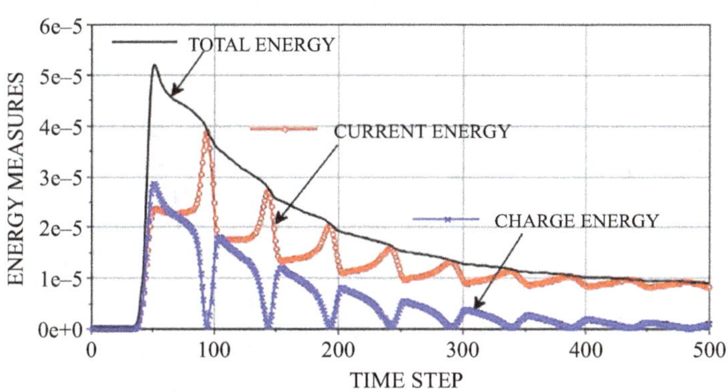

Figure 4.20 The total, current and charge energy measures for a 100-segment circular loop

charge energies are quite different from these quantities for a straight wire. A finite-length straight wire has end-reflected charge maxima while this phenomenon is absent from a loop, which does however have current maxima and charge minima when the counter-propagating *Qc/I* pulses periodically overlap at the feed point and at its conjugate on the opposite side of the loop. This results in the current and charge maxima and minima at 50-segment intervals (or time steps) as the *Qc/I* pulses propagate around the loop.

But there is another effect taking place that makes a closed loop, whatever shape it has, different from a straight wire. After the *Qc/I* pulses have made enough passes around the loop, only a uniform, steady-state current remains because the non-zero energy provided by the impulsive excitation has all radiated away. This effect is more evident where these results are plotted on a normalized log scale as in Figure 4.21. This graph has an absolute scale in Joules because the loop inductance can be determined from the steady-state current and the parameters of the exciting voltage [4,5].

Thus, with the exciting voltage

$$V_{ex}(t) = V_o e^{\left[-A(t-T)^2\right]} \tag{4.19}$$

and with V_o = 1V, A = 4.2 × 10^8/sec and T = 1.5 × 10^{-8} sec, the steady-state current I_o is given by

$$I_o = \frac{1}{L}\int_{-\infty}^{\infty} V_{ex}(t)dt = \frac{\sqrt{\pi}}{AL}. \tag{4.20}$$

Thus, the inductance from the late-time *TWTD* current is simply obtained as

$$L = \frac{\sqrt{\pi}}{AI_o}. \tag{4.21}$$

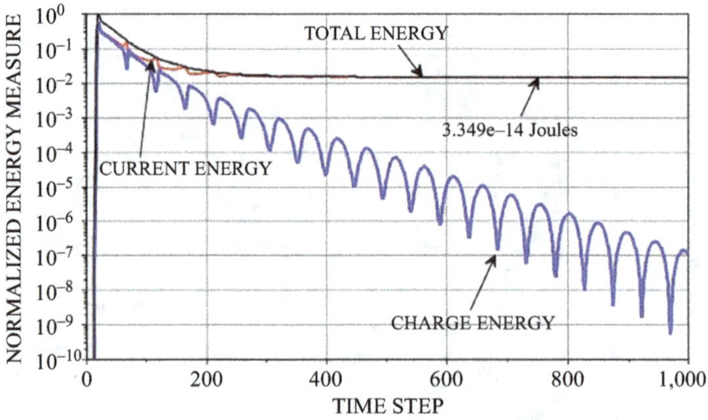

Figure 4.21 A log plot of the energy-value measures in Joules on a 100-segment circular loop

Finally, the energy stored in a loop can be obtained from its inductance and steady-state current as

$$W_T = \frac{1}{2} L I_o^2. \tag{4.22}$$

A possibly unappreciated capability of time-domain computer models such as *TWTD* is thus obtaining the inductance of rather arbitrary–geometry loops of various shapes. Demonstrating the agreement between TWTD-obtained inductance values for loops of various shapes and sizes [4] with published values also provides further validation of TWTD.

The energy measures for an 1,199-segment circular loop and the differentiated total energy are presented in Figure 4.22. The differentiated-energy curve is seen to have an increase at the time where the *Qc/I* pulses overlap, approximately doubling the energy-loss rate, an effect that is not really visible in the total-energy plot. This effect is similar to that of Figures 4.8 and 4.10 for the dipole.

A result for the energy measures of a 300-segment square loop excited as an antenna at segment 100 is presented if Figure 4.23. As for the circular loop of Figure 4.20, the late-time current decays to a non-zero value, becoming uniform around the square loop.

The total energy measures for the square loop are plotted together with the total energy of the 100-segment circular loop for comparison in Figure 4.24. The total energy of the circular loop decreases more rapidly than that of the square loop due to its curvature and the fact that a square loop radiates primarily from its corner 90-degree bends. The effect of the corner bends is emphasized in Figure 4.25 where the square loop total and differentiated energies are plotted. A sequence of larger lobes appears every 75 time steps in the differentiated energy with two smaller intermediate lobes between. All of these lobes are evidently due to *Q/I* pulses interacting with corners of the square loop, due to the asymmetry of the excitation.

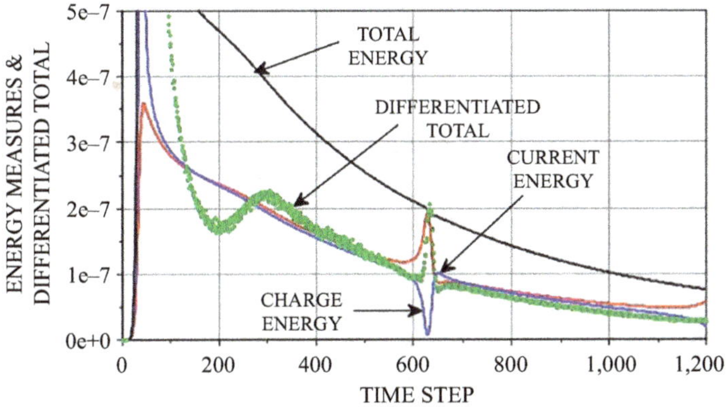

Figure 4.22 The energy measures on an 1199-segment circular loop with the differentiated total energy

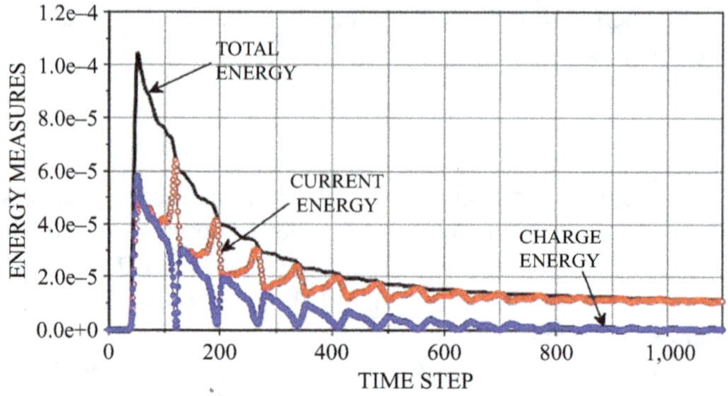

Figure 4.23 The energy measures for a 300-segment square loop excited as an antenna

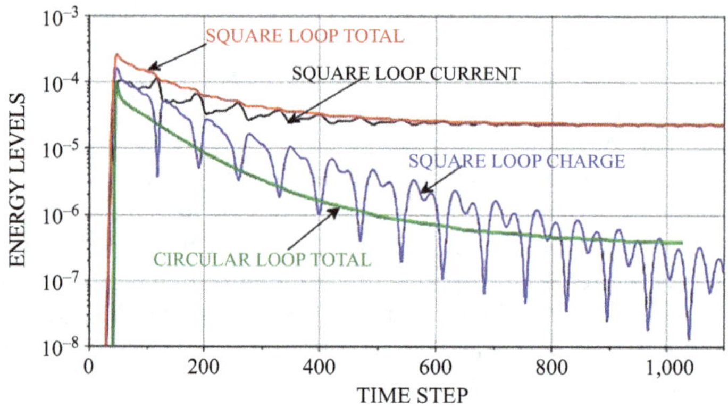

Figure 4.24 A comparison of the energy measures of a 300-segment square loop and the total energy of a circular loop normalized to the square loop, both having a wire radius of 10^{-2} m

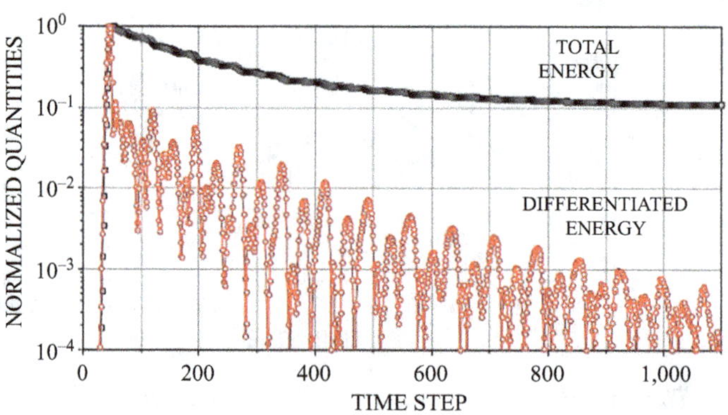

Figure 4.25 The total and differentiated energies for the 300-segment square loop antenna

As time step 1,000 is reached the rate of energy loss has declined by more than three orders of magnitude as the current approaches its steady-state value.

A multiply excited straight wire is next considered to further examine the effects of counter propagating and overlapping *Qc/I* pulses. In this case a 201-segment wire is impulsively excited by positive Gaussian voltages at segments 67 and 135. The energy-measure plot of Figure 4.26 shows a quite different result compared with the center-excited dipole as in Figure 4.6. The first current peak after the total-energy peak occurs when the leftward-propagating current pulse from segment 135 and rightward-propagating positive current pulse from segment 67 meet at the wire center. The following charge peak similarly occurs when the rightward-propagating *Qc/I* pulses from segment 135 and leftward-propagating *Qc/I* pulses from segment 67 reflect from the opposite ends of the wire. Finally, the charge maximum and current minimum occur when the first end-reflected *Qc/I* pulses overlap the oppositely propagating pulses nearing the ends. This pattern repeats periodically over time with a charge minimum occurring when the four *Qc/I* pulses from both sources then meet symmetrically about the wire center. The end reflections that have charge maxima are associated with sharper falloffs in the total energy while at the charge minima there is a slight increase in energy loss that is not evident in the total-energy plot.

The ELRs associated with variations of the excitation for Figure 4.26 are presented in Figure 4.27 for the time range bracketing the latter part of Figure 4.26. The excitation for the single-source cases use 1.414 V to equal the energy input of the 1-V, 2-source excitations. All exhibit an increased radiation loss at about time step 250. For both single-source cases, this is the time at which the once-reflected *Qc/I* pulses first overlap.

Additional extrema of opposite sense occur for the two-source excitations around time steps 217 and 287. The latter are associated with extremes corresponding to increased and reduced spatial radiated energies, respectively. To summarize, the maxima result when charge pulses of opposite sign overlap while the minima occur with overlapping charge pulses of the same sign. This situation will be discussed in Chapter 7 on TDFARS (time-domain FARS).

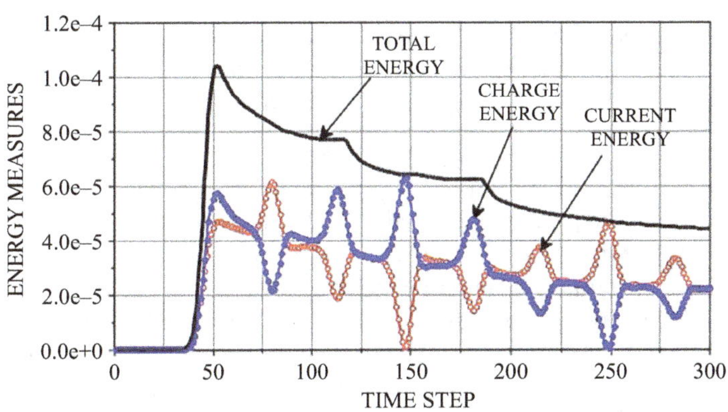

Figure 4.26 The energy measures for a 201-segment straight wire excited by positive impulsive voltage sources at segments 67 and 135

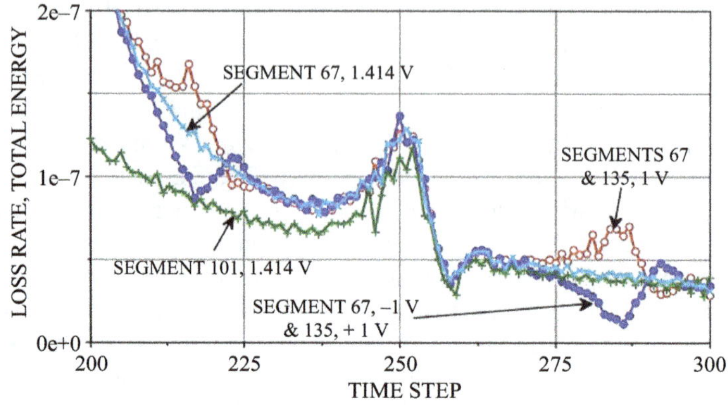

Figure 4.27 The absolute time-differentiated total energies for four different excitations of the 201-segment dipole cases discussed in the text

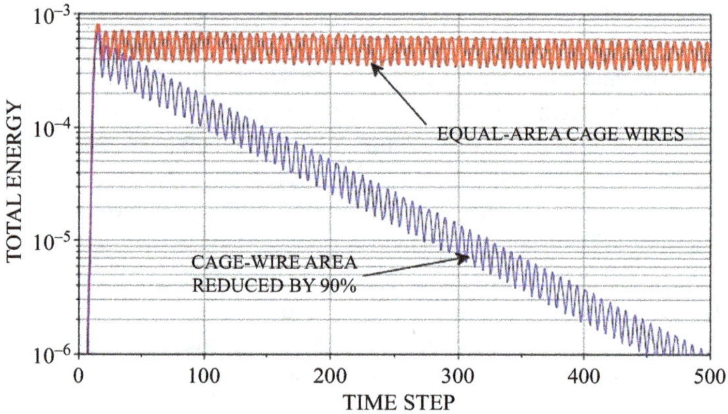

Figure 4.28 Total energy versus time for a short dipole enclosed by an 8-wire cage having an unrolled surface area that matches that of a continuous cylinder and one where the unrolled area is reduced by 90%

The total energy for an impulsively excited dipole located centrally inside an 8-wire cage of parallel wires is plotted for two different cases in Figure 4.28. The 35-segment cage wires and 11-segment dipole wires are 0.89744 and 0.28206 m long, respectively with equal-length segments of 2.564×10^{-2} m. The two plots of Figure 4.28 have wire radii of 3.205×10^{-3} m to form an equal-area cage such that unrolling the cage wires would form a closed cylinder around the dipole, and 0.1 of that. Investigations of various arrangements of parallel wires and wire meshes or grids have determined that if the radii of the wires in such configurations would cover the area that they enclose were they to be "unrolled" that they would approximately function electromagnetically as a continuous metallic surface.

For the electromagnetically "closed" cage, the impulsively excited dipole is effectively shielded from radiating and so the energy input from the exciting source is contained within the cage-dipole region that functions as a cavity, with little loss occurring. What loss does occur is evidently due to some slight amount of leakage as the shield is not perfect. There is a marked contrast in the total-energy oscillations when the radius of the cage wires is reduced, with the energy now decaying exponentially in time via radiation. In the closed system formed by the equal-area-cage shield, the total-energy oscillations must be due to the periodic exchange of energy between the current and charge flowing on the cage wires and that contained in the cage fields.

The final example of time-domain energy measures in Figures 4.29–4.32 is for a dipole antenna having a matched load at its center. A matched load is usually

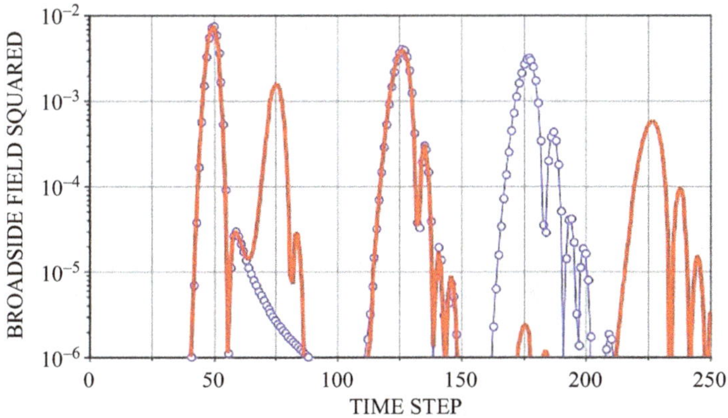

Figure 4.29 The heavy solid curves are the squared radiated field caused by the matched load while the open-circle curves are produced when there is no load

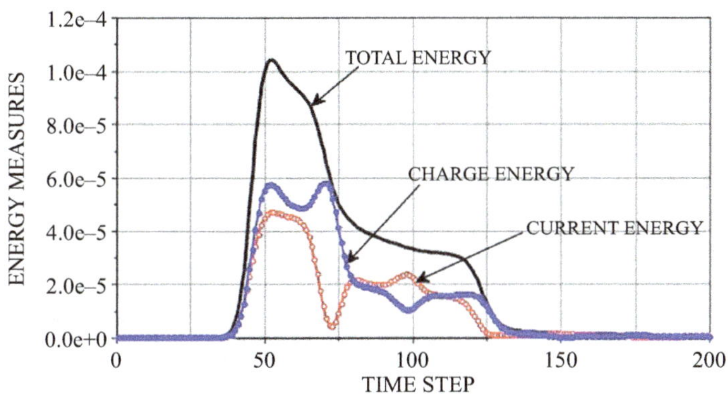

Figure 4.30 The total current and charge energy measures for the match-loaded dipole of Figure 4.29

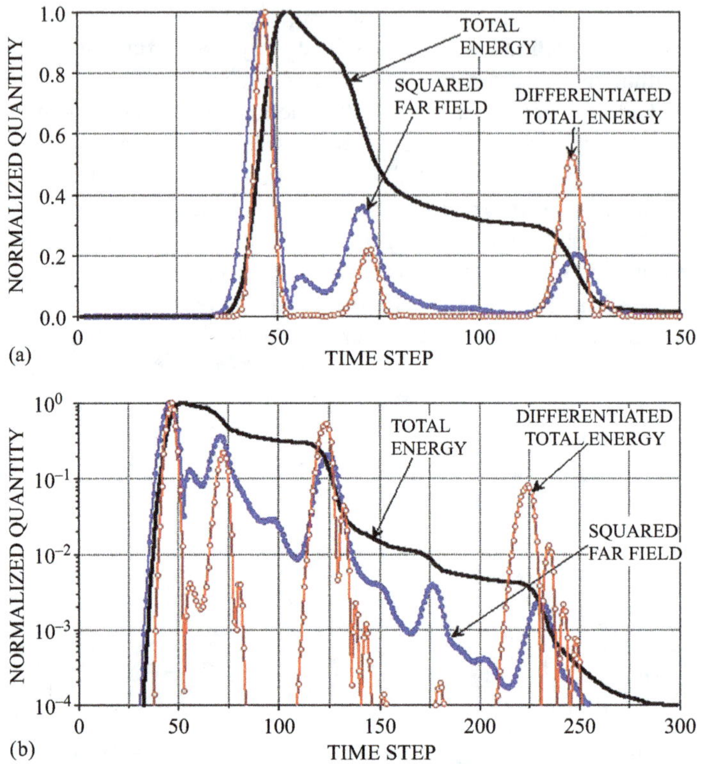

Figure 4.31 *The linear normalized total and differentiated energy measures and the broadside squared far field for the matched-load dipole antenna (a) and for more time steps on an expanded logarithmic scale (b)*

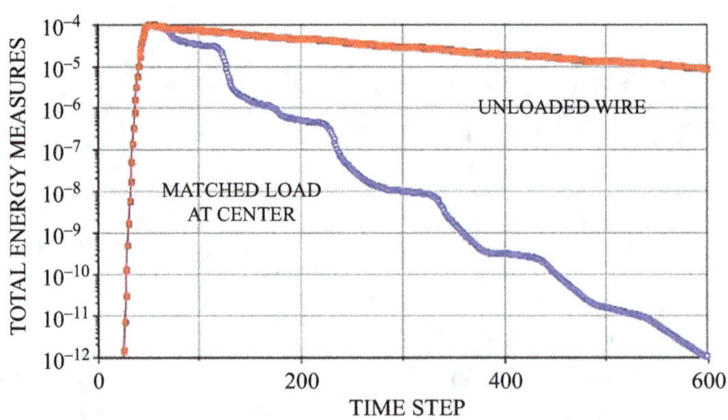

Figure 4.32 *The total energy measures for the unloaded and loaded dipole antennas*

associated with a 2-wire transmission line where a matched condition is one for which there is no reflected current and charge because the incident power (frequency domain) or energy (time domain) is completely absorbed by the load. For this to occur, a positive charge that reaches the load from one wire is balanced by a negative charge that reaches the load from the other wire. The opposite-signed charges cancel in the load that also absorbs their power or energy. This kind of problem has been discussed in Figures 3.14–3.16 of Chapter 3 to demonstrate radiation caused by a matched load.

The square of the broadside radiated field for the two conditions of no center load, and a center load of 1,500 ohms is presented in Figure 4.29. The two fields overlap during the turn-on of the two impulsive sources and at the first end reflections of their respective Qc/I outward-propagating pulses. However, two additional pulses occur when the inward-propagating Qc/I pulses meet at the center load and when their end-reflected pulses arrive back at the matched load with no further load-caused radiation beyond as the wire returns to charge neutrality.

An examination of the energy measures for the matched-load is demonstrated in Figures 4.30 and 4.31. The total-energy and Qc/I energies in Figure 4.30 display a total energy that has decayed to near zero beyond time step 130 on a linear scale. The normalized differentiated total energy and the squared far field in Figure 4.31a demonstrate a correlation between these quantities and the areas of more-rapid decay in the total energy. This correlation is exhibited over a wider range of these normalized quantities on the log scale of Figure 4.31b they span a range of 6 decades.

Finally, in Figure 4.32, the total energy measures for the no-load and matched-load cases of 199-segment dipole are compared. The effect of the load is to increase the ELR by a factor of 10^6 over that of the no-load case on a per-time-step basis. While some of this increase is due to radiative effect of the load, most of it is due to the I^2R loss. In spite of the resistive loss, the goal here has been to demonstrate how an opposite-signed charges meeting in a matched load on an antenna can produce radiation as they combine and stop.

4.5 Summary

The following issues of time-domain energy measures and their applications have been discussed in this chapter:

1. The derivation of time-domain energy measures for the charge and current from $(Qc)^2$ and I^2 integrated over an impulsively excited wire object is first discussed.
2. The sum of the charge energy was found to decay monotonically in time for an open-ended object excited as an antenna except when the charge pulses reflected from the ends where it increased, a nonphysical result.
3. This increase is generally corrected for by using $(Qc)^2/3$ as the charge energy measure of the open segment rather than $(Qc)^2$ to account for a smooth, rather than abrupt, speed change of the charge pulse during the reflection process.

4. Differentiation with respect to time of the total TDEM provides a way to determine when and where electromagnetic energy is radiated by the object, and its associated energy-loss rate (ELR).
5. A demonstration of the major radiation mechanisms of representative wire objects as being due to the exciting source; *Qc/I* reflection from open-ended wires; propagation radiation down the lengths of straight wires; enhanced radiation due to wire curvature; and other features such as bends, impedance loads and changes in wire diameter.

References

[1] E. K. Miller and J. A. Landt, "Direct time-domain techniques for transient radiation and scattering from wires", Invited Paper in Proceedings of the IEEE, **68**, 1980, pp. 1396–1423.

[2] E. K. Miller, "PCs for AP and other EM reflections," *IEEE Antennas and Propagation Magazine*, **40**, 1, February 1998, pp. 96–100, and 41, 2, April 1998, pp. 92–95.

[3] D. Poljak, E. K. Miller, and C. Y. Tham, "Time-domain energy measures for thin-wire antennas and scatterers," *IEEE Antennas and Propagation Society Magazine*, **44**, No. 1, 2002, pp. 87–95.

[4] J. A. Landt and E. K. Miller, "Short-pulse response of a circular loop", *IEEE Transactions on Antennas and Propagation*, AP-**22**, 1974, pp. 114–116.

[5] E. K. Miller, "Time-domain computation of loop inductance," *IEEE EMC Newsletter*, 2011, pp. 62–65.

[6] R. M. Bevensee, Private communication, 1977.

Chapter 5

Radiation-loss dependence of a circular loop antenna on its circumference and wire radius*

Abstract

The Lienard–Wiechert potentials show explicitly that charge acceleration, i.e. a change in charge velocity, causes radiation of an electromagnetic field. The goal of this chapter is to explore the rate of energy loss due to radiation from current and charge flowing on a circular loop as a function of the loop's curvature and wire radius. The results presented are obtained using a thin-wire, time-domain (TWTD) computer model for Gaussian-pulse excitation. Some results for a straight wire are also presented for comparison. Analytical estimates for the curvature and wire-radius effects are developed from best-fit expressions to the computed results using model-based parameter estimation (MBPE).

Keywords: Curvature radiation; Electromagnetic radiation; Farfield analysis of radiation sources; Impulse excitation; Loop radiation; MBPE; Time-domain radiation; TWTD; Wire-radius effect

5.1 Introduction

It has been observed [1–5] and quantitatively demonstrated [6] that charge and current (*Qc/I*) waves propagating on a straight wire attenuate or decay with time and distance due to radiation. The amount of similar radiation-caused decay for a circular-loop antenna due to wire curvature and wire radius is explored in the following. Radiation from a straight wire is also included as a reference.

The results presented here are obtained from the well-validated, time-domain, integral-equation model, thin-wire time domain (TWTD) [7]. A detailed description of the analytical and numerical basis for TWTD can be found elsewhere [8,9]. Briefly, it is based on a thin-wire approximation to an electric-field integral equation solved using the method of moments. The solution is developed by an explicit time-stepping procedure using a 9-term, second-order polynomial basis expansion in space and time and delta-function sampling.

*This chapter is based on "The dependence of time-domain radiation loss on the circumference and wire radius of a circular loop antenna," *Progress In Electromagnetics Research M*, vol. 92, 2020, 1–9, by E. K. Miller.

Modeling an antenna in the time domain usually involves applying an impulsive voltage to one segment of the wire model as is done here. The time-dependent voltage used for the loop results that follow is given by

$$V_{ex}(t) = V_o \exp\left[a^2(t - t_{\max})^2\right] \tag{5.1}$$

The loops and dipoles are modeled using a segment length $dx = 0.05$ m long and a time step $dt = 1.667 \times 10^{-10}$ sec so that $c \times dt = dx$. The half-width amplitude of the time-dependent voltage pulse is about 12 time steps or space segments.

Time-domain results are presented below for loop antennas of circumference $C = 2\pi R$ ranging from 5 to 60 m and radius r, varying from 2×10^{-2} to 10^{-9} m, and for straight wires of comparable length, L, and radii. The primary quantity to be examined is a total energy measure, $W_T(t)$, [6] derived from their time-dependent propagating current, $I(x,t)$, and charge density, $Q(x,t)$, as given by

$$W_T(t) = \int_0^{CorL} [I^2(x,t) + Q^2(x,t)c^2]dx = W_H(t) + W_E(t) \tag{5.2}$$

with c being the speed of light in free space, and the upper limit of the integral is C for loop circumference, or L for dipole length. The quantities W_H and W_E are the respective current (or magnetic) and charge (or electric) energies at the surface of the wire due to the propagating current and charge Results for the loop and dipole antennas are presented respectively in the following.

Example of the impulsive current and charge density for a circular loop having a circumference of 59.95 m (1,199 segments) and a wire a radius, r, of 0.02 m with the feedpoint at segment 600 are shown in Figure 5.1 at intervals of 100 time steps. Note that $I(x,t) = \pm Q(x,t)c$ except where the counter-propagating charge pulses begin to overlap and cancel on the opposite side of the loop. The decay of the current and charge with distance from the feedpoint is clear evidence of the curvature-caused charge acceleration and consequent radiation that occurs.

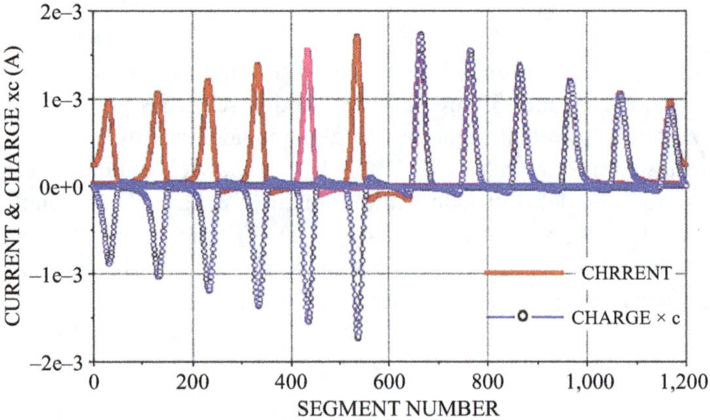

Figure 5.1 The current and charge density times c on one-half of an 1,199-segment loop (59.95 m circumference) with a wire radius r of 0.02 m excited at segment 600 by a Gaussian pulse

5.2 Numerical results for the loop antenna

5.2.1 Varying the loop circumference

The three energy quantities of (5.2) are plotted in Figure 5.2 for a loop 5-m (100-segments) in circumference with a wire of radius 0.02 m as a function of time step. The wide dynamic range of this plot is used to demonstrate that W_E approaches zero with increasing time. The oscillatory decay it exhibits occurs because each time the opposite-signed, oppositely propagating charge pulses meet moving around the loop, they cancel to produce alternating minima. The charge energy eventually becomes zero as the loop radiates the time-dependent energy deposited by the impulsive excitation. This results in a uniform, non-radiating, late-time current I_o and charge neutrality around the loop.

The inductance of a closed loop of any geometry that is excited by a Gaussian pulse can be obtained from the late-time current [10,11]. Knowledge of the loop inductance in turn provides a way to calibrate the current energy plotted in Figure 5.2 as follows. First, determine the loop inductance as given by

$$L = \frac{\sqrt{\pi}}{aI_o} = \frac{1.77245}{aI_o} \tag{5.3}$$

Equation (5.3) yields a value of 4.452×10^{-6} Henries with $I_o = 1.225 \times 10^{-4}$ A for the case of Figure 5.2. Since the energy stored in an indicator carrying a current I_o is given by

$$W_H = \frac{1}{2} L I_o^2 \tag{5.4a}$$

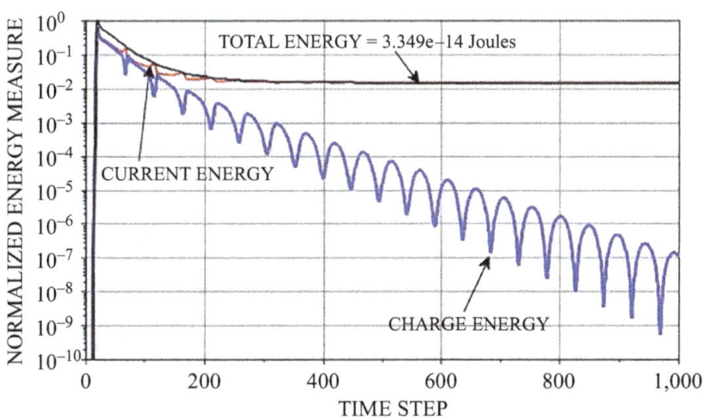

Figure 5.2 The current energy W_H, *charge energy* W_E, *and total energy* W_T *as a function of time step for a loop of 5 m (100 segment) circumference and wire radius of 0.02 m as a function the time step*

it follows that

$$W_H = \frac{\sqrt{\pi}}{2a} I_o \tag{5.4b}$$

Using (5.4b) the energy calibration factor for Figure 5.2 is obtained as 3.340×10^{-14} J = (1.77245) × (1.225 × 10^{-4})/[2 × 3.25 × 10^9].

The time-dependent, normalized total energy on several circular loops of radius 0.02 m and circumferences varying from 5 to 60 m (100 to 1,200 segments) are plotted in Figure 5.3a and on an expanded scale in Figure 5.3b. A normalized energy is plotted for convenient comparison of the various cases included here, but for this particular example, the peak energy is nearly independent of loop size. The decay rates exhibited by the total energy plots in Figure 5.3 are initially independent of the loop size.

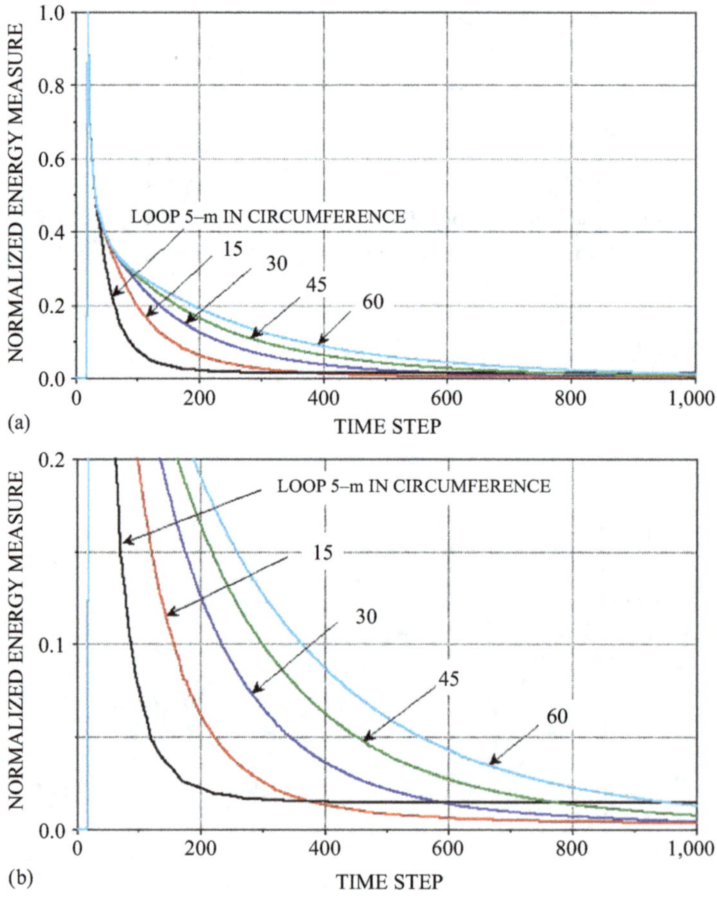

Figure 5.3 Normalized total energy as a function of time for loops having a wire radius of 0.02 m and variable circumferences in m (a) and an expanded scale (b)

Subsequently they vary not only with loop size but with time as exhibited by the decreasing slopes of the $W_T(t)$ curves with increasing time. A numerical estimate of the average decay rate of $W_T(t)$ on a per-meter basis was obtained from the distance traveled by $I(x,t)$ and $Q(x,t)c$ for the normalized, total energy to decrease from 0.2 to 0.1. A plot of these results on a log–log scale is shown in Figure 5.4. A best-fit straight line to the data points of Figure 5.4 yields an estimate for the loop energy-loss rate (ELR_L) as

$$Ln[ELR_L(R)] \approx -2.7 - 0.77Ln(R). \tag{5.5a}$$

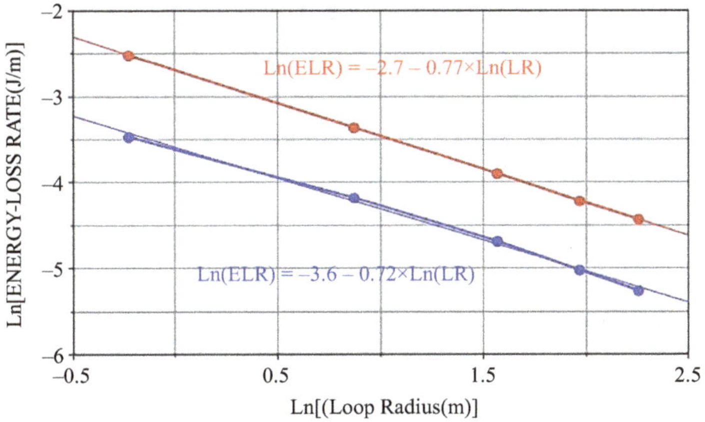

Figure 5.4 *The total energy-loss rate* (ELR$_L$) *over the normalized ranges 0.2–0.1 (open circles) and 0.1–0.05 (solid circles) as a function of the radius of an impulsively excited circular loop having a wire radius of 0.02 m on a log–log scale*

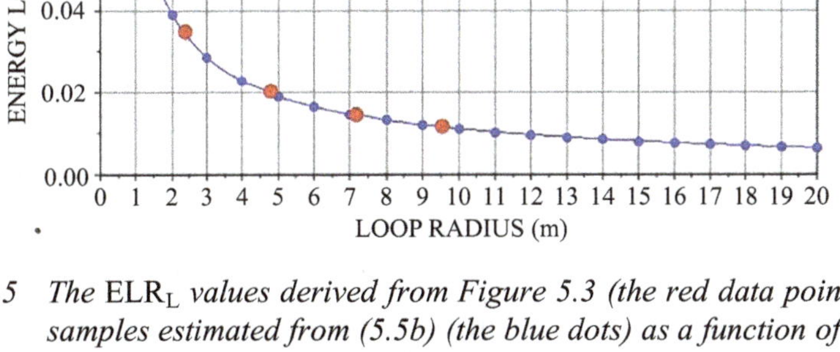

Figure 5.5 *The* ELR$_L$ *values derived from Figure 5.3 (the red data points) and samples estimated from (5.5b) (the blue dots) as a function of the loop radius and a wire radius of 0.02 m*

On simplifying (5.5a) there results

$$ELR_L(R) \approx \frac{6.72 \times 10^{-2}}{R^{0.77}}. \tag{5.5b}$$

or

$$ELR_L(C) \approx \frac{6.72 \times 10^{-2}}{C^{0.77}} \times (2\pi)^{0.77} = \frac{0.277}{C^{0.77}} \tag{5.5c}$$

If instead the average decay rate is estimated from the decay of $W_T(t)$ from 0.1 to 0.05 the best-fit straight line in Figure 5.4 is given by

$$ELR_L(C) \approx \frac{2.75 \times 10^{-2}}{R^{0.72}} = \frac{2.75 \times 10^{-2}}{C^{0.72}} \times (2\pi)^{0.72} = \frac{0.103}{C^{0.72}} \tag{5.5d}$$

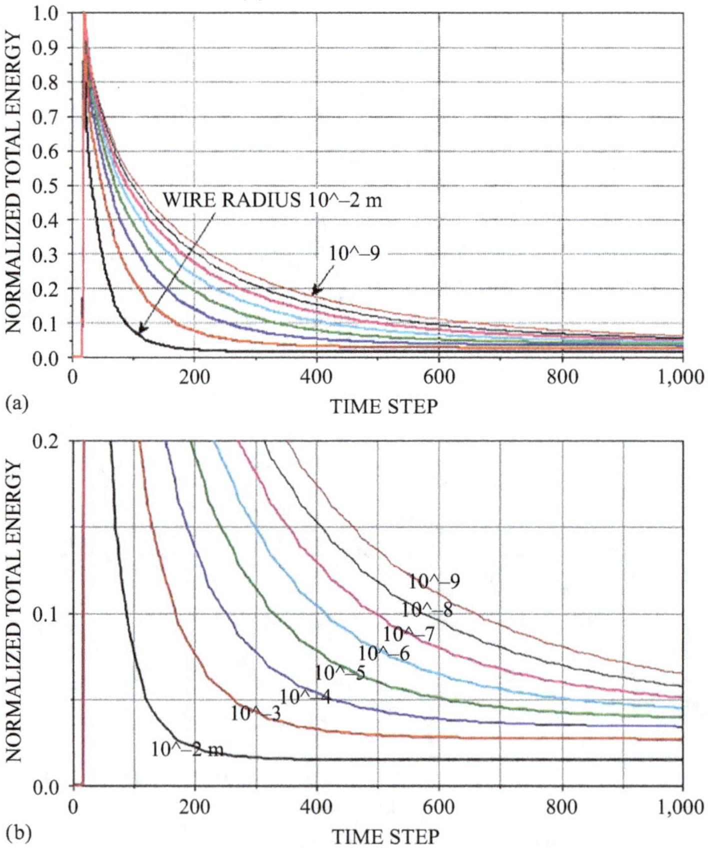

Figure 5.6 Normalized total energy for a 100-segment loop of 5 m circumference with the wire radius a parameter

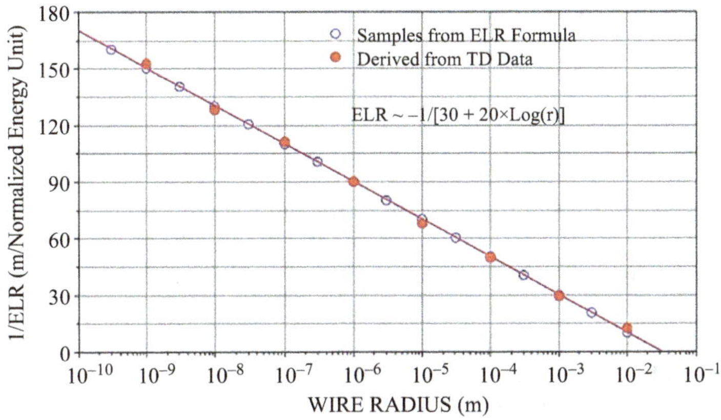

Figure 5.7 Inverse of ELR$_L$ *as a function of the wire radius of a 599-segment (29.95 m) loop compared with the model estimate of (5.6)*

Thus the average ELR_L is found to decrease with increasing loop circumference about as $\sim 1/C^{0.75}$. This behavior occurs because the charge acceleration decreases with decreasing loop curvature. The decreasing slope of $W_T(t)$ with increasing time in Figure 5.3 results in ELR_L becoming smaller as the loop currents approach a steady-state value. Samples of estimated ELR_L values from (5.5b) and the original ELR_L samples from Figure 5.4 are compared in Figure 5.5. They agree within a few percent to validate (5.5).

5.2.2 *Varying the loop wire radius*

The effect on the normalized energy measure of varying the wire radius for a loop of 5 m circumference is plotted in Figure 5.6a and on an expanded scale in Figure 5.6b. The energy decay rate is seen to vary with not only the wire radius but also in time as was found above for varying the loop circumference. The average ELR_L for the results of Figure 5.6 was computed over the energy range 0.2–0.1 as before. In contrast with the results of Figure 5.4, however, plotting the inverse of ELR_L versus $Ln(r)$ now results in a straight-line relationship shown in Figure 5.7. A best-fit straight line to the data of Figure 5.7 is given by

$$ELR_L \approx \frac{1}{-30 + 20Ln(r)} \tag{5.6}$$

5.3 Comparing a loop and straight wire

A comparison of the normalized total energies for loops $[W_L(t)]$ and straight-wire dipoles $[W_D(t)]$ 1,199 segments (59.95 m) long and of variable wire radii is presented in Figure 5.8. The voltage-pulse parameters used for these results are:

$dt = 1.66666 \times 10^{-10}$ sec,
$a = 9.69031 \times 10^{8}$/sec,
and $t_{\max} = 8.33333 \times 10^{-9}$ sec.

The normalized energy results are numerically identical out to about 180 time steps for $r = 10^{-2}$ m while increasing to about 250 time steps for $r = 10^{-9}$ m. These results decrease more slowly than those of the 5-m loop of Figure 5.6 since the loop curvature in this case is almost 12 times smaller. Referring to Figure 5.1, the changing shapes with time of the current and charge pulses show that the higher-frequency content is radiated earlier in their propagation along both the loop and dipole. This may explain the early-time equality of their energy loss in Figure 5.8. At later times, where Figure 5.8 shows that $W_L(t) < W_D(t)$, it can be concluded that the loop radiates the lower-frequency pulse content more effectively, due to its curvature, than does a straight wire.

The normalized energy values at time step 600 are plotted in Figure 5.9 as a function of the wire radius for both the loop and the dipole. The loop energy values are smaller due to the greater loss caused by the loop curvature.

The ELR_D values derived here are averaged over 100 time steps centered at *TS* 400 rather than using only the TDEM difference as done in Chapter 4. This is done so that the loss-rate dependence on the wire radius can be compared. Referring to Figure 5.9, the $ELR_D(400)$is thus estimated using

$$ELR_D(400) \approx \frac{[W_D(350) - W_D(450)]}{100} \tag{5.8}$$

the results of which are shown in Table 5.1. The average of the four ELR_D values in Table 5.1, 2.15/m, demonstrates that the late-time energy loss for the dipole is

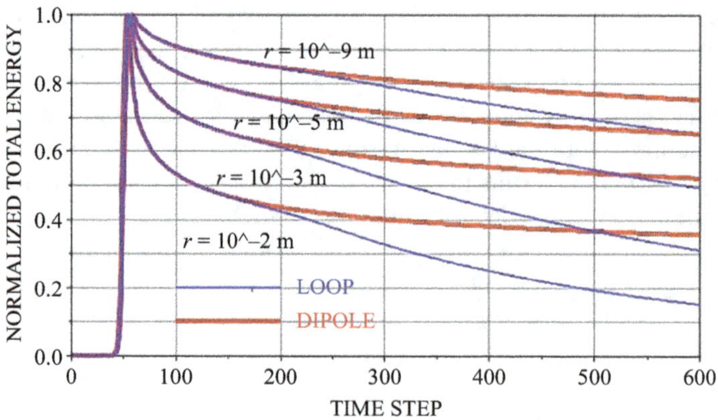

Figure 5.8 The normalized total loop energy $W_L(t)$ *and dipole energy* $W_D(t)$ *as a function of time step for several wire radii and total wire lengths of 59.95 m or 1,199 segments*

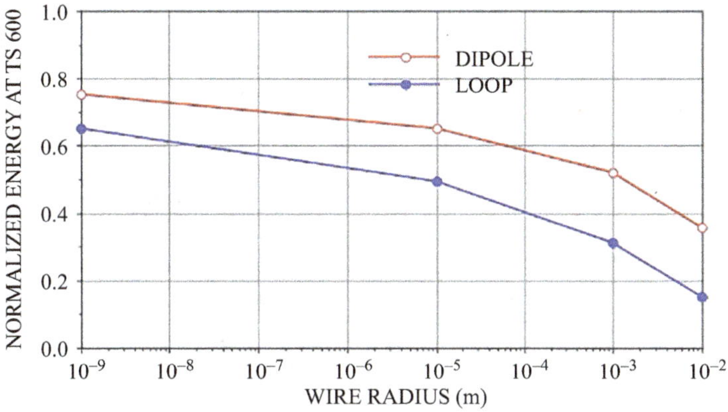

Figure 5.9 The normalized energy at time step 600 as a function of the wire radius for the loop and dipole

Table 5.1 Dipole energy loss rates

Wire radius (m)	W_D (350)	W_D (450)	ELR_D/m
10^{-2}	0.39	0.371	1.9×10^{-4}
10^{-3}	0.565	0.542	2.3×10^{-4}
10^{-5}	0.7	0.678	2.2×10^{-4}
10^{-9}	0.8	0.778	2.2×10^{-4}

Table 5.2 Loop energy loss rates

Wire radius (m)	W_L (350)	W_L (450)	ELR_L/m
10^{-2}	.285	.22	6.6×10^{-4}
10^{-3}	.475	.41	6.5×10^{-4}
10^{-5}	.6355	.585	5.05×10^{-4}
10^{-9}	.765	.716	4.9×10^{-4}

essentially independent of the wire radius over a 7-decade range even though their late-time energy levels differ by a factor of more than 4.

For comparison, the similarly averaged ELR_L values given in Table 5.2 show that the late-time energy loss for the 1,199-segment loop varies by a factor of about 4/3 over the same range of wire radii.

Samples of the current at successive intervals of 50 time steps on a 1,200-segment loop and dipole excited by an impulsive 1-V source at segment 600 are shown in Figure 5.10. The pulses are plotted at successive intervals of 50 time steps over the wire half extending from 600 to 1,200 segments. Besides the greater decay rate of the loop compared with the dipole, the peak of the loop current moves

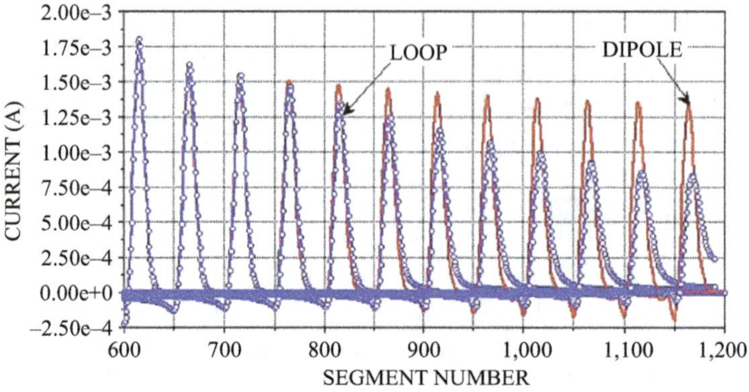

Figure 5.10 The current pulses on 1,200-segment (60 m) dipole and circular loop at 50 time-step intervals beginning at time-step 50

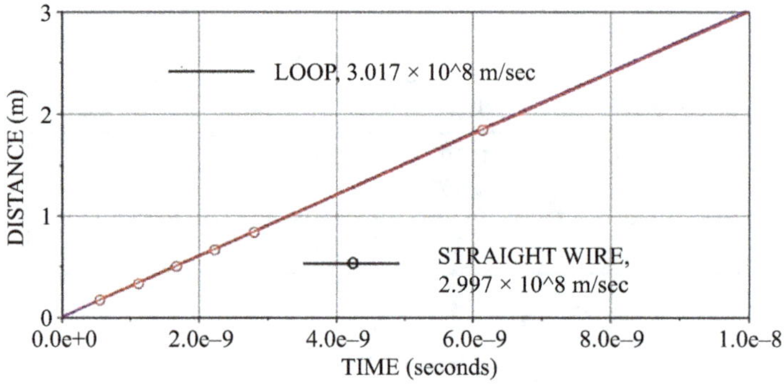

Figure 5.11 Best-fit straight line to the locations of the peaks of the current pulses of Figure 5.10 are shown here as a function of time together with their respective propagation speeds

slightly faster. This is demonstrated in Figure 5.11 where best-fit straight lines are fit to the peak space locations as a function of time. The effective propagation speeds of their respective pulse peaks show the loop speed exceeds that of the dipole by about 2×10^6 m/sec. This is due to the slightly shorter interaction distance along the loop's curvature relative to a straight wire.

5.4 Summary

Examined above is the influence of the size and wire radius on the rate at which an impulsively excited, circular wire loop loses energy due to radiation. Some results for a straight wire for varying wire radius have been included for comparison. A representative analytical estimate for the energy-decay rate, as a function of the loop circumference, C, for a wire radius of 0.01 m was obtained as

$$ELR_L(C) \approx \frac{0.103}{C^{0.75}}. \tag{5.9}$$

The numerator factor, in (5.9), is a variable determined by where in time the decay rate is determined.

Similarly, an estimate of the dependence of the energy-decay rate as a function of the wire radius r has been determined from the time-domain results for a 5-m (100-segment) loop to be given by

$$ELR_L(r) \approx \frac{1}{-30 + 20Ln(r)}. \tag{5.10}$$

Observe that (5.9) was derived specifically for a loop having a wire radius of 0.01 m while (5.10) was developed for a loop 5-m in circumference. The energy-decay rate for the 60-m loop is much less sensitive to the wire radius. Thus, generalizing these results over a wider parameter range, subject to the limits of TWTD, would probably be useful.

As observed previously, the Lienard–Wiechert potentials show explicitly that charge acceleration, i.e. a change in charge velocity, causes radiation of an electromagnetic field. The results presented here have demonstrated the influence of a loop's curvature and wire radius on the rate of energy loss due to radiation from current and charge flowing on it. The results presented are obtained using a thin-wire, time-domain (TWTD) computer model for Gaussian-pulse excitation. Some results for a straight wire are also presented for comparison. Analytical estimates for the curvature and wire-radius effects are developed from best-fit expressions to the computed results.

References

[1] Schelkunoff, S. A., *Advanced Antenna Theory*, John Wiley & Sons, New York, NY, p. 190, 1952.

[2] Jackson, J. D., *Classical Electrodynamics*, John Wiley & Sons, New York, NY, p. 401, 1962.

[3] Shen, L-C., T. T. Wu, and R. W. P. King, A simple formula of current in dipole antennas, *IEEE Trans. AP-S*, AP-16(5), pp. 542–547, 1968.

[4] Anderson, B., Admittance of infinite and finite cylindrical metallic antenna, *Radio Sci.*, 3(6), pp. 607–621, 1968.

[5] Jones, D. S., *Methods in Electromagnetic Wave Propagation*, Clarendon Press, Oxford, p. 295, 1994.

[6] Miller, E. K., The proportionality between charge acceleration and radiation from a generic wire object, *Prog. Electromag. Res.*, 162, pp. 15–29, 2018.

[7] Landt, J. A., E. K. Miller, and M. Van Blaricum, WT-MBA/LLL1B (TWTD): a computer program for the time-domain electromagnetic response of thin-wire structures, *Lawrence Livermore Laboratory Report No. UCRL-51585*, 1974.

[8] Miller, E. K., A. J. Poggio, and G. J. Burke, An integro-differential equation technique for the time-domain analysis of thin-wire structures, 1973. Part I: the numerical method, *J. Computat. Phys.*, 12, pp. 24–48, 1973.
[9] Poggio, A. J., E. K. Miller, and G. J. Burke, An integro-differential equation technique for the time-domain analysis of thin-wire structures. Part II: numerical results, *J. Computat. Phys.*, 12, pp. 210–233, 1973.
[10] Paul, C. R., Partial inductance, *IEEE EMC Soc. Mag.*, pp. 34–42, Summer 2010.
[11] Miller, E. K., Time-domain computation of loop inductance, *IEEE EMC Soc. Mag.*, pp. 34–42, Summer 2011.

Chapter 6

Differentiating the on-surface Poynting vector of a wire to determine its radiation loss

Abstract

This chapter explores use of the Poynting vector (PV) computed from the surface current and charge distribution to investigate the power flow at the surface of various thin, perfect, electric conducting (PEC) wires excited as antennas or scatterers. By differentiating the axial Poynting vector at the wire's surface, the rate of change of power flow in the current and charge along a the wire can be determined. The idea to be explored is whether the loss in power flow along the wire can show where radiation occurs. The differentiated Poynting PV (DPV) results are compared with those obtained using Farfield Analysis of Radiation Sources (FARS), a method briefly introduced here and covered more thoroughly in Chapter 9.

Results are presented for two antenna-excitation models for a straight-wire dipole, the usual tangential E-field model and a two-wire transmission-line feed. Their comparison for a 10-wavelength dipole shows them to agree to within 10% relative to end peaks in the distribution of spatial radiation except in the vicinity of the antenna feedpoint. Both show that radiation occurs not only near the feedpoint and ends of the antenna, but is distributed along its length. Results presented for other antennas such as a bent-wire dipole, and circular and square loops, and for a straight-wire scatterer further demonstrate the effect of shape and excitation on the distributed radiation obtained using the DPV.

Keywords: Current and charge power flow; Charge reflection; Current reflection; Differentiated Poynting vector; Distributed radiation; Electromagnetic radiation; FARS and the Poynting vector; On-surface Poynting vector; Power-flow loss; Poynting vector; Radiation power loss

6.1 Introduction: current decay and propagation radiation in the frequency domain

Various different ways of determining energy loss and the distribution of radiated energy in the time domain have been outlined in the preceding chapters. An alternative way to find the distribution of radiated power in the frequency domain (FDFARS) is introduced here for comparison with the DPV. Further more detailed discussion about FDFARS follows in later chapters beginning with Chapter 8. The basic question being considered is "What is the spatial distribution of radiation emitted by an antenna or scatterer?"

To begin the discussion, it is relevant to mention that the decay of current with increasing distance from the feedpoint of a long-wire antenna has been attributed as being due to radiation by various authors. Examples from frequency-domain analyses include:

- "Manneback's solution for waves on a semi-infinite perfectly conducting (PEC) wire of infinitesimal radius shows that when the radius is finite the current decays, hence energy is slowly diverging from the wire" [1,2].
- "If damping due to the emission of radiation is neglected . . . the current along a (finite dipole) antenna can be taken as sinusoidal [3].
- "Most of the radiation takes place in the exponential decay (of the conical spiral current) region" [4].
- "An outgoing traveling wave (of a finite dipole antenna) . . . decays slowly . . . as a result of radiation" [5].
- "The current wave on the infinite rod decreases slowly with distance and it is damped by radiation losses" [6].
- "The current wave decays slowly with distance along the (infinite cylindrical antenna) rod . . . gradually radiating energy" [7].
- "For the latter case, therefore, the magnitude of the current wave (on an infinitely long cylinder) should decay with distance from the excitation point due to radiation losses" [8].
- "This paper is a didactic discussion of how the electromagnetic energy radiation by an antenna emerges from a localized source, is guided by the antenna's conductors, and ultimately shakes (this author's emphasis) free to form the radiation described by the asymptotic Poynting vector" [9].

Transient and time-domain analyses also discuss this issue, some of which are associated with modeling lightning, including:

- "Does an infinitely-conducting, traveling-wave vertical wire antenna above a perfectly-conducting ground have radiation losses so that current along the antenna is attenuated? . . . The answer is . . . 'yes' for the case of a longitudinal electric field source . . ." [10].
- "In view of the absence of ohmic losses, the observed current attenuation is attributed to 'radiation losses'" [11].
- "The peak of the traveling current wave in the lossless vertical conductor decreases markedly within the lowest 500 m or so, which is ascribable to radiation loss" [12].
- "In this approximation, the apparent current attenuation with height can be attributed to waves reflected back to the source" [13].

Attributing radiation to current decay is plausible since this implies that the on-surface Poynting vector consequently must carry less power or energy as a function of distance down the wire [14]. Of the references above, only one [13] mentions reflection of the current (charge) being carried by an outward-propagating current wave as an explicit cause of current decay. But even then it does not explain how and why charge reflection could cause electromagnetic radiation. These topics are explored here to demonstrate how such an object can radiate along its length.

However, other interpretations have also been proposed, e.g.:

- For a pulse-excited straight wire: " . . . current propagates along the antenna without radiating (this author's emphasis) until it reaches the end of the wire where the charges are decelerated and the radiation of the negative pulse takes place" [15].
- For a pulse-excited straight wire: " . . . the charge-density pulse travels along the monopole at the speed of light, with no dispersion . . ." and " . . . there is no radiation along the monopole (this author's emphasis)" [16].

The preceding two interpretations are incompatible with the results that follow below.

6.1.1 Current decay and charge reflection in the frequency domain

An analysis of the current on an infinite PEC wire antenna developed the following expression for its variation with distance z from the feedpoint as [5]

$$I_{\infty}(z) = \frac{ie^{ikz}}{\eta_o} \log\left[\frac{2\pi i}{2C_w + \log\left(kz + \left[(kz)^2 + +e^{-z\gamma}\right] + \gamma + i\frac{3}{2}\pi\right)}\right] \tag{6.1}$$

where $C_w = \log(1/ka) - \gamma$ and $\gamma = 0.57721566$ (Euler's constant) with k being the wavenumber of free space and a the wire radius. A somewhat simpler, alternate expression for the infinite-antenna current is given in [1,2] as

$$I_{\infty}(z) = Ae^{ikz}[2\log(2z/a) - 1]^{-1/2}. \tag{6.2}$$

The envelopes of maximum current magnitudes from (6.1) and (6.2) for $ka = 0.01$ are plotted in Figure 6.1a normalized to a value at 0.16 wavelengths (as a reference point) from the feedpoint. Both current envelopes decay monotonically with increasing distance from the feedpoint due to what is described here as propagation radiation (see A in Figure 3.1, Chapter 3). Being approximations, these results are slightly different but both exhibit a monotonic decay in the current envelop with distance from the feedpoint. The decay rates of the currents in Figure 6.1 are plotted in Figure 6.1b where they agree more closely than the do the currents themselves.

The only plausible explanation for the current decay shown in Figure 6.1a is that charge is being reflected continuously as the current-charge wave propagates down the wire. Charge does not "stick" to or leak off the wire, as these phenomena are not included in the numerical model because they are not physically feasible for a PEC in a linear regime of operation. Note that reflected charge is accelerated to become of source of radiation.

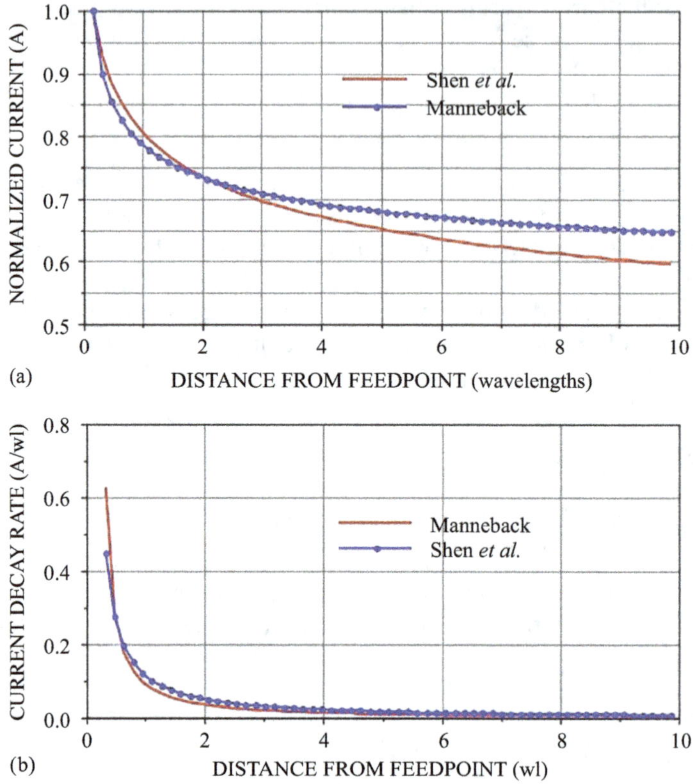

Figure 6.1 The envelopes of maximum current magnitude from Manneback [1,2] and Shen et al. *[5] (a) and the rate of current decay exhibited by these currents (b)*

By contrast, the total current on an infinite biconical antenna [17]

$$I(r) = \int_0^{2\pi} H_\varphi r \sin(\varphi) d\varphi = 2\pi H_\varphi e^{-ikr} \tag{6.3}$$

does not decay with distance because there is no reflected current and radiation from the infinite bicone originates only at its apex, or feedpoint.

The on-surface tangential Poynting vector of a 10-wavelength, center-fed dipole similarly also decays with distance as shown in Figure 5 in [14] and Figure 6.2. This is obtained from the dipole's radial electric field (charge density) and azimuthal magnetic field (current density) using NEC. The power flow is somewhat oscillatory while declining monotonically with increasing distance from the dipole feedpoint. The significance of the current reflection and consequent decrease in power or energy flow is explored next.

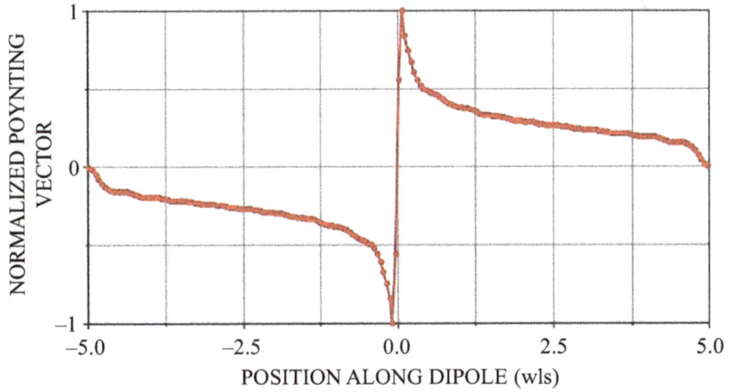

Figure 6.2 *The behavior of the normalized* $PV_{||}(a,z)$ *results as a function of position along a center-fed, 10-wavelength dipole. The outward-directed power flow decreases monotonically with increasing distance in both directions from the center feedpoint, which incorporates two segments of the 200-segment model*

6.2 A preview of frequency-domain FARS

An approach called FARS (Far-field Analysis of Radiation Sources) was initially presented as a way to answer the above question about where radiation is emitted from antennas and scatterers [18]. It is introduced here to provide an independent check on various results to be presented from the DPV. Both frequency-domain FARS (FDFARS) and time-domain FARS (TDFARS) have been developed in the context of the moment method. FDFARS was demonstrated to determine the contribution made to the total power flow at a given far-field observation point by each incremental current length (or segment), s_m, on a PEC boundary or prescribed current [19]. The point-wise FDFARS (denoted for brevity in the following as FARS) incremental power-flow density $p_m^{(FDF)}(\vartheta, \varphi)$ from segment s_m at an observation angle ϑ, φ is given by (for an object of arbitrary geometry)

$$p_m^{(FDF)}(\vartheta, \varphi) = \frac{1}{2\eta}\mathrm{Re}\left[e_{m,\vartheta}(\vartheta, \varphi)E_\vartheta^*(\vartheta, \varphi) + e_{m,\varphi}(\vartheta, \varphi)E_\varphi^*(\vartheta, \varphi)\right],$$
$$m = 1, \ldots, N \tag{6.4a}$$

from which the FARS Linear Power Density (LPD) for the wire objects considered follows as:

$$P_m^{(FDF)} = Lim r \to \infty\left[r^2\int_0^{\pi}\int_0^{2\pi} p_m^{(FDF)}(\vartheta, \varphi)\sin\,\vartheta d\vartheta d\varphi\right] = LPD_m^{(FDF)}. \tag{6.4b}$$

Also, $e_{m,\theta}(\theta,\varphi)$ and $e_{m,\varphi}(\theta,\varphi)$ are the theta and phi components of the incremental, far electric fields due to segment m, the total electric fields for the N segments that comprise the object are given by $E_\theta(\theta,\varphi)$ and $E_\varphi(\theta,\varphi)$ and η is the medium wave impedance. The total *FARS* radiated power, taken to be 1 w for all results below unless stated otherwise, is then obtained from summing over the N $P_m^{(FDF)}$ values to get

$$P^{(FDF)} = \sum_{m=1}^{N} P_m^{(FDF)} = P^{(RAD)} \tag{6.4c}$$

By definition, $P^{(FDF)}$ must equal the usual result for the radiated power $P^{(RAD)}$ since they are both obtained by performing the same overall far-field integration. They differ only in the order in which the integration is performed. $P^{(RAD)}$ is found by first integrating over a source distribution to sample the power-flow at a specified far-field point and then integrating these field samples over the far-field sphere to get the total radiated power. On the other hand, *FARS* first integrates over the far-field sphere to sample the radiated power from a specified source segment or patch, and then integrates these power samples over the source distribution to determine the total radiated power.

The results produced by FARS appear physically plausible as well in agreeing with different, related analyses [18]. For example, for standing-wave current distributions such as the sinusoidal current filament (*SCF*), where the time-average Poynting vector is zero, FARS agrees, within 1% or so, the distributed power determined from the Induced EMF (IEMF) method. FARS exhibits comparable agreement with the concept of a distributed radiation resistance introduced by Schelkunoff and Feldman [20]. And, as is shown here, the differentiated on-surface Poynting vector agrees similarly with *FARS* when appropriately normalized for straight wires.

One less-attractive feature of FARS is that it requires that an accurate integration be performed over the entire far-field sphere of the power flow from the object whose radiating properties are being studied. This can be time-consuming for objects large with respect to the wavelength, especially since it requires the far field to be sampled rather finely to achieve the needed integration accuracy. The accuracy of this integration can be checked in the usual way, i.e., by comparing the input power with the far-field radiated power. Unless otherwise noted, the *NEC* input power and the total FARS power for the results that follow agree to within a percent. But a more efficient way to investigate the origin of radiated power would be desirable.

In the following, the time-average, normal or tangential Poynting vectors ($PV_\perp$) or ($PV_\|$) near or at the surface of a straight-wire dipole are computed to analyze the power-flow density. A subsequent differentiation of $PV_\|$, designated as $DPV_\|$, then provides a measure of the power loss rate due to radiation by the dipole as an alternative to FARS [14]. Additional FARS and $DPV_\|$ results for scattering from a straight wire and some other objects of interest are also presented to demonstrate the utility of this procedure for determining the origin of radiated power. The chapter concludes with some general comments and observations.

6.3 Power flow near a straight-wire antenna

Various methods have been used to investigate and display the fields and power flow near an object including vectors, streamlines, false-color and perspective surfaces. In the following, the time-average Poynting vector parallel and perpendicular to straight wires will be shown as conventional rectangular graphs. This kind of plot is chosen to offer a possibly more quantitatively interpretable display of the physical phenomena to be considered. All computations that follow were performed using the Numerical Electromagnetics Code (NEC) [20], in most cases for a 10-wavelength wire as the nominal configuration and a wire radius of 10^{-3} wavelengths unless stated otherwise. Results included here are for both the time-average $PV_{\|}$ and surface integrals of the *PV* (*IPV*) over annular volumes or "pill boxes" centered on the wire axis. Particular attention is paid to $PV_{\perp}$ and $PV_{\|}$, and their integral values denoted by $IPV_{\perp}$ and $IPV_{\|}$, respectively. Note that the PV at point $\boldsymbol{r}$ in space is normally defined as

$$\boldsymbol{PV}(\boldsymbol{r}) = \frac{1}{2}\mathrm{Re}[\boldsymbol{E}(\boldsymbol{r}) \times \boldsymbol{H}^*(\boldsymbol{r})]. \tag{6.5}$$

Similarly, the power flow at the surface of the wire at axial coordinate z can be obtained from the wire current $I(z)$ and charge $Q(z)$ as

$$PV_{QI}(z) = \frac{1}{2}\mathrm{Re}[Q(z)I^*(z)]. \tag{6.6}$$

Equation (6.6) is the on-surface equivalent for a wire of (6.5) since the on-surface magnetic field is equivalent to the current and the radial, on-surface electric field is equivalent to the charge. Observe that differentiating $PV(z)$ yields the power-flow loss with distance as a spatial measure of radiation emitted by the wire. Observe that when *PV* or *DPV* have no arguments or subscripts that they are generic or on-surface quantities.

6.3.1 The PV

The *PV*, either parallel or perpendicular to the wire, is of interest because, among other reasons, of the insight that it can yield about propagation, radiation and scattering processes, as will be demonstrated in the following and as discussed further in Chapter 11. The discussion here argues that any change in the power flow over an incremental length (segment) at the surface of a PEC wire is due to one or more of five fundamental causes. These are:

1. Excitation power, coming from a tangential incident electric field in a scattering problem, or a localized voltage source for an antenna.
2. Interaction power, representing either a net loss or gain due to tangential electric-field coupling with other parts of the wire, a quantity discussed further in Chapter 11.
3. Dissipative power, an inward power flow due to imperfect conductivity or resistive loading of the wire.

4. Conduction power, coming from neighboring segments due to current and charge flowing along the wire, explicitly included in the *PV*.
5. Radiation power, a perpendicular, outward-power flow to the far field, the quantity of primary concern here.

To put this into some perspective, the parallel, on-surface $PV_{\|}$ for a 10-wavelength, dipole antenna excited at its two center segments, was shown in Figure 6.2. On this plot, power flow to the right is positive and power flow to the left is negative, showing power flowing away symmetrically from the center excitation. This plot illustrates the loss of power due to radiation down the wire as the $PV_{\|}(a,z)$ [obtained from $I(z)$ and $Q(z)$] decays monotonically towards each end of the dipole from the feedpoint. The normalized current distribution from which this result is obtained is plotted in Figure 6.3 where the real component of the current also decays toward zero while the imaginary component is relatively constant.

It is worth noting that using the on-surface $PV_{//}$ and $DPV_{\perp}$ to investigate power-related questions such as where radiation originates from a PEC object may seem inappropriate, since the power that the object creates resides in its fields. But whatever power the object produces is directly due to the current and charge flowing on it. These induced sources are, in turn, caused by a specified excitation without which neither sources nor the power that these sources produce would exist. Thus, it seems reasonable that inferences about the power can be made from the boundary current and charge as an alternative to evaluating the fields that these sources produce. A further advantage of using the induced sources for this purpose is that they are normally obtained in a boundary-value model, whereas computing many additional far-field values other than the principle-plane patterns can be computationally expensive. This possibility is demonstrated below where results obtained from FARS and the on-surface *DPV* are found to be in general agreement.

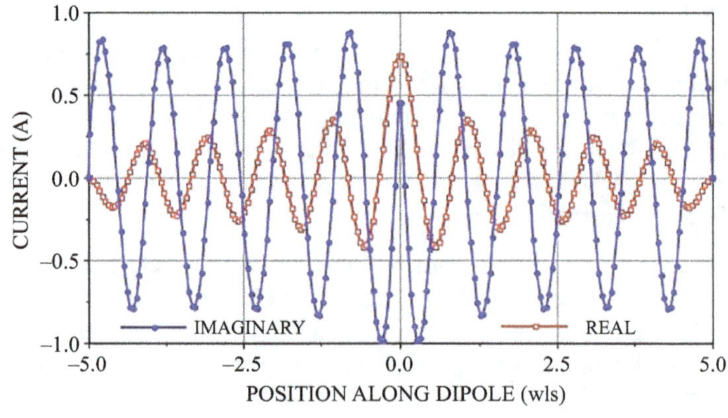

Figure 6.3 The normalized current on a 10-wavelength dipole antenna

Although the *PV* has the dimension of power flow per unit area, it is strictly defined in terms of its normal component integrated over a closed surface to establish the net power flow into, or out of, the enclosed volume, i.e., the integrated *PV* (*IPV*). While it's acknowledged that care must be used in making inferences based upon the *PV*, it has been widely used to investigate spatial power flow in EM fields. Furthermore, the *PV* has some similarity to the ray tubes employed in the Geometrical Theory of Diffraction and the role of rays in Geometrical Optics. In any case, it's reasonable to ask what physical interpretation can be attached, if any, to the *PV* at the surface of a PEC wire, a question to be considered further below.

6.3.2 Power flow near a PEC wire

In evaluating the *PV* and *IPV* for the wire objects to be considered here, an approach analogous to the usual moment-method modeling is being employed. The object being modeled is divided into a number of uniform segments of length Δz, typically 20 per wavelength and the on-surface $PV_{\|}$ is evaluated at the center of each segment at locations z_m, $m = 1, .. N$. The *IPV*s are evaluated over the bounding surfaces of N pillbox-shaped, annular volumes of radius R and length Δz congruent to the wire and whose ends are located at $z_{m\pm} = z_m \pm \Delta z/2$, as illustrated in Figure 6.4.

The *IPV* integrals are given by

$$IPV_{\perp}(R, z_{m\pm}) = 2\pi R \int_{z_m-\Delta/2=z_{m-}}^{z_m+\Delta/2=z_{m+}} PPV_{\perp}(R, z)dz \tag{6.7}$$

and

$$IPV_{\|}(R, z) = 2\pi R \int_{a}^{R} PPV_{\|}(r, z)dz \tag{6.8}$$

Because the wire is taken to be a PEC, the normal power flow at the wire's surface is zero.

Since the normal or perpendicular power flow from the surface of the *PEC* wire is zero, the power flowing out of the segment m annular volume is then given by

$$IPV_{\perp}(R, z_{\pm}) = [IPV_{\|}(R, z_{m+}) - IPV_{\|}(R, z_{m-})] \tag{6.9}$$

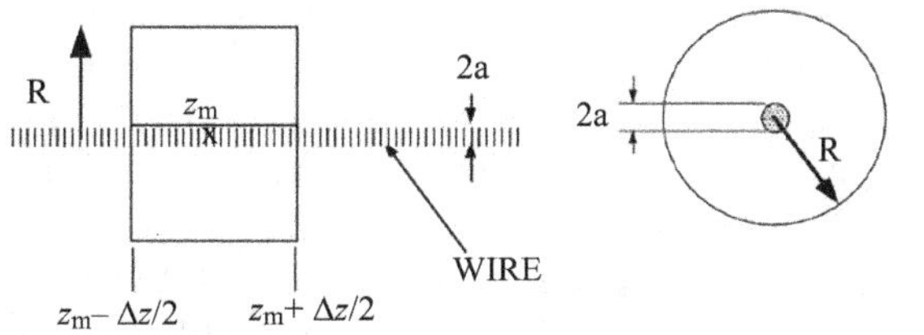

Figure 6.4 The parameters of the pillbox volume related to the IPV *evaluation*

as an alternative to using (6.7). Thus, a change in the power flowing parallel to the wire across annuli defined by $R,z_m \pm \Delta z/2$, will be revealed by the condition that $IPV_{\perp}(R,z) \neq 0$, and vice versa.

Note that it is the net $PV_{\perp}$(a,z), which includes the fields of both the excitation and the current and charge induced by it on the wire, that will always be zero at the surface of a *PEC*. However, $PV_{\|}(a,z)$ would generally not be zero as it represents the flow of power along, or parallel to, the wire. The question arises about how these two observations can be consistent, i.e., if $PV_{\perp}(a,z) = 0$ all along the wire, why then is the $PV_{\|}(a,z)$ not constant?

By way of explanation, note that it is plausible that the net $PV_{\perp}$(a,z) can vanish at a PEC surface if it were to be identically zero, a trivial case not of interest here. Alternatively, the more interesting case is if $PV_{\perp}(a,z)$ consists of inward- and outward-flowing PV components that must be equal at $R = a$. because of the boundary condition on the electric field, a result discussed and illustrated in Figure 11.2, Chapter 11. The inward-flowing component is comprised of whatever power is provided by the excitation and inward-flowing interaction power. Similarly, the outward-flowing component is comprised of whatever power is radiated plus outward-flowing interaction power. Observe that these three components, excitation, interaction and radiation powers, can cause a corresponding increase or decrease in the $PV_{\|}(a,z)$ depending on whether there is a net power gain or loss by the wire at z. A trivially simple non-wire example of a cancellation of incident and radiated (or reflected) power is provided by an infinite *PEC* plane illuminated at normal incidence by a plane wave. The instantaneous and time-average power flows in the incident and radiated fields then are always equal and opposite at the surface. A simple dipole antenna helps to illustrate these phenomena in a more relevant way.

Consider a dipole antenna whose two halves open at right angles from a finite-length, two-wire transmission line having a voltage generator connected across its far end, a case to be considered quantitatively below. The voltage generator is the only location where excitation power is delivered to the system and so the $PV_{\|}(a,z)$ will exhibit an increase there. Along the transmission-line wires themselves, the $PV_{\|}(a,z)$ must be constant since this part of the system is, presumably, non-radiating. Furthermore the net interaction power, $PV_{\mathrm{int}}(a,z)$, will be zero along the transmission line assuming the usual transmission-line equations hold. Along the wires that comprise the dipole arms a decrease in $PV_{\|}(a,z)$ must occur due to radiation from the dipole. These various issues are illustrated further below.

Values were computed for $IPV_{\|}(R,z_{m\pm})$ and $IPV_{\perp}(R,z_{\pm})$ with $a \leq R$, along a center-fed dipole of length $L = 10$ wavelengths modeled using $N = 200$ segments. The pillboxes thus align with the segments of length Δused in NEC to model the dipole. Some results are shown in Figure 6.5 for the right-hand half of the dipole with the two center segments of the dipole excited and a total radiated power equal to 1-w. The total power flowing parallel to the dipole as a function of distance from the feedpoint for various R-values on the right-hand arm of the dipole is presented in Figure 6.5. The power flow at the source segment is entirely perpendicular to the wire at its surface, but decreases with increasing R, which accounts for the growth

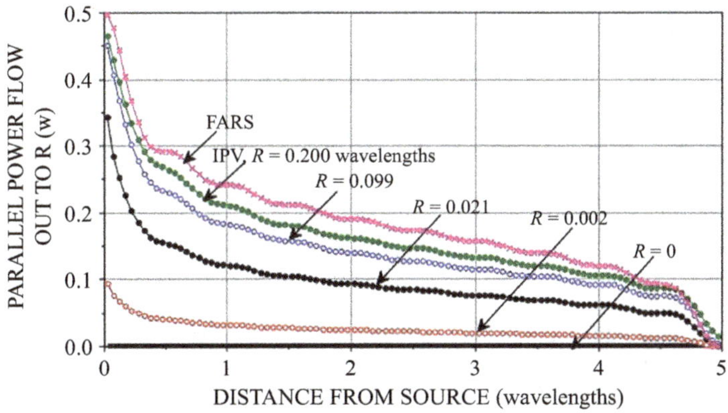

Figure 6.5 *The* IPV$_{//}$(R,z) *on the right-hand half of a 10-wavelength dipole from (6–8) showing the power flow through each annular area of radius* R *for a total radiated power of 1 w divided equally between the two halves of the dipole. These* IPV$_{//}$(R,z) *results asymptotically approach the* FARS *results with increasing* R

of the parallel-flowing power as R increases. For the special case $R = a$, $IPV_{||}(a,z)$ is zero since the annular area of the pillbox is then zero. This does not mean, of course, that the $PV_{||}(a,z)$ is zero on the wire surface as should be obvious from the fact that the normal E- and tangential H-fields (or the charge and current) are nonzero there. This will be discussed further below. Also note that the *FARS* curve in Figure 6.5 is obtained from integrating *FARS* from the dipole end toward its center to accumulate the total radiated power leaving the dipole from along its end to the feedpoint.

For $0 < R < 0.099$ wavelengths in Figure 6.5, the $IPV_{||}(R,z_{m+})$ curves monotonically approach an end value of 0. as z ——> $+L/2$, showing that the same total amount of power passes through the outer surface of an enclosing cylinder whose length equals that of the dipole. Consequently, $IPV_{||}(R,L/2)$ is essentially zero for $R < 0.099$, and no significant amount of power flows out the end of the cylinder. This occurs even though the power flow perpendicular to the dipole axis $IPV_{\perp}(R,z)$, changes substantially along the dipole with increasing radial distance from it as shown in Figure 6.5. For the $R = 0.2$ wavelength case, however, the value of $IPV_{||}(R,L/2) > 0.$, showing that some power is now passing through the end of the enclosing cylinder. If R is increased still further, this effect will continue to grow because more of the power radiating to the far field then passes through the end annulus. A plot of this effect out to $R = 0.2$ is shown in Figure 6.6.

The $IPV_{\perp}(R,z)$ results of Figure 6.7 are presented as an linear-power density (LPD) in w/wavelengths. Also shown are the *FARS* results for this dipole. Note that the $IPV_{\perp}(R = a,z)$ is zero except at the first segment, i.e., at the source, being zero elsewhere as a consequence of the boundary condition on the tangential electric field at the surface of a *PEC*. With increasing values of R, however, $IPV_{\perp}(R,z)$

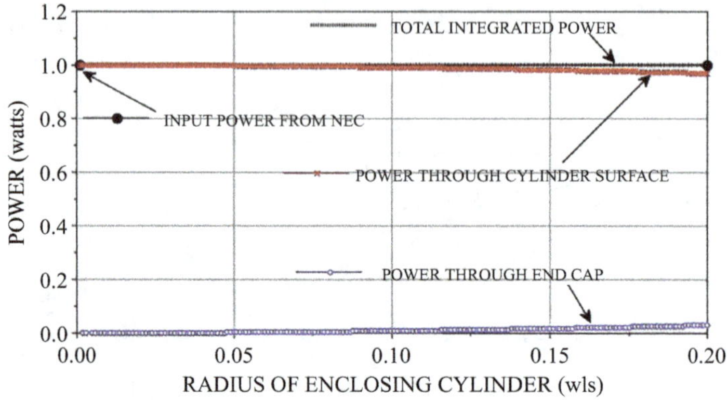

Figure 6.6 Endcap and cylinder-surface powers as a function of the radius of the enclosing annular cylinder. The NEC *input power and the total integrated power flows are shown for comparison*

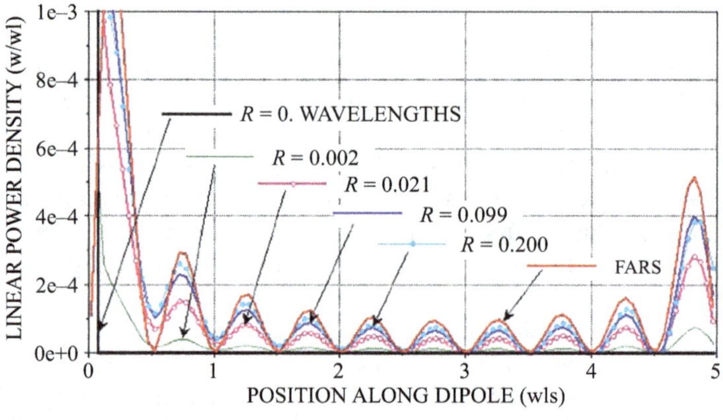

Figure 6.7 The $\mathrm{IPV}_{\perp}(\mathrm{R})$*, in w/wavelength, flowing through the outer surfaces of annular pill boxes having inner and outer radii of a and R and end caps at* $z_m \pm \Delta z/2$*, with* Δz *the length of each of the* m = *1, . . .*N *segments used to model the dipole, for the right-hand half of a PEC dipole*

values grow with increasing *R*. Also, the *FARS* and $IPV_{\perp}(R,z)$ results exhibit the same spatial variation away from the source region, with the result for $R = 0.2$ wavelengths approaching that obtained from *FARS*. The results of Figure 6.7 are equivalent to taking spatial derivatives of the $IPV_{\|}(R,z)$ shown in Figure 6.2, as they are the difference between the power flowing into, and out of, the ends of successive annular pillboxes, and provide the motivation for using the on-surface *PV* and *DPV* shown next.

6.4 FARS and differentiated-PV (DPV) results for a dipole antenna

Referring to Figure 6.2, where the normalized on-surface $PV_{||}(a,z)$ is plotted for the entire 10-wavelength dipole, observe that $PV_{||}(a,z)$ decreases monotonically away from the center feedpoint towards each end of the dipole and that

$$PV_{||}(a, \pm L/2) = 0. \tag{6.10}$$

In the discussion above, the parallel power-flow density at R,z has been obtained from

$$PV_{||}(R,z) \equiv PV_{EH}(R,z) = \frac{1}{2}\mathrm{Re}\left[E_r(R,z)H_\varphi^*(R,z)\right] \tag{6.11a}$$

with $E_r(R,z)$ being the normal electric field and $H_\phi(R,z)$ the tangential magnetic field at R,z relative to the surface of the wire. Instead of using the normal electric and tangential magnetic fields to obtain $PV_{||}(a,z)$ when $R = a$, the charge density, Q, and current, I, on the wire can be used, a more efficient approach since these quantities are the usual outputs of the numerical model.

Thus, with $H = I/(2\pi a)$ and $E = Q/(2\pi a\varepsilon)$,

$$PV_{QI}(z) = \frac{1}{2}\mathrm{Re}[Q(z)I^*(z)]\frac{1}{\left[(2\pi a)^2 2\varepsilon\right]} = PV_{||}(a,z) \tag{6.11b}$$

where ε is the permittivity of the surrounding medium. Either $PV_{QI}(z) = PV_{||}(a,z)$ can be used to compute $DPV_{||}$ on the wire to show the net rate at which the power is changing as a function of position.

The decay in $PV_{||}(a,z)$ or $PV_{QI}(z)$ (the on-surface Poynting vector) that can be seen in Figure 6.2 is due to reflection of the current and charge caused by the spatially varying wave impedance of a constant-radius wire [6,21] and can be inferred as being a loss of power via radiation to the far field. A particular point to observe in Figure 6.2 is that the "incident" or exciting power is clearly shown as an abrupt change in the $PV_{||}$ across the two source segments. While other examples of the time-average power flow near objects such as a PEC dipole can be found elsewhere, e.g. [22–24], the kind of plot shown in Figure 6.2 provides a somewhat different perspective concerning radiation from such an object. Since all *PV* results that follow employ $PV_{QI}(z)$, the "QI" subscript will be dropped from *PV* and *DPV* as will the subscript "||" since only their parallel components will be shown. Also note that, in the results to follow, we use

$$DPV_{||}(a,z) = \frac{d\left[PV_{||}(a.z)\right]}{dz} = DPV \tag{6.11c}$$

Upon differentiating the *PV*(z) curve of Figure 6.2 with respect to z and normalizing the result with respect to the end peak of the 1-w FARS result, the plots for *DPV*(*z*) and FARS as shown in Figure 6.8a are obtained.* Except near the feedpoint, the two curves differ by at most on the order of 5–10%, relative to the normalized end peaks. Part of the feedpoint difference between the two results is due to the fact that FARS does not include the excitation while the *DPV*(*z*) approach automatically does so. This outcome indicates that the power loss

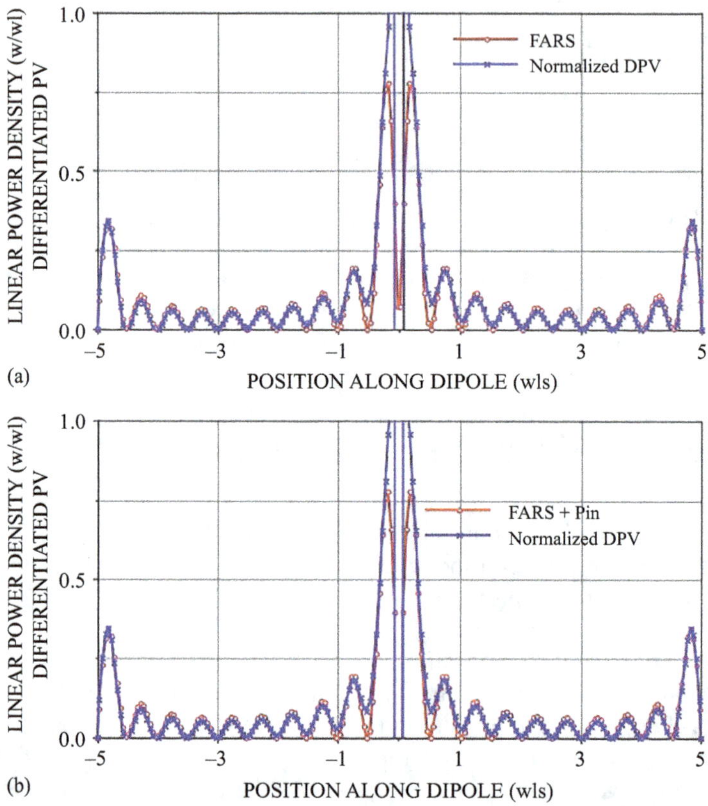

Figure 6.8 *Comparison of* DPV*(z) normalized to the end peak of* FARS *is shown in (a), while the effect of adding to* FARS *the incident power in the feedpoint region is shown in (b). These results demonstrate that, for practical purposes, the* DPV *and* FARS *are in acceptably good quantitative agreement to employ* DPV *as a tool to evaluate the origin of radiation from a simple wire object*

*Initially the differentiation was done using finite differences. Analytical differentiation was later added to NEC to exploit its analytically differentiable basis functions. The accuracy of the former was checked by halving the segment size of the numerical model with the curves for the two results being graphically indistinguishable. Results presented here use a combination of the two approaches.

indicated by the decrease in the on-surface Poynting vector along the dipole is largely consistent with the radiated-power distribution obtained from FARS. From this relatively close agreement between FARS and *DPV*(z) we might also conclude, for the simple case of the straight-wire dipole, at least, that the loss in power with distance from the feedpoint is dominated by radiation with the influence of near-field interaction power not being as important.

The result of adding the incident power density to FARS as provided by the induced EMF (IEMF) method is illustrated in Figure 6.8b. For a wire-like object, $p_{\text{inc}}(z)$ involves the incident tangential electric field at the wire, $E(z)_{inc}^{(\tan)}$, from which the IEMF power density, $p(z)_{inc}^{(IEMF)}$ for a wire current $I(z)$ is then given by

$$p(z)_{inc}^{(IEMF)} = \frac{1}{2}\mathrm{Re}\left[E(z)_{inc}^{(\tan)} I^*(z)\right] \tag{6.12}$$

The two curves of Figure 6.8b exhibit improved agreement in the feed region relative to Figure 6.8a. Since the effect of the incident power is essentially the same for the wire antennas discussed below as seen in Figure 6.8, it will not be considered further for the antenna results that follow.

FARS *LPD*s for the same dipole as shown above are plotted in Figure 6.9 for comparison with the square of the current magnitude on this antenna normalized to the second FARS peak. (The squared current is used since the power is related to a product of the current and voltage or one or the other of these quantities squared.) The total FARS power and the NEC input power agree to within 0.1% for this case, a typical result. The spatial variation in the radiation, as shown by the *LPD*s, is very similar to that of the squared current except at the ends and near the center

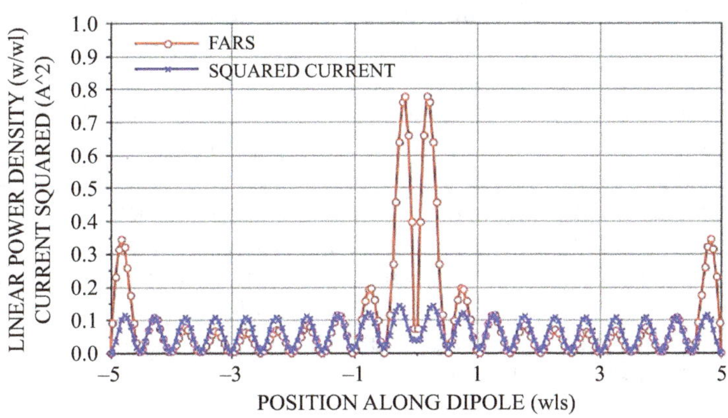

Figure 6.9 *The* FARS LPD *for 1-w radiated power and the square of the current normalized to the* FARS *peak second from the end for a center-fed dipole 10-wavelengths long and 10^{-3} wavelengths radius. The squared current and* LPDs *are closely correlated with respect to locations of their maxima and minima, although the current does not match the FARS excursions near the ends and in the feed region*

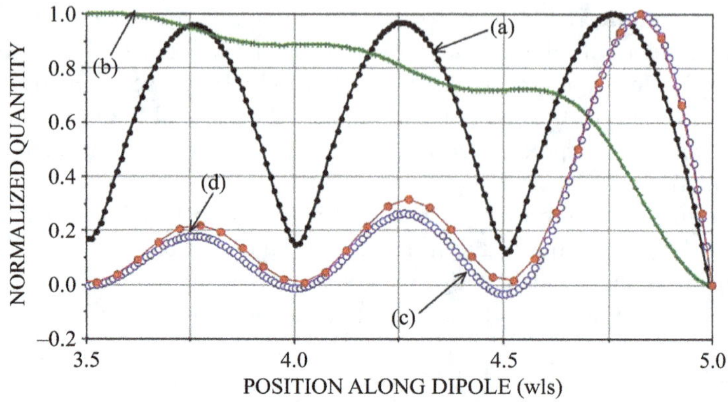

Figure 6.10 Details of (a) the normalized current magnitude, (b) the on-surface PV, PV*(a,z), (c) the differentiated PV,* DPV*(a,z), and (d) the* FARS *linear power density, near the end of a 10-wavelength, center-fed dipole antenna*

feedpoint where the *LPD*s exhibit substantially larger maxima. The *LPD*s indicate that reduced, but significant, intermediate radiation evidently occurs all along the antenna. This phenomenon is discussed in detail elsewhere [2].

The behaviors of the current magnitude and surface PV, together with the *DPV* and FARS results, are displayed in more detail near the right end of the dipole in Figure 6.10. While an expected maximum in the current magnitude occurs about 0.25 wavelengths from the end, the maximum in both the *DPV* and FARS results is found instead at about 0.17 wavelengths from the end. That *FARS* and *DPV* indicate that there is zero radiation right at the end of the dipole is consistent with the fact that the parallel *PV* must be identically zero there. This being the case, a smooth increase to a peak some distance from the end should be expected when the combined effects of current-maxima and end-reflection radiation are considered [25].

Peaks of the current magnitude in Figure 6.10 occur at approximately 1.235, 0.735, and 0.235 wavelengths from the wire end, or at about 5/4, 3/4, and 1/4 wavelengths, respectively. The maximum in the slope of the *PV* and corresponding peaks in the *DPV* and *LPD* nearest the end of the wire, on the other hand, are located measurably closer to the end at about 0.165 wavelengths away. This shift toward the end of the end-radiation peak relative to the current peak is evidently due to the effect of end-reflection radiation.

6.5 FARS and *DPV* results for a transmission-line driven dipole antenna

Since the greatest difference between FARS and the *DPV* occurs in the vicinity of the feedpoint, it may be worthwhile to examine a problem with a more-realistic source model. Results for the 10-wavelength antenna excited by a two-wire transmission line are included in Figures 6.11–6.13 for a total FARS power of 1 w. The transmission line is 4.2 wavelengths long and has a characteristic impedance of 300 ohms. The difference between the *FARS* and input powers in this case is about 7%.

The plots of Figure 6.11 exhibit the *PV* and the *DPV* for the transmission-line-driven dipole. The *PV* is seen to be essentially constant along the transmission line, as should be expected, since the transmission line is non-radiating and lossless. As the vicinity of the dipole connection is approached and beyond that bend, the *PV* falls monotonically to the end of the antenna and the *DPV* exhibits the periodic lobing of the FARS results previously seen for the straight-wire dipole.

The *FARS* and *DPV* results for the transmission-line driven dipole in Figure 6.12, for an input power of 1-w, agree to within 10%, relative to their normalized end peaks, except in the vicinity of the bend between the transmission line and dipole at ±5 wavelengths from each end. Differences of comparable magnitude also occur near the feedpoint of the transmission line, at 0 on this plot. It is probably significant that the 10% agreement between FARS and the *DPV*

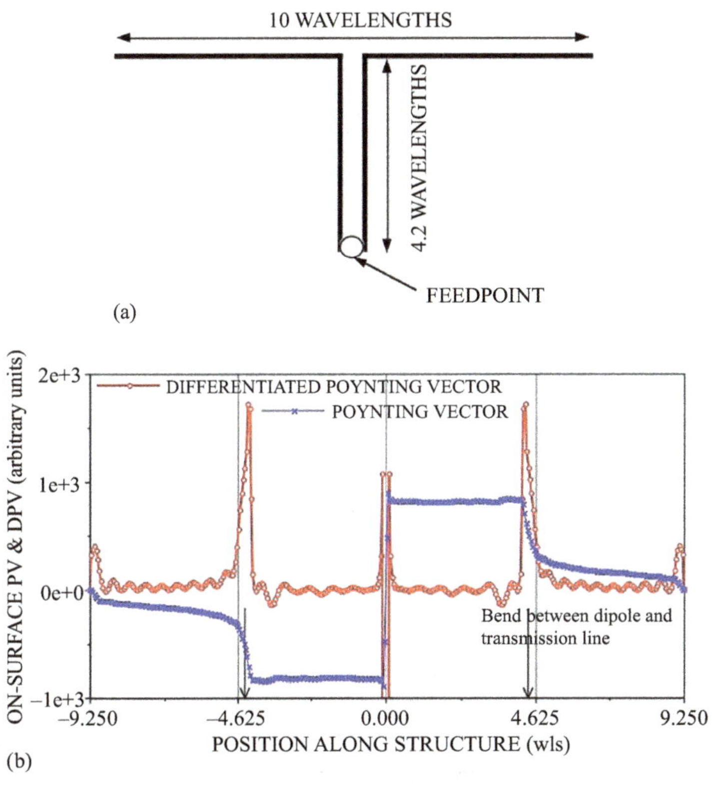

Figure 6.11 *The transmission line and dipole (a) and (b) the* PV *and* DPV *results for this structure. The power flow changes from being essentially constant along the transmission line to monotonically decaying beyond at the bend between the dipole and transmission line*

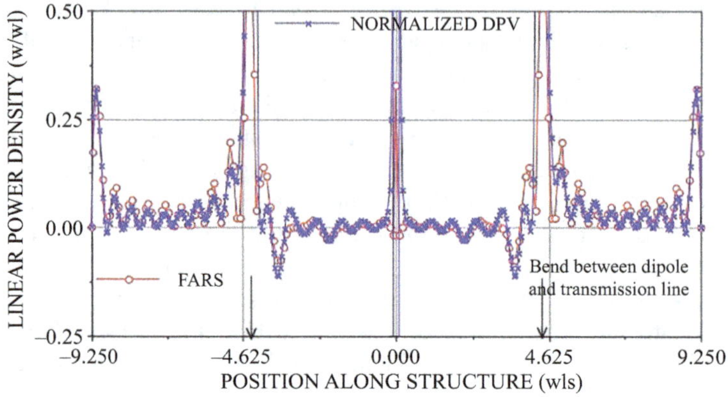

Figure 6.12 The linear power density from FARS and the DPV *result normalized to the FARS end peak for a 10-wavelength dipole driven by a 4.2-wavelength, 300-ohm transmission line*

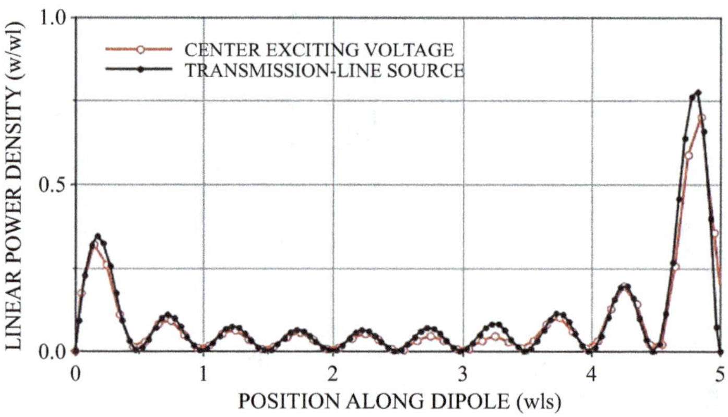

Figure 6.13 FARS linear power densities on the left arm of a 10-wavelength dipole excited by a voltage source at its center and by a 4.2-wavelength transmission line, the radiated power in each case being 1 w

extends even into the negative region just before the bend as well as along the transmission line where no power should be radiated.

The negative *LPD*s may appear to be non-physical as the radiated power must be a positive quantity, but this is simply explained. Observe that there will be zero field outside a closed PEC envelope within which is located an electromagnetic source, e.g. an antenna. This is because the antenna will induce a source distribution on the envelope that cancels the external fields that the antenna would otherwise produce, i.e., there is zero field outside the envelope. A FARS analysis of this problem would find that whatever external radiated fields the antenna would otherwise produce are exactly cancelled by the induced source distribution on the PEC envelope. An example of this situation is included in Chapter 7 for a dipole enclosed by a wire cage, where the cage wires have LPDs the sum of which numerically cancels the dipole fields. This is not exactly analogous to the situation here since there is no enclosing envelope, but the principle is the same. That is, the fields coming from some parts of an object can reduce the overall far-field power (or energy) flow from that object and thus exhibit negative LPDs. Also observe that the non-zero values on the transmission line are not compatible with transmission-line theory. The latter, however, excludes radiation whereas this possibility is included, of course, in a model based on Maxwell's equations.

A comparison of two FARS results on one arm of the dipole is shown in Figure 6.13 for the transmission-line-driven dipole and for a dipole center-excited by a voltage source, where the dipole left end is located at 0 in this plot and the feedpoint is at 5 wavelengths. The radiated power in each case is 1 w. These two FARS results agree overall to within 5% relative to the peak near the feedpoint. Observe that, in contrast to comparing FARS and the *DPV*, any differences between these two curves are due to the antenna model itself since each result has been obtained using only FARS.

6.6 FARS and DPV results for a straight-wire scatterer

6.6.1 FARS and the DPV *for a 10-wavelength wire scatterer at normal incidence*

Results from the *PV*/*DPV* approach and *FARS* for a 10-wavelength straight-wire scatterer are shown for normal incidence in Figures 6.14–6.16 and for near-axial

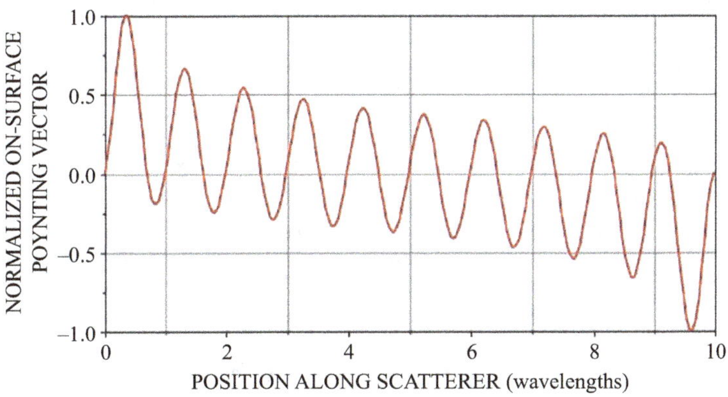

Figure 6.14 The normalized PV *as a function of position along a 10-wavelength wire excited by a normally incident plane wave with negative power flow to the left and positive to the right. The near-end power flow is predominantly towards the center but oscillates as a function of position, being odd-symmetric about the center*

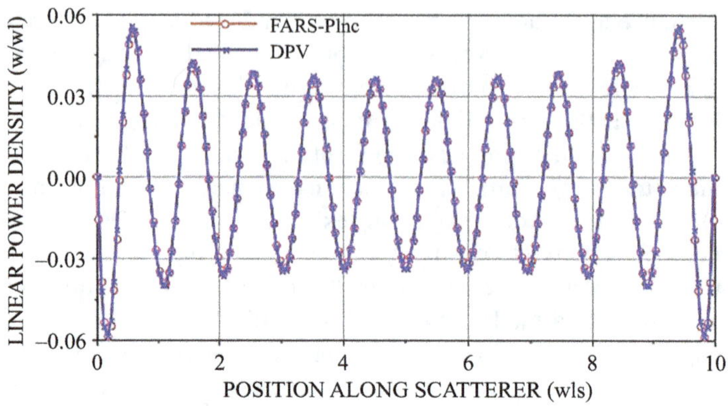

Figure 6.15 *The* DPV *normalized to the end peak of* FARS *for the wire of Figure 6.14 where the incident IEMF power of 1-w is subtracted from the* FARS *results. These results agree within few percent over the entire wire*

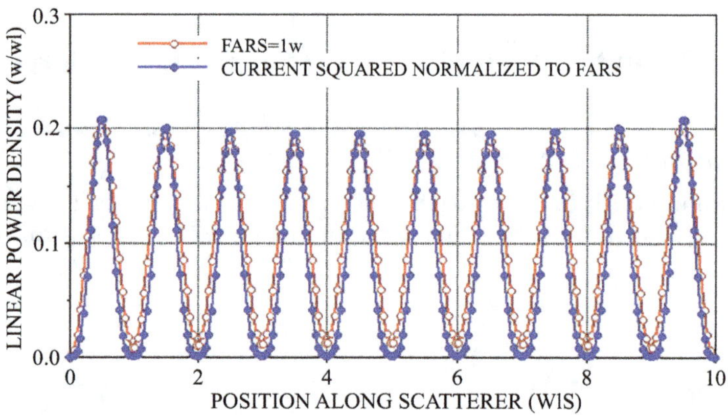

Figure 6.16 *The FARS LPD for 1-w scattered power together with the square of the current normalized to the FARS end peak for a 10-wavelength wire having a radius of 10^{-3} wavelengths. The squared current and* LPDs *are more closely correlated here than for the dipole antenna of Figure 6.9*

incidence in Figures 6.17 and 6.18. The *PV* for a plane wave normally incident on a 10-wavelength wire and 1-w scattered power shown in Figure 6.14 is negative for leftward-flowing power and positive for rightward-flowing power. The *PV* is seen to be odd symmetric about the wire's center with the largest peaks near each end. The *DPV* results normalized to FARS from which the incident power density is subtracted are contained in Figure 6.15. There, the two results agree to within a percent or so over the entire wire as contrasted with the dipole-antenna results above. The closer agreement between FARS and *DPV* for the scattering case is evidently due to the fact that the input power is distributed along the length of the wire whereas it is concentrated over a small area for the antenna. In Figure 6.16, the FARS data is compared with the current squared, normalized to the end peak of the FARS LPDs. As seen previously for the dipole antenna, the squared current and LPDs are closely correlated. These results also might be interpreted to imply an approximate validation of the *DPV* in so far as indicating where, and in what amounts, radiation occurs.

6.6.2 FARS and the DPV for a 10-wavelength scatterer at near-axial incidence

Application of the *PV*/*DPV* approach to a plane wave incident at 10 deg from end-on and propagating right-to-left on a 10-wavelength scatterer is illustrated in Figures 6.17 and 6.18. The *PV* shown in Figure 6.17 demonstrates that the power flow is now asymmetric, being predominantly towards the left, consistent with the plane wave arriving from the right. The FARS and normalized *DPV* results minus the incident IEMF power are compared in Figure 6.18 where they exhibit agreement comparable for the case of normal incidence and show that the largest peak in the radiated (scattered) power occurs at the left end. The predominance of this end-peak radiation in Figure 6.18 seems compatible with the well-known result from the geometrical theory of diffraction in showing that the ends are the primary scattering centers for oblique illumination of a straight wire.

6.6.3 The near-interaction field on a straight-wire scatterer

For a PEC object, the total $PV_{\perp}(a,z) \equiv 0$ on a local basis over the entire surface. It follows then that the sum of the incident, interaction, and radiated power flows

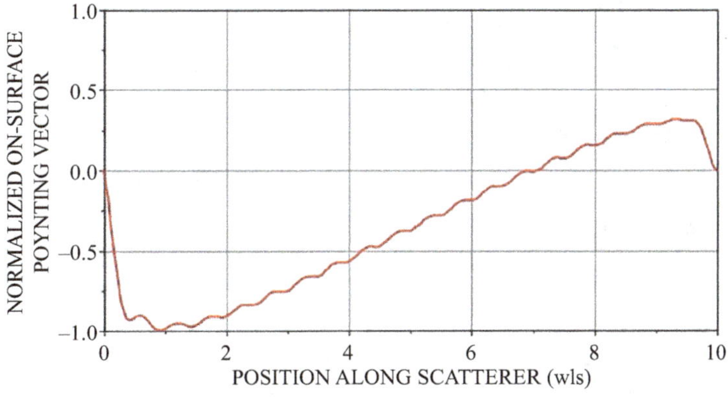

Figure 6.17 The normalized PV *for a 10-wavelength wire excited by a plane wave incident from the right-hand side at 10 deg from the wire axis. The power flow is predominantly toward the left end beyond three wavelengths or so away from the illuminated end, and is much less oscillatory than for normal incidence*

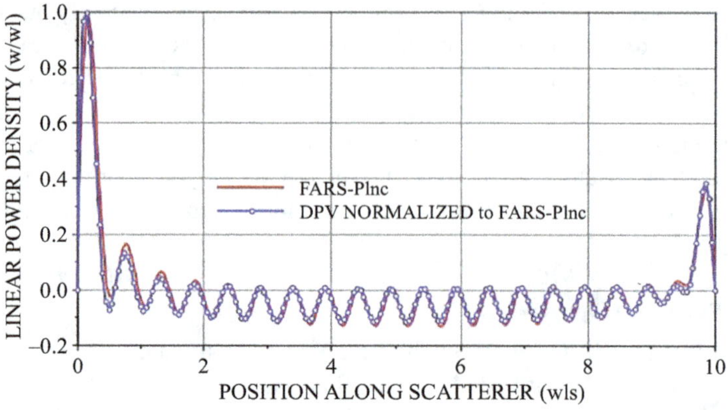

Figure 6.18 *The* DPV *normalized to the far-end peak of FARS for the 10-wavelength* PEC *wire whose* PV *results are shown in Figure 6.17. These results agree within few percent relative to the peak over the entire wire, again providing validation for the* DPV *approach. It is coincidental that the left-hand-end peak is about unity, as the incident and scattered powers are set to 1 w*

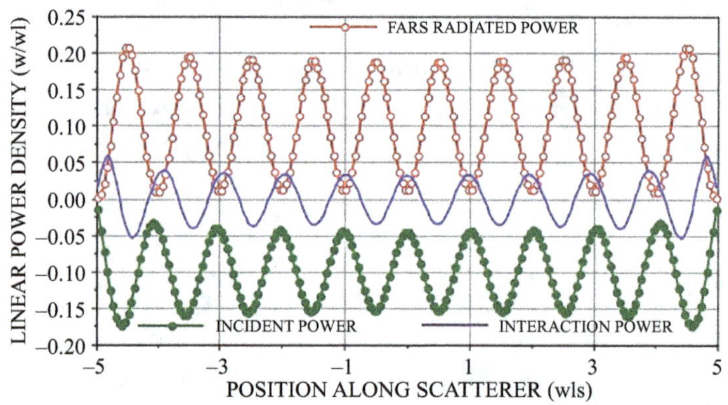

Figure 6.19 *The linear densities of incident, interaction and radiated powers for the 10-wavelength, straight-wire scatterer at normal incidence*

must be 0. Since the first and the third of these quantities are available from the IEMF method and FARS, it is possible to obtain the interaction power, an example of which is given in Figure 6.19 for the straight-wire, 10-wavelength scatterer at normal incidence for 1-w of collected, or incident, power. The oscillations in the interaction power follow those of the incident power, and since

$$p_{inc}(z') + p_{rad}(z') + p_{\text{int}}(z') \equiv 0 \tag{6.13a}$$

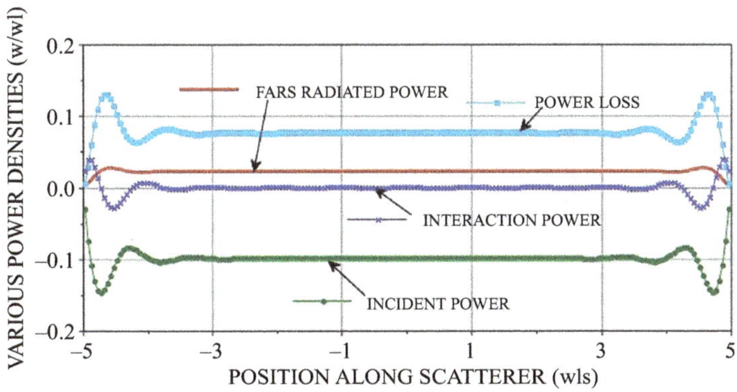

Figure 6.20 The linear densities of incident, interaction, loss, and radiated powers for normal incidence on a 10-wavelength, resistively loaded wire scatterer at normal incidence

then

$$p_{\text{int}}(z') = -[p_{inc}(z') + p_{rad}(z')]. \tag{6.13b}$$

The effect of adding 100-ohm resistors to each of the 200 segments of the straight-wire scatterer of Figure 6.19 is shown in Figure 6.20 to smooth out the oscillatory behavior of the unloaded case. The self-interaction power is essentially zero away from the wire ends and the radiated power is nearly uniform over the entire wire.

6.7 FARS and DPV results for some other thin-wire antennas

6.7.1 A dipole with two right-angle bends

Results for a 10-wavelength antenna having two right-angle bends two wavelengths from the ends are included in Figures 6.21 and 6.22. The bent-wire *PV* is compared with that of a straight-wire dipole in Figure 6.21 where the influence of the bends appears as an enhanced radiation-induced loss in their vicinity. The FARS and normalized *DPV* results for the bent dipole in Figure 6.22 for 1-w radiated power agree qualitatively in that the locations of their respective maxima and minima track well. They differ in magnitude relative to the normalized end peaks by 30% or so as well as in the source region. The effect of the bends in Figure 6.22 can be seen as a "notch" in the FARS and *DPV* values at the bends. In this respect, the bends produce a result similar to the end peaks that occur on the open end of a wire, and the local radiation peaks exceed those at the ends of the bent dipole.

6.7.2 A circular loop antenna

The *PV*s for a 10-wavelength dipole and circular loop are compared in Figure 6.23 and FARS and *DPV* results for the loop are shown in Figure 6.24. The loop *PV*

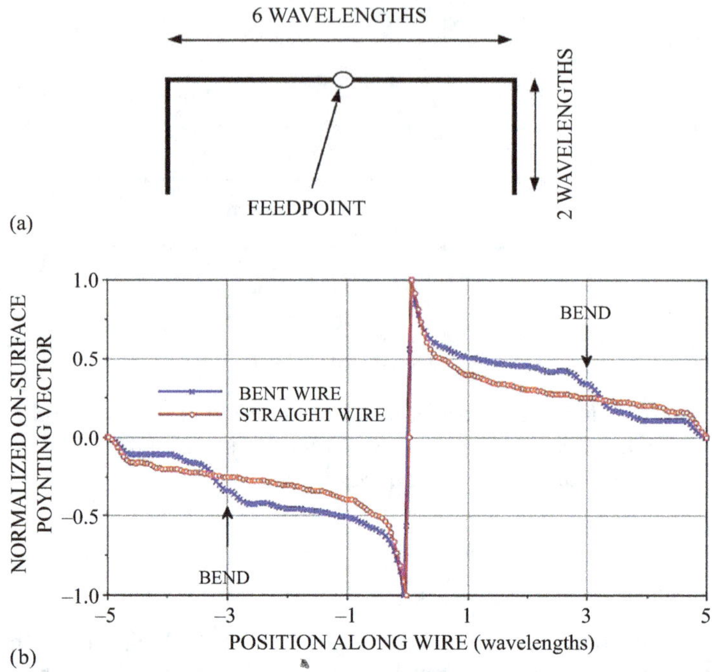

Figure 6.21 *Geometry of the bent wire (a) and (b) the normalized* PV *for this structure compared with that for the straight dipole. The bend locations are indicated by the arrows in the vicinity of which changes in the* PV *slope are seen*

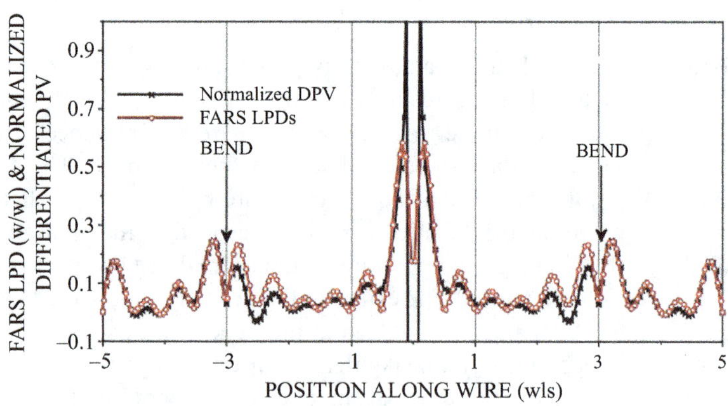

Figure 6.22 *The* DPV *normalized to the far-end peak of* FARS *for the 10-wavelength bent dipole with the total radiated power 1 w. These results are qualitative agreement but with differences of 30% or so, and are not as close as for the straight-wire antenna and scatterer*

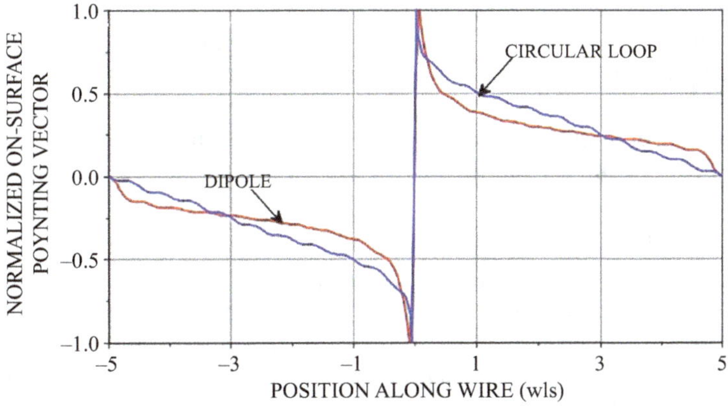

Figure 6.23 *The normalized* PV *for a 10-wavelength circular-loop antenna, excited at 5 wavelengths on this plot, compared with that for the center-fed dipole*

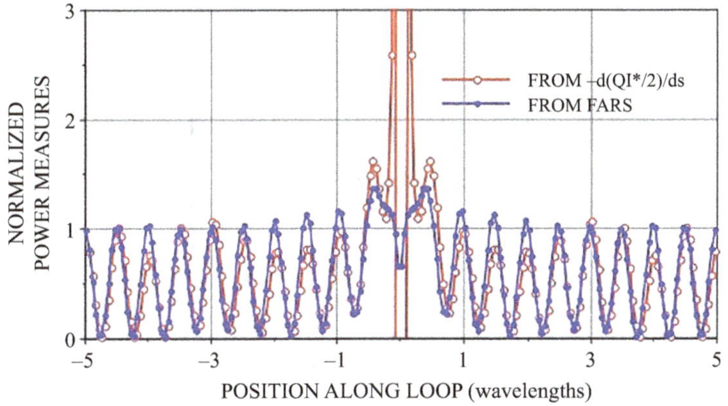

Figure 6.24 *The* DPV *normalized to the far-end peak of* FARS *for the 10-wavelength circular loop. These results are qualitative agreement but, with differences of 20% or so, again not as close as for the straight-wire antenna and scatterer*

displays a rather constant rate of decrease from the feedpoint, contrasting with that for the dipole where the decay rate is noticeably larger near the feedpoint and the dipole ends. A comparison of the *FARS* and *DPV* results in Figure 6.24 shows that while they are qualitatively similar in their patterns of minima and maxima, they do not agree as well as for the dipole. Their maximum differences relative to the normalized peaks, excluding the vicinity of the feedpoint, are 20% or so. It is not clear why FARS and *DPV* for the circular loop differ so much relative to the dipole, but it is worth noting that field coupling, or mutual interaction, on a circular loop might be expected (and near the bends of the bent dipole) to be greater than for a straight wire. This is because the tangential E-field of the straight wire, i.e., the interaction field, falls as $1/r^2$ along the wire, while that across the diameter of the loop it falls as $1/r$. This interaction field would represent an input power that the *DPV* may not represent well. The *DPV* approach does not explicitly include these coupling effects that are automatically included in FARS.

6.7.3 A square loop antenna

Results equivalent for the loop antenna above are repeated for a 10-wavelength square loop antenna in Figures 6.25 and 6.26. The effects of the successive right-angle bends of the square loop are seen as enhanced decreases in the *PV* of Figure 6.25 and increased radiation loss in the *DPV* and FARS results of Figure 6.26. As for the bent dipole, the FARS and normalized *DPV* results are in

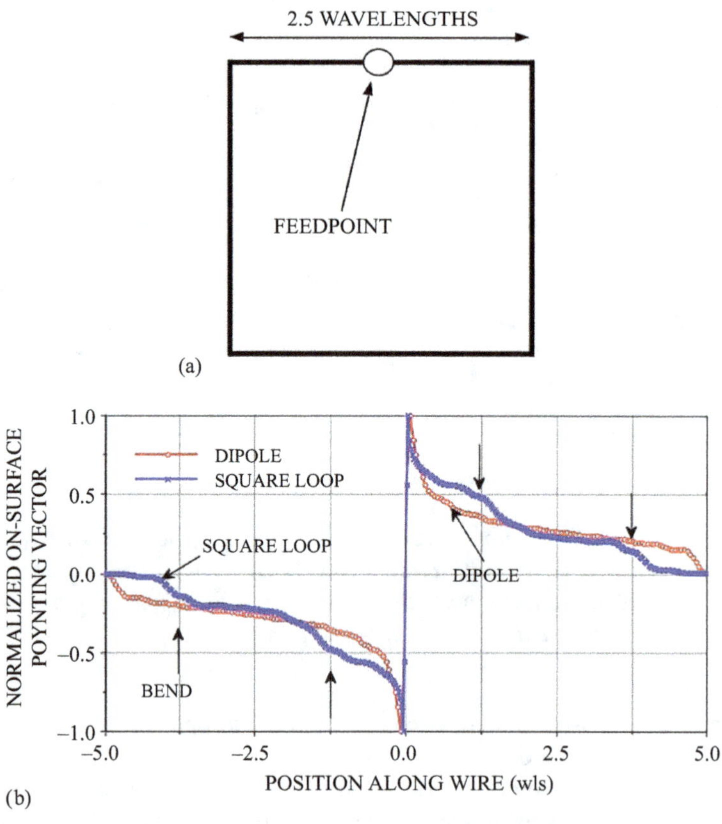

Figure 6.25 The square-loop antenna (a) and (b) the normalized PV *for this structure, excited at 0 on this plot, compared with that for the center-fed dipole. The loop's* PV *exhibits slope changes in the vicinity of each right-angle bend, whose locations are shown by the arrows*

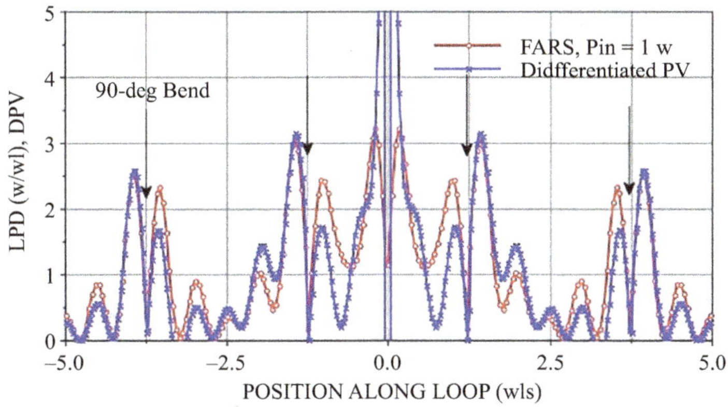

Figure 6.26 The DPV *normalized to the far-end peak of FARS for the 10-wavelength square-loop antenna. These results are qualitative agreement but, with differences of 40% or so, again not as close as for the straight-wire antenna and scatterer*

qualitative agreement in terms of their maxima and minima and differ on the order of 40% in their magnitudes.

6.7.4 A zigzag antenna

Zigzag antennas of various lengths and geometries were modeled using NEC and to investigate their *PVs* and *DPVs*. A generic example of a Zigzag antenna is shown in Figure 6.27. All segments of the Zigzag structure are 0.05 wavelengths long, with the total number of segments varied from 200 (10 wavelength) to 1,000 (50 wavelengths), and the included angles varied from 180 (a straight wire for reference) to 40 deg.

The FARS and normalized *DPVs* for a 10-wavelength Zigzag antenna and included angle of 100 deg are included in Figure 6.27. The two results agree to within 10% or so relative to the end-lobe maximum and, as in previous examples, the largest difference occurs in the source region. Also, over the region between the feedpoint and end lobes, negative powers again occur but are exhibited by both approaches with the FARS results in this case being somewhat more erratic.

Several Zigzag antennas having included angles varying from 180 deg (a straight wire), to 60 deg were run having 200 segments and an overall wire length of 10 wavelengths, to compare the FARS radiated powers with the NEC-computed input power. These results are plotted in Figure 6.28 where the two power values agree to within 1% to validate the two results.

In Figures 6.29 and 6.30, are shown results for a 50-wavelength Zigzag and straight-wire antennas excited by 1-V sources at segments numbered 6 and 7 from their left ends. The envelopes of their *PVs* are shown in Figure 6.29 where the characteristic decay of a long wire is displayed for the straight wire, while that for the Zigzag is nearly uniform over more than 60% in its center. Upon differentiating

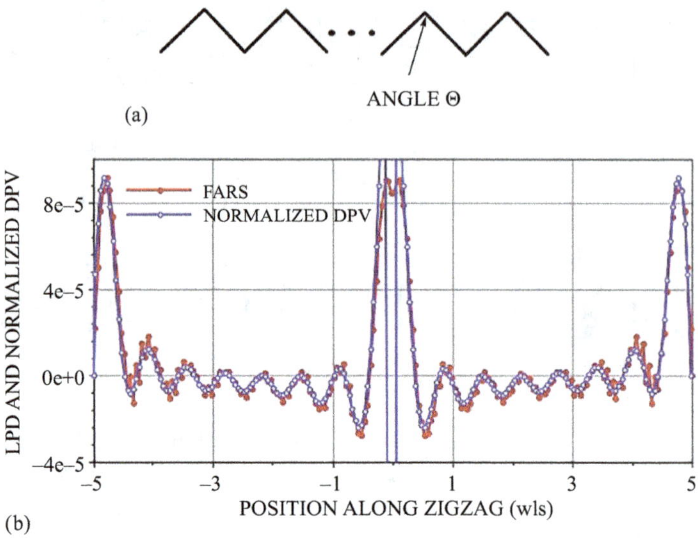

Figure 6.27 The Zigzag antenna (a) and the FARS and normalized DPV *results for a 10-wavelength antenna for 1-V excitation at the two center segments*

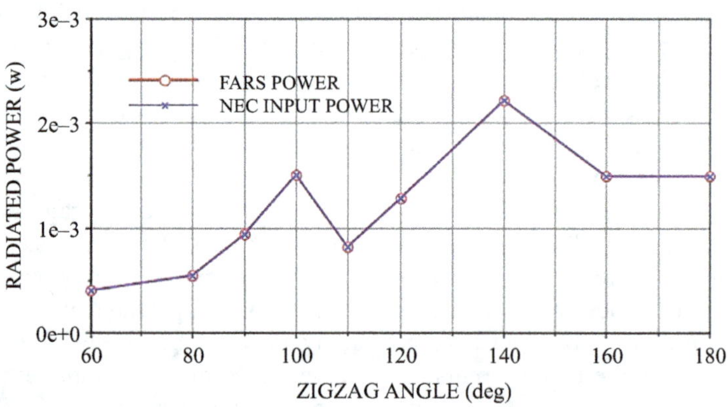

Figure 6.28 Comparison of the radiated power from the Zigzag antenna of Figure 6.26 obtained using FARS and the input power from NEC

these *PV* plots the *DPV* results of Figure 6.30 are obtained. For both antennas, radiation is primarily concentrated near their ends, but the Zigzag exhibits substantially lower *DPV* levels in its center compared with the straight wire, an interesting result. The Zigzag also has a greater amount of negative *DPV* values near the source region and the end.

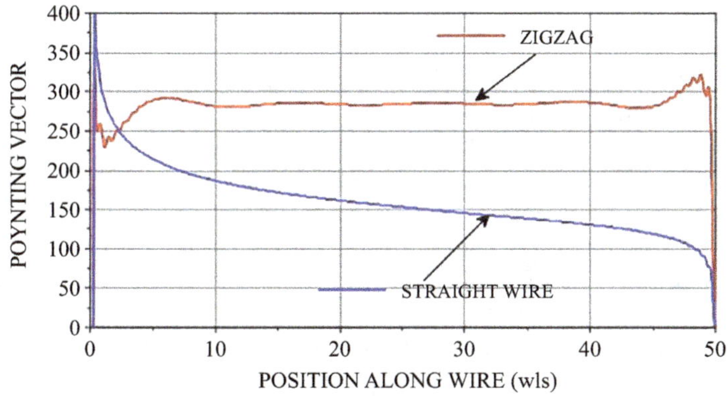

Figure 6.29 *The envelopes of the on-surface* PVs *for an end-excited straight wire and Zigzag antennas having 50 wavelengths of wire*

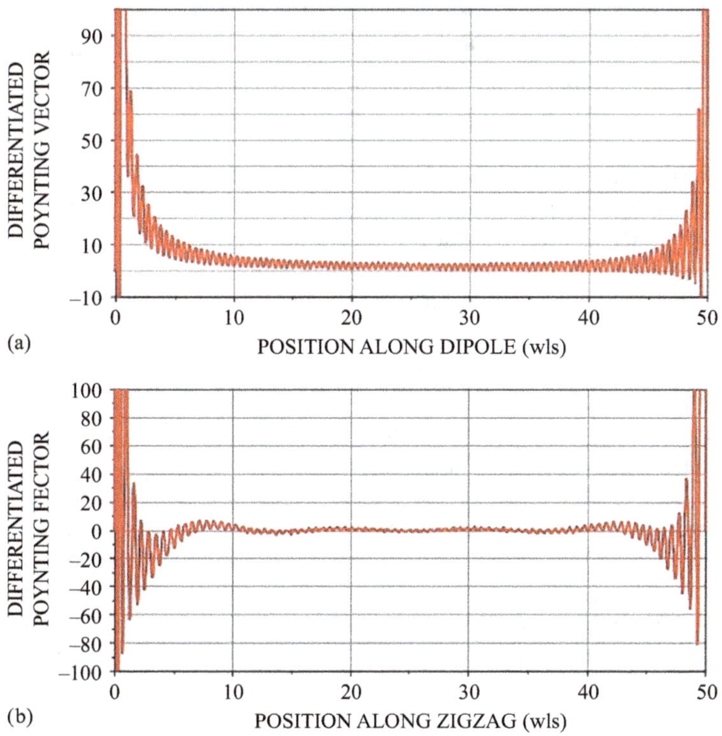

Figure 6.30 *The* DPVs *derived from the* PVs *of the antennas as shown in Figure 6.29*

6.8 Summary

The motivation for the approach described above has been to investigate the possibility that the on-surface, *DPV* as a means to qualitatively, preferably quantitatively, determine the amount of power being radiated per unit length from *PEC* wire objects excited as antennas or a scatterers. Use of the *DPV* begins by first computing the axial power flow represented by the on-surface *PV* along the object using either the normal E- and tangential H-fields at its surface, or equivalently, the current and charge distributions. For wire antennas, the *PV* is found to be a monotonically decreasing function from the feedpoint, reflecting the fact that all of the power deposited on the wire at its feedpoint is radiated away along its length.

The results obtained for straight-wire antennas and scatterers demonstrate that the *DPV* approach is generally in agreement with those obtained from *FARS* in terms of the spatial distribution of radiation from such objects. For antennas, the largest difference between the two approaches is confined to within ±0.5 wavelengths of the feedpoint, while otherwise differing, on a normalized basis, by 10% or so relative to the near-end peaks. For scatterers, the agreement is much better, within a few percent, over the entire object. For some simple problems, however, the *DPV* will be zero even though radiation does occur. Standing current waves, such as the sinusoidal current filament and constant-amplitude traveling-wave currents, are such examples. This is because the time-average power flow along the two counter-propagating, equal-amplitude current waves that comprise a standing-wave current is zero. For a constant-amplitude traveling wave, the *DPV* is zero because the time-average *PV* is a constant independent of position.

The results presented above show that the *DPV* does not fully agree with FARS even for the simple straight wires. As was noted, the *DPV* includes the effect of the excitation to which an antenna or scatterer is subject. The *FARS* approach, on the other hand, accounts only for the radiated, or far field. Furthermore, the *DPV* involves the near field and whatever interactions that might occur when power flowing from one part of an object is partially intercepted by another part of it. Such coupling would affect the *FARS* evaluation only to the extent that the interaction power becomes radiated, far-field, power. It is this phenomenon that is most likely the cause of significant differences that are found between FARS and the *DPV*. This kind of coupling should be a minimum for straight wires, but can be expected to be more pronounced for shapes such as curved and bent wires and closed loops, for example. In spite of its possible limitation in quantifying the spatial distribution of radiated power, a distinct advantage that the *DPV* has over FARS is that it is much less computationally intensive, since FARS requires accurate integration of the radiated power over the entire far-field sphere. In spite of the limitations to which the *DPV* can be subject, it has been found qualitatively to indicate where the objects investigated do radiate.

Acknowledgment

The author thanks G.J. Burke for adding FARS and the *DPV* approach to NEC.

References

[1] S. A. Schelkunoff, *Advanced Antenna Theory*, John Wiley & Sons, New York, NY, p. 190, 1952.

[2] C. Manneback, "Radiation from transmission lines," *AIEE Journal*, Vol. 42, pp. 95–105, 1923.

[3] J. D. Jackson, *Classical Electrodynamics*, John Wiley & Sons, New York, NY, p. 401, 1962.

[4] Y. S. Yeh and K. K. Mei, "Theory of conical equiangular-spiral antennas. Part II – current distributions and input impedances," *IEEE Transactions of AP-S*, Vol. AP-16, No. 1, pp. 14–21, 1968.

[5] L-C. Shen, T. T. Wu, and R. W. P. King, "A simple formula of current in dipole antennas," *IEEE Transactions of AP-S*, Vol. AP-16, No. 5, pp. 542–547, 1968.

[6] J. Bach Anderson, "Admittance of infinite and finite cylindrical metallic antenna," *Radio Science*, Vol. 3, No. 6, pp. 607–621, 1968.

[7] D. S. Jones, *Methods in Electromagnetic Wave Propagation*, Clarendon Press, Oxford, p. 295, 1994.

[8] A. V. Kudrin, E. Yu, Petrov, G. A. Kyriacou, and T. M. Zaboronkova, "Insulated cylindrical antenna in a cold magnetoplasma," *Progress in Electromagnetics Research,* Vol. 53, pp. 135–166, 2005.

[9] J. D. Jackson, "How an antenna launches its input power into radiation: the pattern of the Poynting vector at and near an antenna," *American Journal of Physics*, 74, 2006, pp. 280–288.

[10] B. Kordi, R. Moini, and V. A. Rakov, "Comment on 'Return stroke transmission line model for stroke speed near and equal that of light,' by R. Thottappillil, J. Schoene, and M. A. Uman," *Geophysical Research Letters*, Vol. 29, No. 10, 2002, p. 1369.

[11] B. Kordi, R. Moini, and V. A. Rakov, "Comparison of lightning return stroke electric fields predicted by the transmission line and antenna theory models," in *Proc. 15th Int. Zurich Symp. Electromagn. Compat.*, 102P2, Zurich, 2003, pp. 551–556.

[12] Y. Baba and M. Ishii, "Characteristics of electromagnetic return-stroke models," *IEEE Transactions of Electromagnetic Compatibility*, Vol. 45, No. 1, 2003, pp. 129–135.

[13] Y. Baba and V. A. Rakov, "On the mechanism of attenuation of current waves propagating along a vertical perfectly conducting wire above ground: application to lightning," *IEEE Transactions of Electromagnetic Compatibility*, Vol. 47, No. 3 August 2005.

[14] E. K. Miller, "The differentiated on-surface Poynting vector as a measure of radiation loss from wires," *IEEE Antennas and Propagation Society Magazine*, Vol. 48, No. 6, 2006, pp. 21–32.

[15] R. Gómez Martin, A. Rubio Bretones, and S. Gonzalez Garzia, "Some thoughts about transient radiation by straight thin wires," *IEEE AP-S Magazine*, Vol. 41, No. 3, 1999, pp. 24–33.

[16] C. C. Bantin, "Radiation from a pulse-excited thin wire monopole," *IEEE AP-S Magazine*, Vol. 43, 2001, pp. 64–69.

[17] C. Balanis, *Antenna Theory Analysis and Design*, Harper and Row Publishers, New York, NY, 1982.

[18] E. K. Miller, "PCs for AP and Other EM Reflections," *IEEE AP-S Magazine*, Vol. 41, No. 2, 1999, pp. 82–86; "Far-field analysis of radiation sources," IEEE AP-S International Symposium, Renaissance Orlando Resort, Orlando, FL, June 21–26, 1999, p. 2008; "PCs for AP and Other EM Reflections," *IEEE AP-S Magazine*, 42, 3, June 2000, pp. 84–89.

[19] S. A. Schelkunoff and C. B. Feldman, "On radiation from antennas," *Proceedings of the I.R.E.*, Vol. 30, 1942, pp. 511–516.

[20] G. J. Burke, Numerical Electromagnetics Code – NEC-4 method of moments, Part II: program description – theory, Lawrence Livermore National Laboratory, Rept. USRL-MA-109338, January 1992.

[21] E. K. Miller, "The proportionality between charge acceleration and radiation from a generic wire object," *Progress in Electromagnetics Research*, Vol. 162, pp. 15–29, 2018.

[22] F. Landstorfer, "Energy transport within the nearfield of rod antennas," *Archiv der elektr. Ubertragung*, Vol. 26, 1972, pp. 189–196.

[23] F. Landstorfer and R. R. Sacher, *Optimisation of Wire Antennas*, Research Studies Press, Letchworth, Hertfordshire, England, 1985.

[24] E. K. Miller and F. J. Deadrick, "Visualizing near-field energy flow and radiation," *IEEE Antennas and Propagation Society Magazine*, Vol. 42, No. 6, 2000, pp. 46–54.

[25] E. K. Miller, "Comparison of the radiation properties of the sinusoidal current filament and a PEC dipole of near-zero radius," *IEEE AP-S Antennas and Propagation Society Magazine*, Vol. 8, No. 4, pp. 37–47, 2006.

Chapter 7

Time-domain far-field analysis of radiation sources and TWTD

Abstract

The procedure called FARS (Far-Field Analysis of Radiation Sources ha been briefly introduced in earlier chapters of this book. It is a technique for determining the quantitative distribution of power or energy radiated from a perfect electric conductor. FARS is based on a source-integral expression for the fields of an object of interest for which both time-domain (TDFARS) and frequency-domain (FDFARS) versions have been developed. These two versions provide complementary perspectives of general electromagnetic phenomena. While a frequency-domain result explicitly demonstrates the effects of standing waves, for example, a time-domain result has the advantage of separating various contributions to the far field due to their different time delays. The motivation for both FDFARS and TDFARS is to provide information about radiation from a perfect electric conductor. A brief description of TDFARS is given here and demonstrated for some simple wire geometries.

Keywords: Antennas; Dipole antennas; FARS; FDFARS; Electromagnetic radiation; Linear energy density; Loop antennas; NEC; Propagation radiation; Radiation; Reflection radiation; TDFARS; Time-domain radiation; Time-domain scattering; TWTD

7.1 Introduction

From a numerical perspective, far-field radiation from a perfect electric conductor (PEC) can be readily and accurately determined. Near-field coupling between such objects and their far-field patterns are routinely obtained for increasingly complex and larger geometries using well-validated computer models. Consequently, ever better understanding of radiation is being developed from computing such observables.

A motivation for developing Far-field Analysis of Radiation Sources (FARS) [1–3] is the goal of determining the quantitative distribution of radiated power from over the surface of a PEC. Although not mandatory from the perspective of solving the kinds of problems just mentioned, accomplishing this goal would seem to be a valuable tool for better understanding electromagnetic radiation. Given the above observation concerning the far field it would appear to be straightforward to do this. A variety of objections have been raised about the perceived impracticality of achieving this goal.

7.1.1 The perfect electric conductor as a radiator

For the most part, a metallic object modeled either as antenna or scatterer is assumed to be a PEC, resulting in the total electric field tangential to its surface being zero. That is assumed to be the case throughout what follows. A consequence of this boundary condition, of course, is that the Poynting vector normal to the object's surface is everywhere zero. It has been claimed that it then follows there can be no radiation from the surface of such an object since the net Poynting vector is zero normal to its surface. While the observation concerning the Poynting vector is unquestionably valid, it is reasonable to ask, "Can the claim that the object doesn't radiate possibly be correct?"

An alternative observation is that the total tangential electric field is zero because its positive and negative components must exactly cancel to satisfy the assumed boundary condition. Associated with these canceling positive and negative electric fields are normally directed inward and outward Poynting vectors. As shown in [3], these Poynting vectors comprise two kinds of real power flow. One is the near-field interaction between an incremental part of the object, a wire segment in a moment-method context, and the rest of the object while the other accounts for power loss due to radiation. Examples of this phenomenon are shown in Figures 11.2 and 11.6.

Scattering from a PEC that has been impulsively excited provides a counter-example to the non-radiating claim. After the exciting field returns to zero, so that there is no further energy being collected by the object, there remain a non-zero, or late-time, current and charge distributions on the object. Only after enough time has passed will the object return to charge neutrality. The energy associated with the late-time current and charge is converted to radiation through the interaction of the electric field produced by these induced sources with the PEC object.

An additional claim about a PEC not radiating is that since all of the radiated power has to originate at the exciting source for an antenna, it is self-evident that no radiation comes from anywhere else. This claim seems to be based on the fact that streamlines of the Poynting vector can all be traced back to the source region, seeming to support this claim, as illustrated in Figure 7.1 [4]. But these Poynting-vector streamlines exist only because of the magnetic and electric fields from which they are derived are due to the current and charge (sources) that flow on the antenna. Without these sources, there would be no fields, no Poynting vectors and no streamlines. But, considering the inhomogeneous Maxwell curl equations, electric current and charge are the forcing functions on which the electric and magnetic fields depend. The latter would not exist without the former.

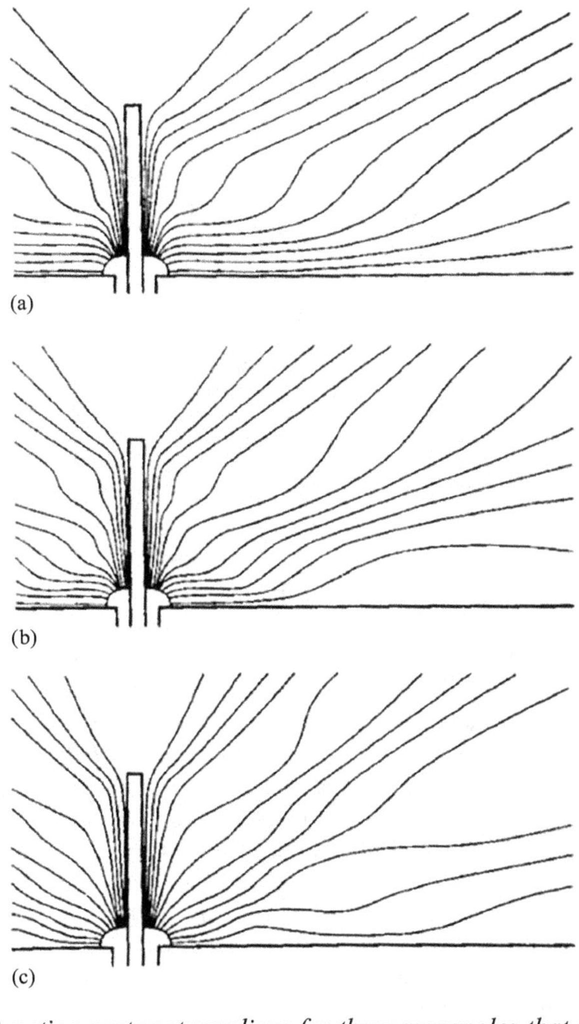

Figure 7.1 Poynting-vector streamlines for three monopoles that are (top to bottom) 1.0, 1.5, and 2.5 wavelengths long. After Landstorfer and Sacher [4] Figure 3.22 (a)–(c)

Observe in Figure 7.1 that, proceeding along the monopole, the streamlines appear to successively "bend" away from it. This bending process implies the development of a small radial component of the Poynting vector being traced by a particular streamline. What might cause such a radial Poynting-vector component? As shown in Chapter 2, an accelerated charge develops a radiating, outward-propagating electric field parallel to that acceleration. An example of the broadside far electric field caused by an impulsive, feedpoint excitation applied to a straight-wire antenna is demonstrated in Figure 7.2. The radiated field is closely correlated with the feedpoint current and associated charge set into motion, or accelerated, by the excitation.

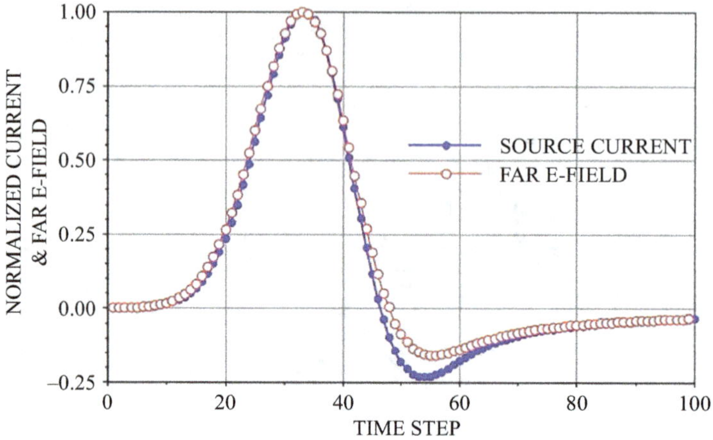

Figure 7.2 The broadside far electric field compared with the feedpoint current for a wire 10^{-3} m in radius

It has been previously established, in Chapter 3 for example, that there is a small but continuous reflection of current and charge propagating along a constant-radius straight wire because the wire's effective wave impedance varies with distance [5]. In the time domain, this reflection results in a slowly decaying current and charge pulses, an example of which is shown in Figure 7.3a, where the charge density Q times light speed c is seen to match the current I. The partially reflected propagating Qc/I pulses, undergoing acceleration, produce an accompanying Poynting vector normal to the wire to produce the intermediate or propagation radiation shown as the square of the broadside electric field of Figure 7.3b. While the total tangential electric field on the wire surface must remain zero, this acceleration-caused Poynting vector would seem to be how the wire "steers" or "sheds" power to the far field as described by Jackson [7].

On successively increasing the wire lengths from 0.33 (99 segments) to 2.00 m (599 segments), end-reflection radiation pulses result as shown as a function of the time step in Figure 7.3b. The end-reflection radiation is of decreasing magnitude because of the continuous reduction in the Qc/I pulse amplitudes with distance. Note that the peaks of the end-reflection radiation pulses occur coincident with the time that the Qc/I pulse peaks reach the wire ends.

Observe that there are two first-principles means of determining the fields produced by an object. One approach is the defining Maxwell curl equations. The other is a Green's-function formulation. In either case, the object serves as a terminating boundary for the fields. The current and charge distributions, in particular, terminate the tangential magnetic field and normal electric field on the object. How EM radiation leaves the object depends entirely on these sources. The explicit working assumption in this discussion, where a Green's-function or source-integral approach is followed, is that sources comprised of current and charge on an object cause the radiation it produces.

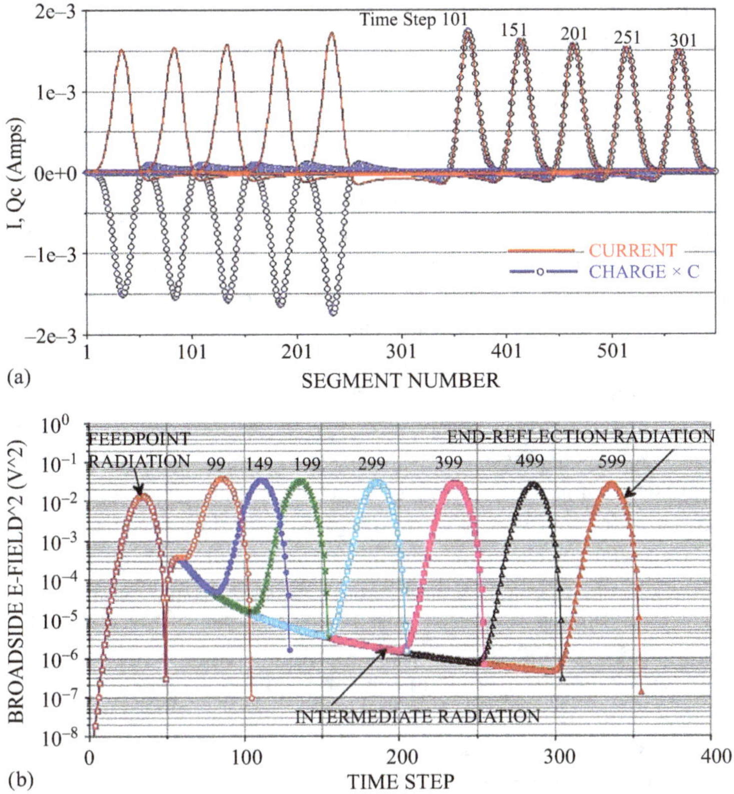

Figure 7.3 Charge density Q *times light-speed* c *and the current* I *for a center-fed 599-segment wire at several time steps (a) and the square of the far electric field for dipoles of increasing lengths from 0.33 to 2.00 m (99–599 segments) and radii of 10^{-3}-m radius*

7.1.2 *Why does radiation occur?*

That the physical cause of radiation from free charge is acceleration is shown by the Lienard–Wiechert potentials as discussed beginning in Chapter 2. However, this fact is not explicitly considered in the solution of boundary-value problems where it is essentially irrelevant. By contrast, acceleration and relativistic effects must be included in analyzing problems such as particle accelerators. A pertinent question to ask is how does free charge on a PEC object compare with equivalent sources? These sources result from formulating a boundary-value problem using the equivalence principle [8]. Equivalent sources might be argued to be non-physical by someone who wants to model charge and current flow within an imperfect, i.e., a real-metal, conductor. But measurements of observables for physical objects developed from models based on the equivalence principle and equivalent sources, ranging from radiation patterns to near-field coupling and source distributions, are found to agree to within measurement and modeling uncertainties. Thus, results derived from an equivalence-based model have been proved to be physically valid.

Where charge acceleration occurs on such a simple object as a straight wire, excited as an antenna or scatterer, may not be entirely clear however. Application of an exciting electric field, whether local as for an antenna, or global as for a scatterer, causing charge to move in response and thereby to be accelerated, is obvious. The end of a wire, where the current goes to zero due to charge reflection, is as well. But what happens elsewhere on the wire seems still to elicit disagreement.

Some insight regarding this issue may be provided by a source-integral approach. Consider that the fields anywhere in space can be found by integrating the source distribution over the antenna using a Green's function [9]. Evaluating electric and magnetic fields by integration over a source distribution is one of the most basic mathematical operations of EM theory. It explicitly incorporates the concept that electric current and charge are the cause of such fields. It is true that the fields outside some arbitrary closed surface within which current and charge sources are confined can be found by integrating the tangential electric and magnetic fields over that surface. But since these surface fields exist only because of the enclosed sources, it is clear that those sources are the cause of the fields on and external to that enclosing surface. In the particular case of the far field, it follows that these sources must radiate.

At a given observation point in the far field the current integral, except for some special geometries, requires numerical evaluation. This necessarily involves the use of a piece-wise procedure where the current distribution is divided into small increments. All non-zero incremental current samples will be found to affect the far field to a greater or lesser extent. The claim that parts of the current distribution do not radiate is inconsistent with this fact. The issue instead is to ask how much a given incremental current affects the total radiated power that the antenna produces over the entire far-field sphere? It is the answer to this question that motivated the development of FARS. From the perspective of the Lienard–Wiechert potentials, a fundamental issue seems to be whether or not the acceleration of equivalent charge needs to be explicitly included in trying to answer this question.

A qualitative connection between equivalent charge acceleration and radiation is provided by the electric field "kink" model [10] discussed in Chapter 2 that is independent of the Lienard–Wiechert potentials. If equivalent sources experience charge acceleration, it would seem to follow that radiation occurs as a result even though the Lienard–Wiechert potentials appear not to be directly applicable in this situation. This issue is discussed in some detail elsewhere [11] where the electric field "kink" model is demonstrated. The kink model is not strictly analytically based, but instead exploits two facts: (1) EM fields propagate at a finite speed; and (2) electric field lines are continuous. The kink model helps explain radiation from equivalent charge on a PEC as determined from FARS for which the basis is a source-integral description of the far fields. Reference made hereafter to current and charge is with respect to their equivalent forms as discussed above.

7.1.3 Charge acceleration on a perfect electric conductor excited as an antenna

Since charge acceleration causes radiation, it is relevant to consider where and how charge acceleration occurs on simple wire objects. It has been noted that energy to produce charge acceleration on antennas and scatters is provided by an exciting electric field. Various other kinds of acceleration on wires also occur as previously demonstrated in Chapter 3, including reflections that occur at open ends of wires and discontinuities such as impedance loads, changes in wire diameter and bends, and continuously as a result of a spatially varying wave impedance. Shape-caused acceleration takes place on objects like circular loops, helices and conical-spirals. Some examples of these effects are shown below.

For the dynamic case of an impulsively excited, antenna, a situation somewhat analogous to the kink model holds. The initially charge-neutral antenna responds to the impulsive excitation with outward-propagating, opposite-signed charge pulses and accompanying electric and magnetic fields around the antenna. After a sufficient time has passed, all of the energy delivered by the excitation to charge separation and the resulting electric and magnetic fields consequently produced will have been lost in radiation and charge neutrality will be re-established. The source of the impulsive excitation could be simply an applied electric field across one or a few segments, or some transmission line at the end of which is a generator. Examples of both will be shown below. The radiated power or energy is typically obtained by integration of the far-field power flow or by evaluating the input power supplied by the excitation whether the problem at hand is an antenna or scatterer.

A classical method for obtaining the power radiated by an arbitrary current such as a sinusoidal current filament is the induced electromotive force (IEMF) method. Assuming a current has some specific variation such as the sinusoidal current filament implicitly implies the cause to be whatever input electric field is necessary to maintain that current. In essence, the IEMF method determines the input power due to the effect of the electric field of the entire current acting on itself. The IEMF method can also be implemented in an incremental sense, the incremental IEMF method described in [3] and covered in Chapter 11, to account for the interaction energy on a piece-wise basis as in the moment method. Note that the IEMF method employs the tangential electric field and current, or equivalently the tangential magnetic field, at the surface of a PEC. For a wire, this is identical to integrating the normal Poynting vector at the wire's surface around its circumference.

Although such electric field interactions provide the force to produce charge motion and a resulting current, explicit incorporation of charge acceleration is not involved in Maxwell's equation as normally written. In assuming an object is a PEC, mechanical force equations are not needed and the mass of real charge, e.g. an electron, does not enter the picture. Instead, it is implicitly assumed that the equivalent charge is essentially massless and so responds instantaneously to any applied field. In spite of such apparently non-physical assumptions, analysis based on equivalent sources agrees to high accuracy with real-world measurements.

The radiation caused by local excitation is demonstrated in Figure 7.2 where the feedpoint current and broadside far electric field of a straight-wire dipole excited by a Gaussian voltage pulse, aligned in time to remove the time delay to the far field, are plotted on a normalized basis. It can be seen that these two quantities agree to within a few percent relative to their peak values. Charge acceleration due to the impulsive voltage produces a feedpoint current pulse, resulting in the observed radiation. The current undershoots because of reflection from the outgoing pulse and extraction of opposite-signed charge from the feedpoint [11]. The time-dependence of the Gaussian exciting pulse used here is $\exp[-a^2(t-t_{max})^2]$, with $a = 7.467 \times 10^9\ sec^{-1}$ and $t_{max} = 4 \times 10^{-10}$ sec, with the time step $\Delta t = \Delta x/c$, typical parameters for the TWTD computations included here.

A more detailed example of end-reflection radiation is illustrated in Figure 7.4. Here, a wire antenna 0.33-m long (99 segments) is impulsively excited near its right end, at segment 80, to produce a more complex *Qc* pulse incident on the left end of the wire. The broadside far electric field is shown over 100 time steps beginning at time step 100. The spatial *Qc* distribution on the wire is plotted for time step 100 just prior to its beginning reflection at the left end. The negative maxima in the charge distribution comes from the feedpoint itself while the positive maxima comes from reflection at the right end nearest the feedpoint. Agreement between the normalized radiated field and the charge distribution is within a few percent relative to the pulse peaks.

The propagation speed of the peak of the *Qc/I* pulses on the 599-segment wire of Figure 7.2 is shown at a sequence of time steps in Figure 7.5. A best-fit straight line to this data shows that the charge pulse has a speed of 2.9868×10^8 m/s, essentially the speed of light.

The magnitude of the radiated field as a function of the magnitude of the end-reflected charge pulse for dipoles of increasing lengths is shown in Figure 7.6. There the magnitude of the peak of the broadside far electric field is plotted as a

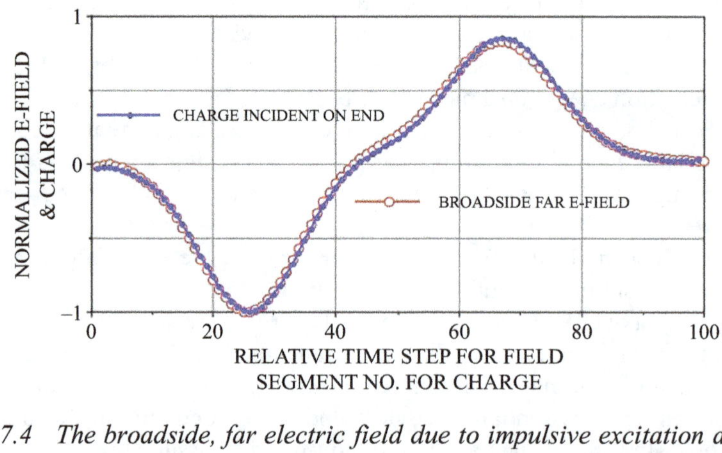

Figure 7.4 The broadside, far electric field due to impulsive excitation applied near the right end at segment 80 and reflection of the Qc/I pulses reflecting from the right end of a 99-segment wire of 10^{-3}-m radius

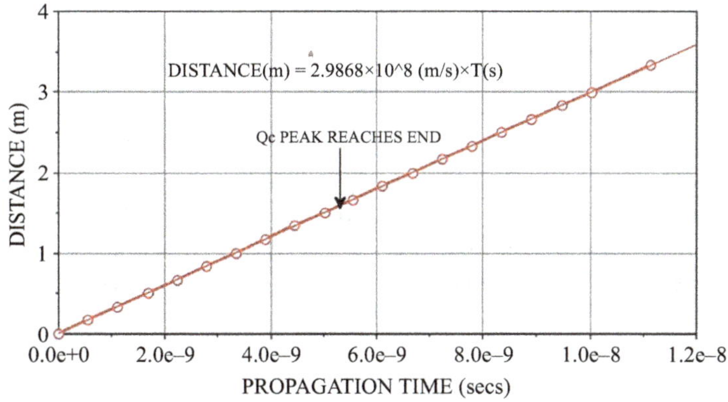

Figure 7.5 The position of the equivalent charge peak on an asymmetrically excited 2-m (599-segment) dipole whose radius if 10^{-3} m is shown as a function of time. A straight-line fit shows the pulse to be propagating at a speed of 2.9868 m/s

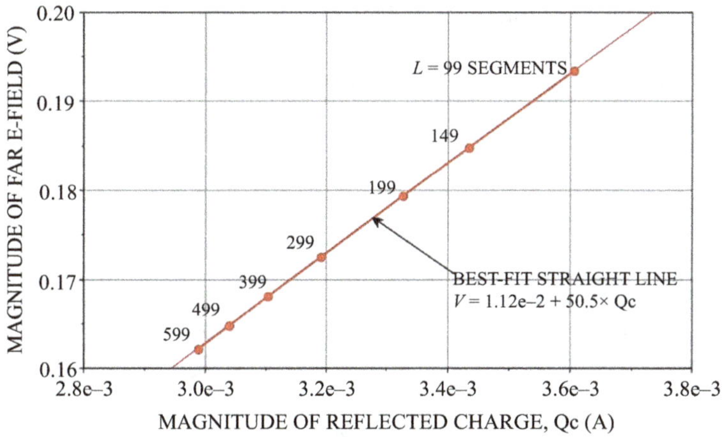

Figure 7.6 The absolute magnitudes of the peak broadside far electric field as a function of the magnitude of the peak charge upon its first end reflection for center-excited dipoles of lengths varying from 0.33 m (99 segments) to 2.00 m (599 segments) in length and 10^{-3} m radius

function of the peak in the incident *Qc* pulse for center-fed, impulsively excited wires of lengths ranging from 0.33 to 2.00 m. Since the outward-propagating *Qc/I* pulses decreases with distance due to the continuous charge reflection as a result of the spatially varying wave impedance, the peak of the charge pulse that reflects from the wire end also decreases. There is a clear linear relationship between the reflected charge and radiated field.

7.2 Time-domain FARS (TDFARS)

TDFARS works in essentially the same way as in the frequency domain (see Chapter 8 for example), but while tracking the time variation of both the power and energy radiated by the object being studied. In either case, all that is needed are the total far electric field from the object of interest together with the incremental field from each of the segments or patches of its moment-method model.

In the time domain, the total far electric field is obtained as a function of observation and source times t and t' and observation and source locations $\boldsymbol{r}$ and $\boldsymbol{r}'$ for a wire of contour $C(\boldsymbol{r}')$ from [13,14]

$$\mathbf{E}(\mathbf{r},t) = \frac{\mu_0}{4\pi}\int_{C(\mathbf{r}')}\left[\frac{\mathbf{s}'\partial}{R\,\partial t'}I(s',t') + c^2\frac{\mathbf{R}}{R^2}\frac{\partial}{\partial s'}I(s',t') - c^2 Q(s',t')\frac{\mathbf{R}}{R^3}\right]ds' \tag{7.1}$$

where $\mathbf{R} = \boldsymbol{r} - \boldsymbol{r}'$, $R = |\mathbf{R}|$, s' is a coordinate along the wire axis, $\mathbf{s}'$ is a unit vector in the direction of the current density $I(s',t')$, and $Q(s',t')$ is the charge density. The retarded time, given by $t' = t-R/c$, is the time delay between a source at $\mathbf{r}',t'$ and its effect at $\mathbf{r},t$. The incremental far electric field from segment m, $\boldsymbol{e}_m(\boldsymbol{r},t)$, is given by (7.1) with $C(\boldsymbol{r}')$ replaced by $C_m(\boldsymbol{r}')$. The incremental and total far-field power-flow densities $P_m^{(TDF)}(\boldsymbol{r},t)$ and $p(\boldsymbol{r},t)$, thus become

$$p_m^{(TDF)}(\mathbf{r},t) = Lim(r\rightarrow\infty)\left[\frac{r^2}{\eta_0}\mathbf{e}_m(\mathbf{r},t)\cdot\mathbf{E}(\mathbf{r},t)\right] \tag{7.2}$$

and

$$P(\mathbf{r},t) = Lim(\mathbf{r}\rightarrow\infty)\left[\frac{r^2}{\eta_0}\mathbf{E}(\mathbf{r},t)\cdot\mathbf{E}(\mathbf{r},t)\right] \tag{7.3}$$

respectively, with η_o being the free space wave impedance. By integrating over 4π steradian observation angles, the total FARS linear power density, i.e. the power radiated from segment m, $P_m^{(TDF)}(t)$ at time t, is given by

$$P_m^{(TDF)}(t) = \int_0^{2\pi} p_m^{(TDF)}(\mathbf{r},t)\sin\,\theta d\theta d\varphi \tag{7.4}$$

with the total *TDFARS* power coming from summing over all segments so that

$$P^{(TDF)}(t) = \sum_{m=1}^{N} P_m^{(TDF)}(t) = P_{rad}(t) \tag{7.5}$$

with $P^{(RAD)}(t)$ the usual radiated power at time t. The per-segment linear energy density, $W_m^{(TDF)}(t)$, and total energy radiated up to time t, $W^{(TDF)}(t)$, can then be written as

$$W_m^{(TDF)}(t) = \int_{\infty}^{t} P_m^{(TDF)}(t')dt' \tag{7.6}$$

and

$$W^{(TDF)}(t) = \sum_{m=1}^{N} W_m^{(TDF)}(t) = W^{(RAD)}(t). \tag{7.7}$$

Observe that while $P^{(TDF)}(t)$ and $W^{(TDF)}(t)$ must be positive, $P_m^{(TDF)}(t)$ and $W_m^{(TDF)}(t)$ can be negative. This can happen if the contribution of segment m reduces the total far field. A specific example of this possibility is discussed in Appendix A of this chapter. The above quantities can be converted to a linear power or energy density on a per-unit-length basis by dividing by the segment length, Δ_m. Unless otherwise indicated, this is the quantity presented in the figures that follow, with the amounts expressed on a per-meter basis. An alternative derivation of TDFARS was suggested and included in [12] where this material was first developed.

7.3 Validating time-domain FARS

The validity of TDFARS follows that of FDFARS and differing only in being implemented in the time domain. As evidence of the validity of TDFARS two examples are included here. One is a comparison with independent results. A second is a quantitative comparison of TDFARS and FDFARS. Both FARS versions were first implemented in the Thin-Wire Time-Domain code [13,14] and the Numerical Electromagnetics Code [15] respectively. Unless otherwise stated, frequency-domain results are for a straight wire 10-m long, 1-mm in radius, and a wavelength of 1-m. The same wire geometry is used for the time-domain results in Sections 7.3 and 7.4 with Gaussian-pulse excitation having a time variation given by:

$V_{ex}(t) = \exp[-a^2(t-t_{\max})^2]$,
$a = 4.2 \times 10^8 \text{ sec}^{-1}$,
$t_{\max} = 1.5 \times 10^{-8}$ sec,
and a time step $\Delta t = \Delta x/c = 3.333333 \times 10^{-10}$.

This makes the 0.1 power width of the exciting pulse about 15 time steps or 1.5 m.

The linear energy density for a straight wire obtained from TDFARS is shown Figure 7.7 together with an analytically derived result from Smith and Hertel [16]. The results from [16] show the normal time-integrated power flow, or energy, out of a cylindrical surface near to, and centered on, the axis of a filamentary current. The filamentary current consists of a centered Gaussian current/charge (Q/I) pulse that splits into two outward-propagating, time-constant-amplitude pulses that reflect from the filament ends and are absorbed upon arriving back at the feedpoint. The observing-cylinder radius $\ll$ the filament length.

The linear energy density results and near-field energy flow are similar qualitatively, but differ in detail. This is likely due to the difference in the current/charge flow in the two cases. The TDFARS linear energy density come from solving a boundary-value problem for a wire modeled as a PEC, involving multiply

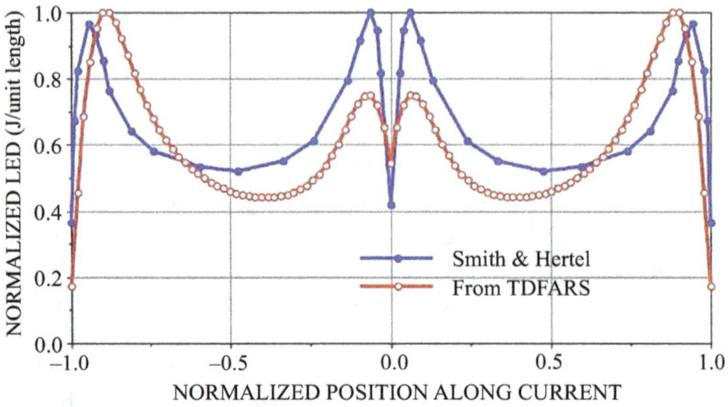

Figure 7.7 Results for the normal energy flow near a Gaussian current filament excerpted from Smith and Hertel [16] and for time step 999 of a Gaussian-pulse-excited straight wire as obtained from TDFARS

reflected *Q/I* pulses whose amplitudes decrease in time as energy is lost due to radiation. The multiple reflections account for the larger linear energy density values at the ends of the wire relative to the feedpoint. The analytic model of [16], on the other hand, exhibits a slightly larger peak at the center feedpoint than at the ends of the filament. This is due to the *Q/I* pulse being "turned on" or started, and then stopped, at the center, with only a single end reflection. The fact that these two different results agree so well qualitatively provides a degree of mutual validation for both. Based on the above and various other confirming examples, it is reasonable to interpret the following TDFARS results as providing valid results for the energy density radiated from along a wire object.

A comparison of the linear power density for a 10-m, center-fed dipole 10^{-3} wavelengths in radius is presented in Figure 7.8 with the total power normalized to 1 w. The time-domain results were obtained using a Gaussian-step, time-harmonic excitation and are shown after a steady-state response was reached. These two results again agree to within a few percent except in the source region and near the wire ends, to support the validity of both FARS methods.

As a further FDARS validation, a frequency-domain counterpart of Figure 7.7 is shown in Figure 7.9. Plotted there are FDFARS results (see Chapter 8) and the radial power flow at a radial distance of 10^{-3} wavelengths for a 10 wavelength-long dipole of 10^{-20} wavelengths radius modeled using the Numerical Electromagnetics Code. This small radius was used to simulate a sinusoidal current filament. The normal Poynting vector, $\boldsymbol{E} \times \boldsymbol{H}$, is zero at the surface of the PEC because the positive and negative E-field components due to a specified excitation and the induced source distribution that results must cancel there to satisfy the PEC boundary condition. However, the radial power flow only 10^{-3} wavelengths away is within a few percent of the FDFARS linear power density results, except near the feedpoint and wire ends.

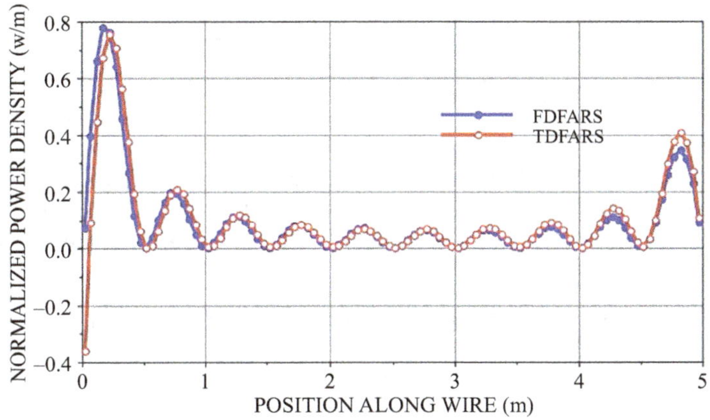

Figure 7.8 A comparison of the linear power-flow densities from the Numerical Electromagnetics code using FDFARS and from the Thin-Wire Time-Domain code using TDFARS for the right half of a 10-m dipole 10^{-3} m radius, for 1-w total input power with each model having 200 segments and a wavelength of 1-m

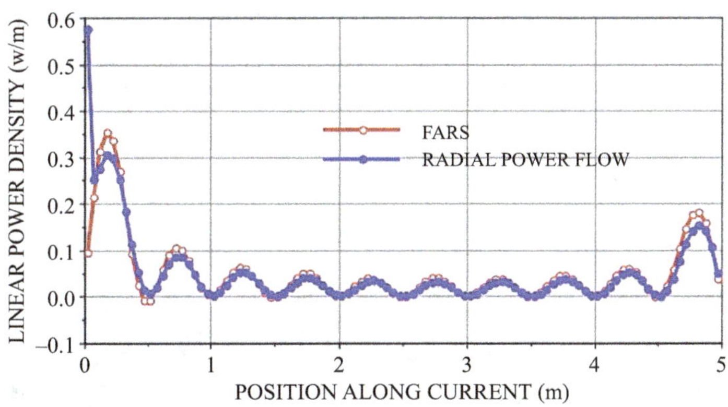

Figure 7.9 A comparison of FDFARS and the radial power-flow density in watts/m for the right half of a 10-m dipole 10^{-20} m in radius for a wavelength of 1-m radiating a total power of 1-w, modeled with 200 segments using the Numerical Electromagnetics Code

While the inward- and outward-directed Poynting vectors cancel at the wire surface, just a little distance out from that surface, a net outward-directed radial flow is already being established. This implies that the balancing inward-directed flow from near-field interaction and the exciting field is confined relatively near the wire's surface. With increasing distance from the dipole, however, a transition to the usual far-field radiation pattern will take place, which will no longer correlate with the near-field power flow in Figure 7.9. Note also that the radial power flow explicitly includes the effect of the source near the dipole, which is not the case in FARS. These results are obtained by integrating the normal Poynting vector over a closed cylindrical surface defined by the dipole, a surface 10^{-3} m in radius and annular "washers" at each end. The radial power flow is equal to the difference between the input power and the longitudinal power flows through the ends of that same cylindrical surface since the power flow at the wire's surface is zero.

7.4 Some representative applications of TDFARS

7.4.1 Radiation from straight wires

Perspective plots of the linear power density and linear energy density from TDFARS for the dipole of Figure 7.7 are presented in Figures 7.10 and 7.11. The radiated power in Figure 7.10 is seen to peak at the center source as is expected. This is followed by continuing radiation along the wire length as the *Q/I* pulses propagate back and forth along the wire. Further center peaks of radiation occur as the *Q/I* pulses meet again at the center. This phenomenon is found whenever propagating *Q/I* pulses overlap in similar situations and is discussed further in [18].

The linear energy density in Figure 7.11 increases essentially monotonically with time. These are replotted in an *x–y* graph in Figure 7.12 at several intermediate time steps for a more quantitative display. In this result and those that follow, the linear power densities are for a total power of 1-w and the linear energy densities

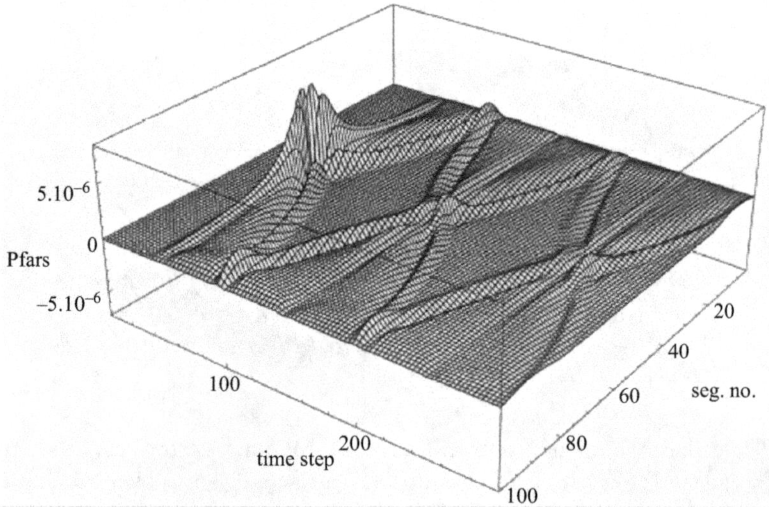

Figure 7.10 A perspective plot of the linear power density per segment for an impulsively excited dipole as a function of time [17]

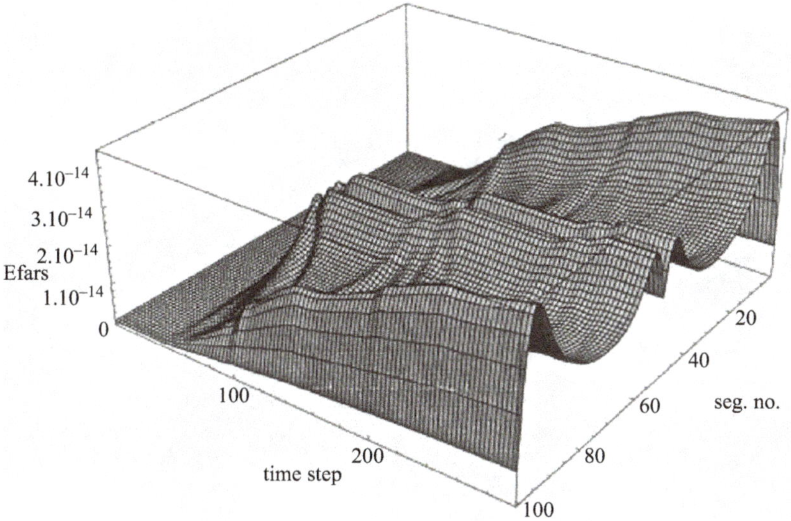

Figure 7.11 A perspective plot of the linear energy density per segment for an impulsively excited dipole as a function of time [17]

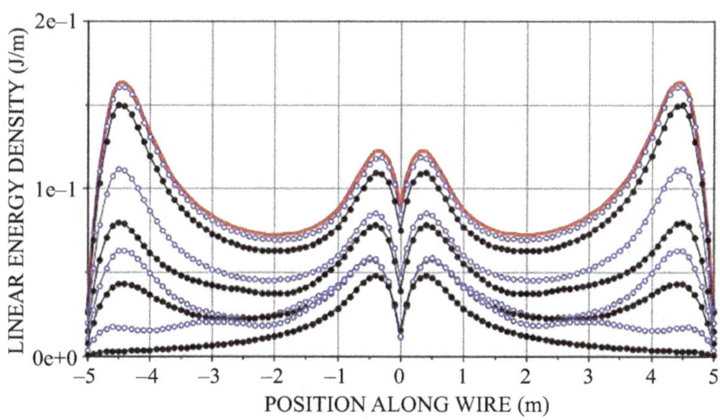

Figure 7.12 Time domain linear energy densities at various time steps 50, 75, 100, 125, 175, 250, 500, 750, and 999 for a 10-m wire of 10^{-3}-m radius excited at its center by a Gaussian pulse and radiating a total energy of 1 J

are for a total energy of 1-J. Nearly all the energy has been radiated away by the last time step shown on this figure. The end-radiated energy increases by a factor of about 8 between time steps 75 and 999. This seems consistent with the expectation that once charge and current (*Q*/*I*) pulses have been set into motion, an obvious place for radiation to occur is at the wire ends due to reflection of the *Q*/*I* pulses

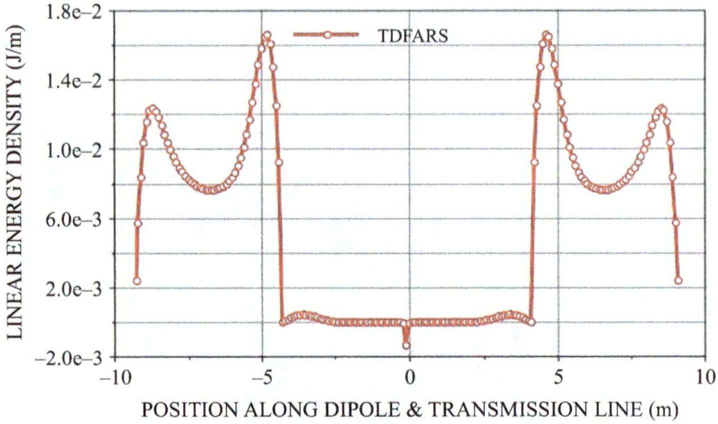

Figure 7.13 The linear energy density from TDFARS for a 10-m dipole fed by a two-wire transmission line for 1-J total energy

there. The source-region maxima increase by a factor of about 2.5. The linear energy density for this problem varies smoothly along the wire, showing that radiation takes place along its entire length. They are otherwise similar to an envelope of the FDFARS linear power density for the straight wire in Figure 7.9, with their largest values occurring near the ends and feedpoint.

The energy densities that result from exciting a dipole with a two-wire transmission line is shown in Figure 7.13. The transmission line is 4.2 wavelengths long with a spacing of 0.1 m. The dipole arms are each 5-m long and the wire radius is everywhere 8.153569×10^{-3} m so that the line impedance is 300 ohms with segment lengths of 0.1 m. The total radiated energy is 1-J. The time-domain radiation is confined almost wholly to the dipole itself.

The linear energy density for the transmission-line excited dipole, but shown only for the dipole itself, and the same-length dipole having a one-segment source are presented in Figure 7.14. The details of the linear energy density are qualitatively similar. The larger peak of the one-segment source is at the end while that of the transmission-line feed is near the connection point. The effect of the transmission-line impedance evidently increases reflection-caused radiation there as compared with the unloaded one-segment source.

7.4.2 Scattering from straight wires

Shown in Figure 7.15 for the same wire used for the antenna of Figure 7.12 is the linear energy density at time step 999 when the excitation is a broadside-incident plane wave. For comparison, the linear energy density for the antenna is also included, with both curves depicting results for a total energy of 1-J. The ends of the wire for the scattering case are seen clearly to be less predominant than for the antenna. The center peak in the linear energy density for scattering is due to the nature of the current for broadside-incidence as shown in Figure 7.16. The current

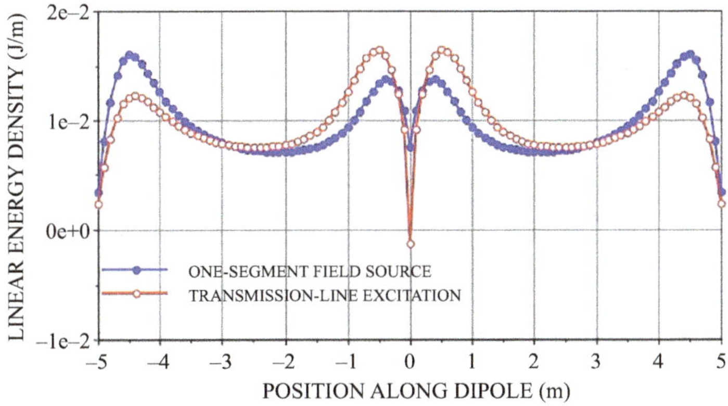

Figure 7.14 A comparison of the linear energy densities for the 10-m dipole excited by a transmission line but shown for the dipole only, and by a one-segment source for 1-J total energy

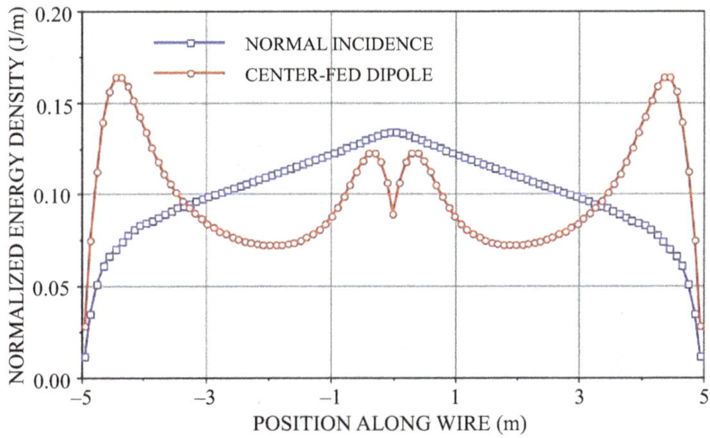

Figure 7.15 The time-domain linear energy densities for a 10-m wire excited at its center as an antenna and by a broadside-incident plane wave as a scatterer for a total energy of 1 J

builds up uniformly along the length of the wire, collapsing inward from the ends towards the center. The current flow persists proportionately longer with increasing distance from the ends to produce a linear energy density having a monotonic increase towards the center of the wire as seen here.

Results for the linear energy density are shown in Figure 7.17 for both broadside incidence and 10 deg from axial incidence and a total scattered energy of 1-J. For near axial incidence, the ends of the wire are now seen to be predominant, with larger radiation at the end away from the incident wave than that on the near

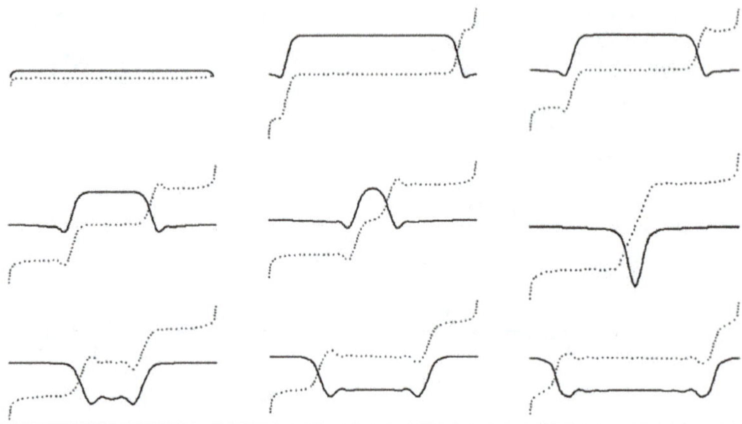

Figure 7.16 The current (solid line) and charge density (dashed line) from left to right and top to bottom on a straight wire for broadside incidence of a plane wave

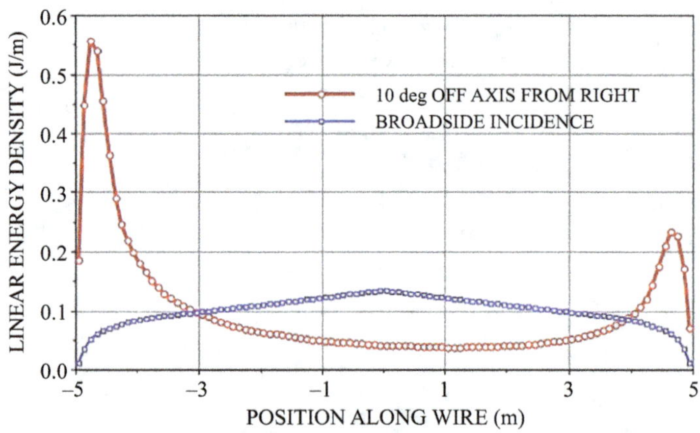

Figure 7.17 The linear energy densities per m for a 10-m wire scatterer excited by plane waves incident from broadside and 80 deg from broadside (the right end in this plot) for a total scattered energy of 1 J

end. This result seems reasonable when considering the oblique scattering characteristics of a wire from the perspective of the geometrical theory of diffraction where the wire ends are the dominant scattering centers.

7.4.3 Radiation from loops

The time-domain linear energy density for a circular loop antenna is shown in Figure 7.18. The time-domain LEDs smoothly decline from the feed region while having a large "bump" on the opposite side of the loop. Radiation along the loop, as contrasted

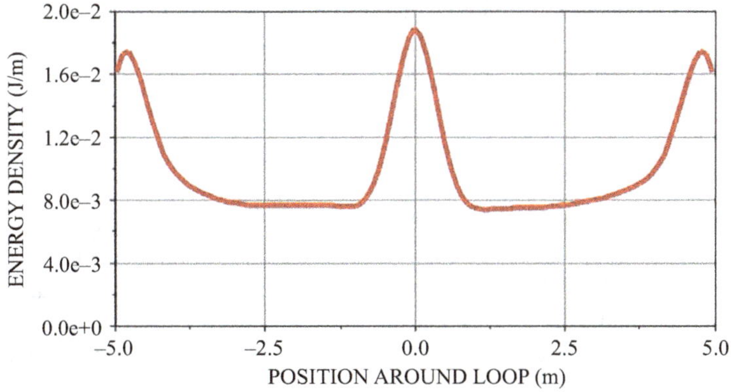

Figure 7.18 TDFARS linear energy density for a circular loop antenna excited at ±5 by an input energy of 1-J with the "bump" in the energy distribution at 0 opposite the feedpoint

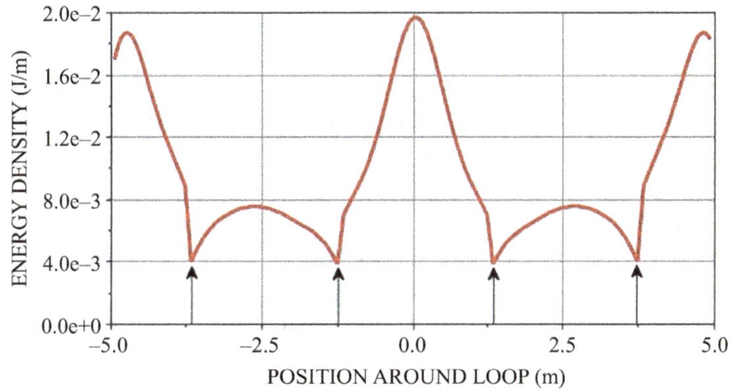

Figure 7.19 TDFARS linear energy density of a square loop excited at ±5.0 as an antenna by an input energy of 1 J with the "bump" in the energy distribution at 0. opposite the feedpoint, with the corners of the loop shown by the arrow

with a straight wire of Figure 7.12, is more nearly uniform between the feedpoint and opposite side. This radiation is due to acceleration of the charge pulses moving around the curved path formed by the loop in addition to radiation due to continuous-reflection acceleration. The bump opposite the feedpoint is due to reflection acceleration of oppositely signed, oppositely propagating pulses meeting there. Their radiated fields are consequently everywhere additive in the far field, thus increasing the resulting power from this location. This phenomenon is discussed in more detail elsewhere [18] and in Chapter 4. The pulse and model parameters are the same as those used previously.

As a last result, time-domain linear energy density for a square-loop antenna is presented in Figure 7.19. The power-density distribution (not shown) is found to have clearly demarcated cusps at the locations of the corners in the loop. These also appear in the energy-density distribution of Figure 7.19, and as for the circular loop there is a bump opposite the feedpoint.

7.5 The effect of varying the source location

The results for straight-wire antennas presented thus far have been limited to center-fed excitation. Examined now are the effects of placing one or two sources at other locations. It will be shown that different radiation phenomena result from this apparently insignificant change.

Consider the two energy-density plots in Figure 7.20 where the excitations are 1.414 V to match the input energy of the two-source examples that follow in Figure 7.21. That for a 201-segment dipole excited impulsively at its center is similar to other results previously shown. However, when an offset, or asymmetric source is used, while some features of the center-fed results are preserved in the end peaks and the energy-density variation around the feedpoint, there is a new "bump" on the right half of the dipole. This feature is caused by the same phenomenon as the bump that occurs opposite the feedpoint for the circular and square loops in Figures 7.18 and 7.19.

The source at segment 67 produces the same outward-propagating *Q*/*I* pulses as for a center-located source, but after their first end reflections these counter-propagating pulses overlap at segment 135 rather than at the dipole center as for center excitation. Observe that segment 67 is 66 segments from the left end of the wire while segment 135 is 66 segments from the right end. In other words, segments 67 and 135 are at what might be termed conjugate locations. The opposite-signed *Q* pulses cancel at overlap and their partial reflections in opposite directions produce additive radiated fields that are coherent over the far-field sphere [18].

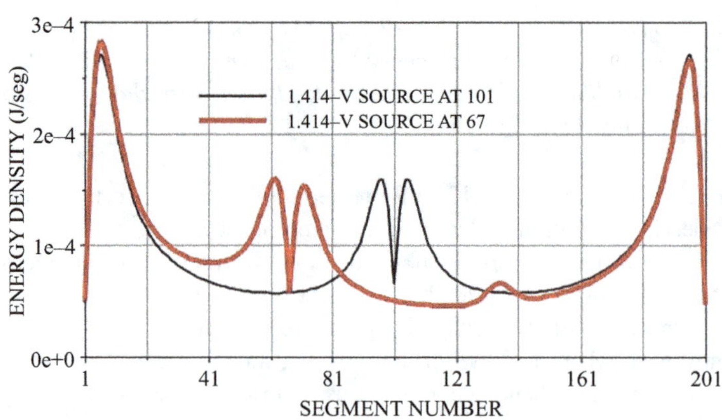

Figure 7.20 The energy densities for a 201-segment wire excited at its center at segment 101 and at segment 67 by a 1.414 V source

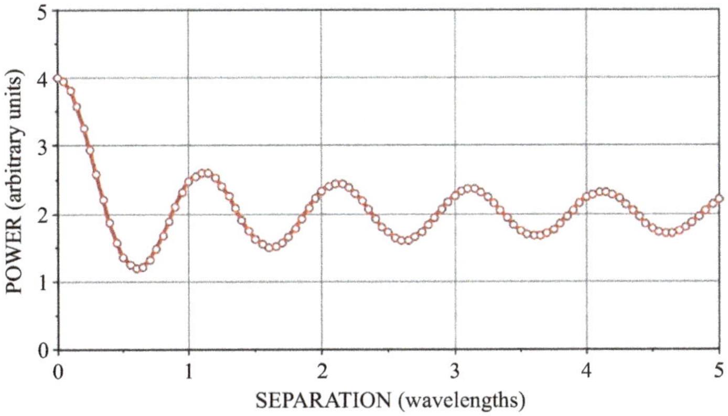

Figure 7.21 *The power radiated by two unit-amplitude sources as a function of their separation*

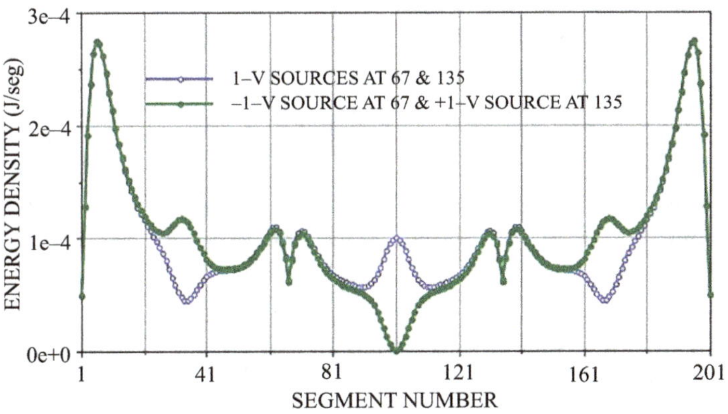

Figure 7.22 *The 201-segment wire excited at segments 67 and 135 by two positive 1 V sources and by segment 67 by −1 V and at segment 135 by +1 V*

This results in increased radiation over their region of overlap and the bump in Figure 7.20, or what might be termed pulse-overlap radiation.

This phenomenon can perhaps be more clearly illustrated in the frequency domain as in Figure 7.21. The far-field radiated power from two unit-amplitude point sources is plotted as a function of their separation in wavelengths. At separations of 5 or more wavelengths, the total radiated power asymptotically approaches 2 units, showing that the radiated fields are becoming uncorrelated, or incoherent, over the far-field sphere. But as the source separation decreases to less than a half wavelength, the total power increases to 4 units at zero separation.

Energy-density plots for two-source excitation of 201-segment straight wires are shown in Figure 7.22. For positive 1-V excitations at segments 67 and 135 a

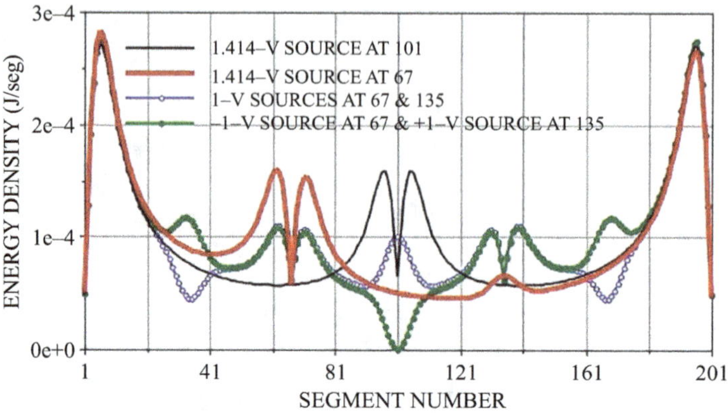

Figure 7.23 A combination plot of Figure 7.20 and 7.22 to clarify their spatial-variation relationships

lobe in the energy density occurs at the center of the wire, while for ± excitations at these same locations, there is a zero-amplitude lobe at the center. Two other symmetric maxima and minima occur midway between segments 67 and 135 and the wire ends. The explanation for both is the same as for the center lobes. The overlapping of same-signed charges pulses propagating in opposite directions results in minima in the energy density while overlapping opposite-signed charge pulses results in maxima. The plots of Figures 7.20 and 7.22 are combined in Figure 7.23 for comparison.

7.6 How charge reflection causes the FARS energy distribution

The TDFARS results presented here are generally smoothly varying functions along a wire except at the feedpoint of an antenna where a sharp minimum occurs. This minimum is coincident with the point where charge separation takes place as the outward-propagating pulses are set into motion. As the Q/I pulses repeatedly propagate across the feedpoint, the minima gradually increases because of the pulse-overlap radiation. Both FDFARS and TDFARS exhibit this kind of feature. Elsewhere, however, the time- and frequency-domain FARS results are very different as displayed in various figures of Chapter 9.

While the FDFARS power-density plots have two lobes per wavelength, the TDFARS energy-density plots are monotonically varying where propagation radiation predominates. The lobed structure of the FDFARS results are due to the fact that the frequency-domain currents are basically standing waves. Thus, while propagation radiation due to continuous partial charge reflection is common to both, their spatial distributions are very different. Why the lobed structure of the FDFARS power density occurs is illustrated conceptually in Figures 7.24 and 7.25.

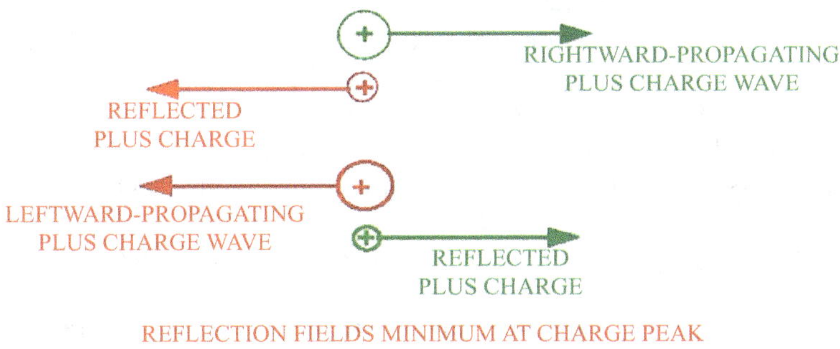

Figure 7.24 Co-located, reflected same-sign charges minimize reflection radiation

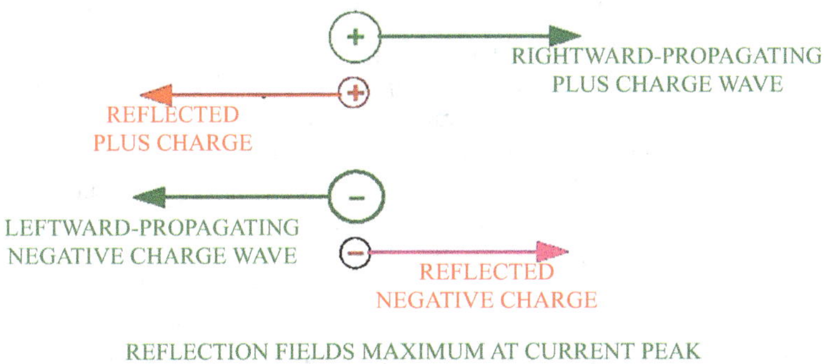

Figure 7.25 Co-located opposite-signed charges maximize reflection radiation

The reflection of oppositely propagating same-signed charges generates canceling electric fields. Conversely, the reflection of oppositely propagating opposite-signed charges generates additive electric fields. Thus, FDFARS maxima are correlated with current standing-wave maxima.

7.7 Summary

Radiation from simple wire antennas and scatterers in the time domain has been explored here using a procedure called FARS. The derivation of the time-domain version of FARS (TDFARS) was briefly outlined and validation examples were given throughout the chapter and Appendix. Application of TDFARS to radiation and scattering from straight wires and for circular- and square-loop antennas was demonstrated.

In connection with simple wire geometries, various kinds of charge-acceleration radiation were discussed. The acceleration mechanisms are: (1) an exciting electric field; (2) continuous reflection due to a spatially varying wave impedance; (3) abrupt reflection at a wire end or bend; and 4) a curved geometry. Excitation of a PEC as an antenna by a localized source is demonstrated in Figures 7.2 and 7.3, and as a scatterer by an incident plane-wave in Figures 7.16–7.18. Continuous reflection radiation from along a constant-radius straight wire is shown in Figures 7.3, 7.7, and 7.10. Abrupt-reflection radiation from the open end of a wire is shown in Figures 7.3, 7.4, and 7.11, and from a square loop in Figure 7.19. Radiation from a curved geometry radiation, a circular loop, is shown in Figure 7.18.

Observe that the mathematical implementation of FARS is straightforward. Interpreting results obtained from FARS might be somewhat challenging. References to electromagnetic radiation are voluminous in the electromagnetic literature, but the origin of radiation from even simple objects has remained rather unclear. Most work in the literature seems to be in agreement that radiation from a simple dipole antenna occurs at or near its feedpoint and at its ends. However, radiation from along the length of a simple, straight wire as has been shown in this and previous chapters seems to raise varying degrees of skepticism. This is in spite of the fact that there seems to be general agreement that the decay of current along a straight wire provides evidence for radiation. The results presented above are not incompatible with the interpretation that all radiation from the kinds of wire objects considered can be concluded to occur because of charge acceleration. A representative sampling of other authors' views relating to some of the issues raised in this chapter can be found in [5–7,19–28].

Appendix A: The occurrence of negative linear-power and -energy results from FARS

As mentioned in Section 7.2, negative values can naturally occur in FARS analysis. These should not be interpreted as being invalid, as can be demonstrated in a situation where they must occur. Negative values signify that some parts of a structure act to reduce the total power that it radiates. For TDFARS, the configuration used to demonstrate this phenomenon is a circular cage of 8 parallel wires centered on the axis of which is located a dipole antenna. The cage wires are 0.9 m (35 segments) long, the dipole is 0.28 m (11 segments) long and the cage radius is 2,564 $\times$ 10^{-2} m. All cage wires are 3.205 $\times$ 10^{-3} m radius, to satisfy the "equal-area" rule, i.e., if unrolled the cage wires would form a continuous surface around the dipole. The rationale here is that the cage then acts as an envelope to effectively shield the dipole from radiating externally.

Results for the TD linear energy densities (LEDs) of this problem are shown in Figure 7.A1. The sum of the cage wire LEDs is seen to be negative, while those for the dipole are positive, so that the negative LEDs of the cage wires nearly cancel the positive LEDs of the dipole. Perfect shielding would produce zero external field. In this particular example, some energy can be expected to "leak" out the open ends of the cage. Besides providing confirmation for the validity of the

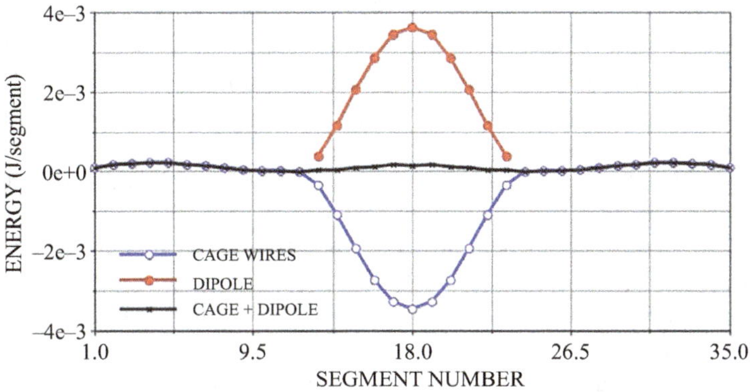

Figure 7.A1 *The energy densities of a dipole the wire of a cage within which it is located*

TWTD model, this example confirms the logical validity of negative linear energy densities, due to the necessity of the far fields partially or, in a situation where they must be zero, completely canceling the net radiation. Similar results are obtained using FDFARS.

Acknowledgment

The implementation by Jerry Burke of TDFARS in the Thin-Wire Time-Domain code and FDFARS in the Numerical Electromagnetics Code and for helping in numerous other ways in the modification and use of these codes is gratefully acknowledged.

References

[1] E. K. Miller, PCs for AP and other EM reflections, *IEEE AP-S Magazine*, 41(2), pp. 82–86 and 41(3), pp. 83–88, 1999.

[2] E. K. Miller, Some further results from FARS: far-field analysis of radiation sources, in *Proceedings of 16th Annual Review of Progress in Applied Computational Electromagnetics*, Naval Postgraduate School, Monterey, CA, pp. 278–285, 2000.

[3] E. K. Miller, Development of the incremental IEMF method to validate FARS (far-field analysis of radiation sources), *IEEE AP-S Magazine*, 52(6), pp. 46–54, 2010.

[4] F. M. Landstorfer and R. R. Sacher, *Optimization of Wire Antennas*, Research Studies Press LTD., Letchworth, Hertfordshire, England, p. 54, 1985.

[5] J. B. Anderson, Admittance of infinite and finite cylindrical metallic antenna, *Radio Science*, 3(6), pp. 607–621, 1968.

[6] L. Shen, T. T. Wu, and R. W. P. King, A simple formula of current in dipole antennas, *IEEE Transactions on Antennas Propagation,* 16, pp. 542–547, 1968.

[7] J. D. Jackson, How an antenna launches its input power into radiation: the pattern of the Poynting vector at and near an antenna, *American Journal of Physics,* 74, pp. 280–288, 2006.

[8] Roger F. Harrington, *Time Harmonic Electromagnetic Fields*, McGraw-Hill Book Company, New York, NY, 1961

[9] J. Van Bladel, *Electromagnetic Fields*, McGraw-Hill Book Company, New York, NY, 1964.

[10] G. S. Smith, *Classical Electromagnetic Radiation*, Cambridge University Press, Cambridge, 1997, Fig. 6.14 *et seq.*

[11] E. K. Miller and G. J. Burke, A multi-perspective examination of the physics of electromagnetic radiation, *Applied Computational Electromagnetics Society Journal*, 16(3), pp. 190–201, 2001.

[12] E. K. Miller, Time-domain far-field analysis of radiation sources, *IEEE Antennas and Propagation Society Magazine*, 53(5), pp. 82–97, 2011.

[13] E. K. Miller, A. J. Poggio, and G. J. Burke, An integro-differential equation technique for the time-domain analysis of thin-wire structures. Part I: the numerical method, *Journal of Computational Physics*, 12, pp. 24–48, 1973.

[14] A. J. Poggio, E. K. Miller, and G. J. Burke, An integro-differential equation technique for the time-domain analysis of thin-wire structures. Part II: numerical results, *Journal of Computational Physics*, 12, pp. 210–233, 1973.

[15] G. J. Burke, Numerical electromagnetics code – NEC-4, method of moments, Part I: user's manual, and Part II: program description – theory, *Lawrence Livermore National Laboratory*, UCRL-MA-109338, January,1992.

[16] G. S. Smith and T. W. Hertel, On the transient radiation of energy from simple current distributions and linear antennas, *IEEE AP-S Magazine*, 43(3), pp. 49–62, 2001.

[17] G. J. Burke, Private Communication, 2008.

[18] E. K. Miller, Time-domain far-field analysis of radiation sources and point-source coherence, *IEEE Antennas and Propagation Society Magazine*, 54(2), pp. 100–108, April, 2012.

[19] J. D. Jackson, *Classical Electrodynamics*, John Wiley & Sons, New York, NY, p. 401, 1962.

[20] Y. S. Yeh and K. K. Mei, Theory of conical equiangular-spiral antennas part ii–current distributions and input impedances, *IEEE Transactions on AP-S*, AP-16(1), pp. 14–21, 1968.

[21] A. M. Abo-Zena and R. E. Beam, Electromagnetic fields at points near pulse-excited linear antenna, *IEEE Transactions on Antennas Propagation*, 19, pp. 129–131 1971.

[22] G. Franceschetti and C. H. Papas, Pulsed antennas, *IEEE Transactions on Antennas Propagation*, 22, pp. 651–661, 1974.

[23] D. S. Jones, *Methods in Electromagnetic Wave Propagation*, Clarendon Press, Oxford, p. 295, 1994.

[24] G. S. Smith, On the interpretation for radiation from simple current distributions, *IEEE Antennas Propagation Magazine*, 40, pp. 39–44, 1998.
[25] J. C. Bogerd, A. G. Tijhuis, and J. J. A. Klaasen, Electromagnetic excitation of a thin wire: a traveling-wave approach, *IEEE Transactions on Antennas Propagation*, 46, pp. 1202–1211, 1998.
[26] R. G. Martin, A. R. Bretones, and S. G. Garcia, Some thoughts about transient radiation by straight thin wires, *IEEE Antennas Propagation Magazine*, 41, pp. 24–33, 1999.
[27] J. Shaeffer and B. Cooper, Electromagnetic visualization for antennas and scattering, in: *1999 IEEE Int. Antennas Propagat. Symp. Dig.* vol. 37, pp. 74–77, June 1999.
[28] Colin C. Bantin, Radiation from a pulse-excited thin wire monopole, *IEEE Antennas Propagation Magazine*, 43, pp. 64–69, 2001.

Chapter 8

Frequency-domain far-field analysis of radiation sources and NEC

Abstract

Time-domain far field analysis of radiation sources (TDFARS) were introduced in the previous chapter and its implementation is illustrated by a variety of results obtained using the computer code thin-wire time domain (TWTD). This order of presentation has been used because initial discussion in this book was devoted to the time domain as a more intuitive way of introducing the physical basis for electromagnetic radiation. This following chapter introduces the complementary version of frequency-domain far field analysis of radiation sources (FDFARS), development of which actually came first. FDFARS was developed in an attempt to determine how much power is contributed to the far field from a PEC object on a per-unit-length or per-unit-area basis. Examples considered here include straight, curved, and bent wires, with the far field analysis of radiation sources (FARS) results appearing to be physically consistent with other expectations about radiation.

Keywords: Distributed radiation; Electromagnetic radiation; FDFARS; Frequency-domain FARS; FARS; NEC; Radiation; Sinusoidal current filament; Spatial radiation

8.1 Introduction

Frequency-domain far field analysis of radiation sources (FDFARS) evolved from some discussion that appeared in a column by the author of this book called "PCs for AP and Other EM Reflections" [1]. One topic that generated some ongoing debate was the rather fundamental question about where radiation originates from a sinusoidal current filament (SCF). The fact that an analytical formula for the far field of the SCF explicitly involves only its ends and possibly its center depending on its length might be interpreted to mean that it only radiated from these points. This formula is [2]

$$E_{rad,\theta}^{(SCF)}(\vartheta) = j\eta_o \frac{I_o e^{-jkr}}{2\pi r} \frac{1}{\sin(\theta)} \left[e^{(jkL/2)\cos(\theta)} + e^{-(jkL/2)\cos(\theta)} - 2\cos(kL/2) \right] \quad (8.1)$$

where the far electric field seems to be due to three-point sources, two due to the first two terms in (8.1), and a third in the last term whose magnitude varies cyclically with the length. But an alternate form for the far field as given by

$$E_{rad,\theta}^{(SCF)}(\theta) = j\eta_o \frac{I_o e^{-jkr}}{2\pi r \sin\theta}\left[\cos\left(\frac{kL}{2}\cos\theta\right) - \cos\left(\frac{kL}{2}\right)\right] \tag{8.2}$$

has a quite different appearance.

Were it not for the $1/\sin(\theta)$ term in (8.1) and (8.2), the SCF radiated field would indeed be that of three-point sources. The effect of that term is to cause radiation from the SCF to occur in a lobe-like fashion all along its length. This effect is discussed in more detail in Chapter 14 and [3] where the degrees of freedom of the far field radiated by a given source distribution provides yet another way to assess the effective radiation sources of a given distribution.

This effect was explored in some detail in [1] using the induced electromotive force (IEMF) method. The tangential electric field was computed along the SCF at a radial distance of 10^{-6} wavelengths to estimate the effective input power required to maintain a unit-amplitude sinusoidal current. (Note that the results were found to be relatively insensitive to radial distance, with some other IEMF results being obtained using 10^{-3} wavelengths.) Because the SCF is a standing wave there is zero time-average propagating power along it, the local input power as deduced using the IEMF method must match the distribution of power radiated by it, i.e., the radiated power density.

Recall that a motivation for FARS was to address the question:

What is the quantitative contribution to the radiated power from the surface of a PEC object on a per unit length (wires) or per unit area (3D surfaces) basis?

This approach that was first described by Miller [1,4] as FDFARS, development of which preceded that of TDFARS. However, TDFARS has been discussed here first in Chapter 7 since the initial part of this book has dealt with time-domain electromagnetics as being a more intuitive introduction to radiation physics. However, similarities between the two versions of FARS will become apparent since they both necessarily involve the same kind of moment-method treatment of the far fields and the piecewise contributions to the power and energy flows from incremental parts of a radiating object.

8.2 Some background

When posed to electromagneticists, quite a wide a variety of responses to the radiation question above is obtained, but with most falling into two broad categories. One is that this is not really a sensible question to ask, or even if so, it is one that probably cannot be answered. The other is that since solving electromagnetic problems of interest has not seemed to be hindered by not considering this question, what's the point of worrying about it? The first response is essentially a non-answer. If EM radiation can be quantitatively evaluated in terms of radiation patterns and radar cross sections, what can be wrong with asking where the radiated power originates? The geometrical theory of diffraction provides just one example where the contributions of edges, discontinuities, and surface curvature to radiation are used explicitly and quantitatively in formulating a solution technique. Physicists who design particle accelerators and other devices involving various charge configurations must deal with this issue routinely. Furthermore, a literature search indicates that some in the physics community regard this as an open question [5]. As for the latter response, it is one that seems to display a singular lack of intellectual curiosity. While the response may be superficially true, not having an answer to the radiation question seems to leave a significant gap in our understanding of electromagnetic physics.

It is relevant to briefly describe how FARS was developed. Since the motivation was to determine where radiation originates from an object, it seemed obvious that this would have to involve investigating the far field. One approach proposed to do this [6] by taking the difference between the total power radiated by an object and the power obtained by omitting a single segment or patch from the calculation. This difference would be taken as the amount of power due to that segment or patch. It was assumed that the spatially distributed radiated-power density originating from the object would then be derived by systemically scanning over the entire object in this way. This approach produces a term like

$$\begin{aligned}
&\frac{1}{2}\mathrm{Re}\left[\left(\mathbf{E}_{d,m}+\mathbf{e}_m\right)\cdot\left(\mathbf{E}_{d,m}+\mathbf{e}_m\right)^*-\mathbf{E}_{d,m}\cdot\mathbf{E}^*_{d,m}\right]\\
&=\frac{1}{2}\mathrm{Re}\left[\mathbf{E}^*_{d,m}\cdot\mathbf{E}^*_{d,m}+\mathbf{e}_m\cdot\mathbf{E}^*_{d,m}+\mathbf{E}_{d,m}\cdot\mathbf{e}^*_m+\mathbf{e}_m\cdot\mathbf{e}^*_m-\mathbf{E}_{d,m}\cdot\mathbf{E}^*_{d,m}\right]\\
&=\frac{1}{2}\mathrm{Re}\left[\mathbf{e}_m\cdot\mathbf{E}^*_{d,m}+\mathbf{E}_{d,m}\cdot\mathbf{e}^*_m+\mathbf{e}_m\cdot\mathbf{e}^*_m\right]
\end{aligned} \tag{8.1}$$

with $\mathbf{E}_{d,m} = \mathbf{E} - \mathbf{e}_m$ where e_m is the incremental electric field of wire segment m. On the other hand, FDFARS instead involves only the term

$$\frac{1}{2}\mathrm{Re}\left(\mathbf{E}\cdot\mathbf{e}^*_m\right). \tag{8.2}$$

FDFARS came about from conducting various computer experiments involving the far field of a sinusoidal current filament (SCF); it was not derived analytically. The last of these experiments was found to produce a distributed radiated power that matches numerically, within a few percent, the IEMF results presented below. Further investigation yielded convincing evidence that it provides an answer to the question raised above.

8.3 The FARS approach

FARS is based on the bistatic far fields of a radiating object. However, instead of imaging using the bistatic far fields, as described in [7], however, FDFARS instead determines the contribution made to the total power flow at a given far field observation point by each incremental source, s_m, on the PEC boundary. This incremental FDFARS power, $p_m^{(FDF)}(\theta,\varphi)$ for segment m in observation direction θ,φ is

$$p_m^{(FDF)}(\theta,\varphi) = \lim r \to \infty\left(r^2\frac{1}{2\eta}\mathrm{Re}[\mathbf{e}_m(\theta,\varphi)\cdot\mathbf{E}^*(\theta,\varphi)]\right) \tag{8.3}$$

where $\mathbf{e}_m(\theta,\varphi)$ is the electric field due to source (or segment) m and η is the medium impedance and FDF is an abbreviation for FDFARS. The total power contributed by each s_m, denoted as $P_m^{(FDF)}$, comes from integrating the incremental result $p_m^{(FDF)}(\theta,\varphi)$ over all θ,φ to obtain

$$P_m^{(FDF)} = \int_0^{\pi}\int_0^{2\pi} p_m^{(FDF)}(\theta,\varphi)\sin\,\theta d\theta d\varphi \tag{8.4}$$

yielding the power coming from segment m.

The total electric field for N sources or segments, denoted by $\mathbf{E}(\theta,\varphi)$, is then given as

$$\mathbf{E}(\theta,\varphi) = \sum_1^N \mathbf{e}_m(\theta,\varphi). \tag{8.5}$$

Finally, the total radiated power is obtained from integrating the $P_m^{(FDF)}$ overall N sources, or

$$P^{(FDF)} = \sum_1^N P_m^{(FDF)}. \tag{8.6}$$

Note that the FARS computation differs from a conventional evaluation of the total far-field power only in defining the intermediate quantity $P_m^{(FDF)}$, the individual contribution of each incremental source, to the total radiated or scattered power. Although there is no constraint that each $P_m^{(FDF)}$ be positive, the physical significance of negative values may be unclear at this point and is discussed elsewhere in Chapters 7 and 9.

8.4 The induced electromotive force method

Because it provides a basis for comparison with *FARS* for a standing-wave current, the induced electromotive-force method (IEMF) for computation of boundary power is also considered here. Further results obtained from the IEMF method are included in Chapters 10 and 11. The differential IEMF power, $dp^{(IEMF)}(x)$, for a current distribution $I(x)$ whose tangential near-electric field, $E_{tan}(x)$, is given by

$$dp^{(IEMF)}(x) = 0.5\mathrm{Re}\left[E^{(\tan)}(x)I^*(x)dx\right]. \tag{8.7a}$$

and the per-segment IEMF power can then be approximated as

$$p_m^{(IEMF)} = 0.5\mathrm{Re}\left[E_m^{(\tan)}\Delta_m I_m^*\right] \tag{8.7b}$$

where $E_m^{(\tan)}$ is the tangential electric field at the center of segment m whose length is Δ_m and on which the current is I_m. The total *IEMF* power is then

$$P^{(IEMF)} = \sum_{1}^{N} p_m^{(IEMF)} \tag{8.8}$$

where, for a PEC object it is required that

$$P^{(FDF)} = P^{(IEMF)} \tag{8.9}$$

since all input power is radiated. Note that for boundary-value problems $p_m^{(IEMF)}$ is non-zero only on segments that are excited by some applied field, whereas for a current such as the *SCF*, it will generally be non-zero everywhere. However, a modified version of the IEMF method is found to work for a PEC as shown in Chapter 11.

8.5 Tangential electric fields of the sinusoidal current filament

Some results for the tangential electric field of the SCF are presented in Figure 8.1. It is these fields that are used together with their respective currents to implement the IEMF method results which are shown in Figure 8.2.

8.6 Initial FDFARS results for the sinusoidal current filament

The FDFARS results in Figures 8.3–8.7 are for sinusoidal current filaments of lengths varying from 20 to 21 wavelengths in quarter-wavelength steps, for a radiated power of 1-W. The FARS power density for the SCF in Figures 8.3 and 8.5 agree within a few percent of the IEMF result used to compute the "input" power as presented in Chapter 10. It was the close agreement obtained between the IEMF method and FARS for the SCF that provided the initial demonstration for the validity of FARS. Note that the only connection between the two procedures is that both exploit knowledge of the current distribution, but FARS uses the far field and the IEMF method uses the near field.

Since there is zero *net* power flow along the SCF, its spatial distribution of input power as determined by the IEMF results must also equal its spatial distribution of radiated power as determined by FARS and as shown in Figures 8.3 and 8.5. As a reference in this and the following three figures, the current in the center wavelength or so (the source region) of the SCF is included in the top graph of these results. The power density in the source region will be seen to vary substantially with the current there. Results are presented in Figure 8.4 for a SCF length of 20.25 wavelengths, where only the FARS result is plotted. The FARS power in the source-region has changed substantially from the

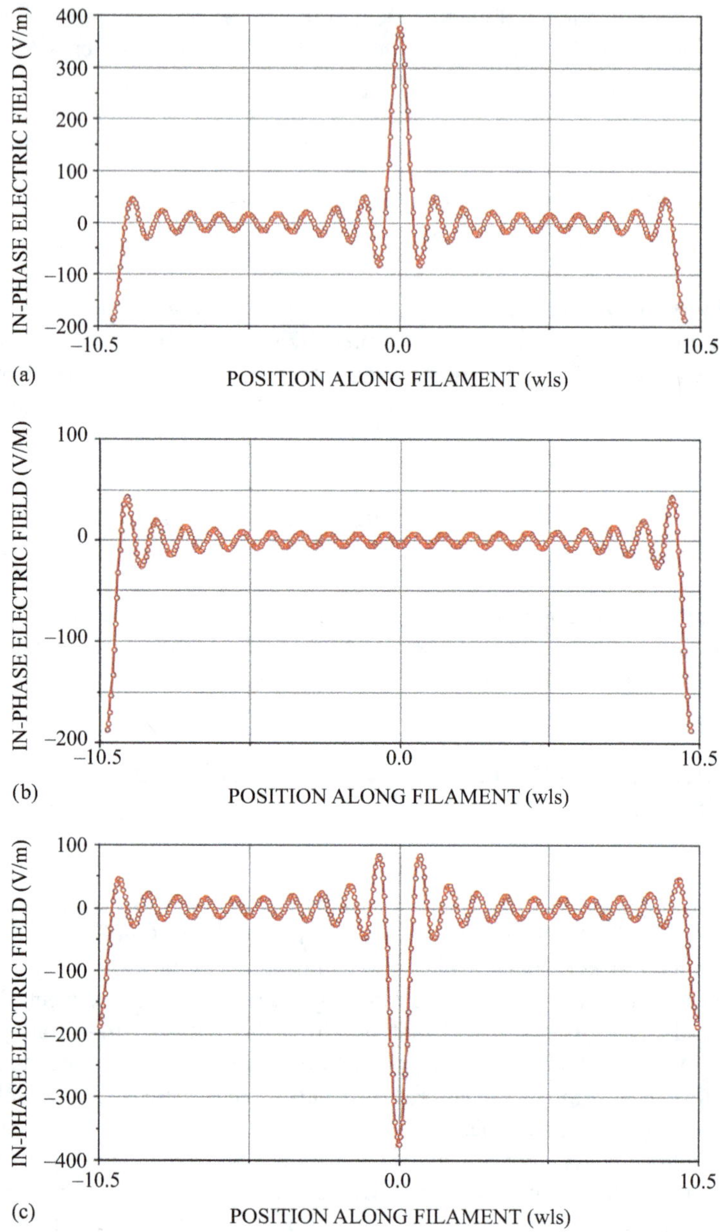

Figure 8.1 The tangential electric fields at a radial distance of 10^{-6} wavelengths from a 20 (a), 20.5 (b), and 21 wavelength (c) sinusoidal current filaments

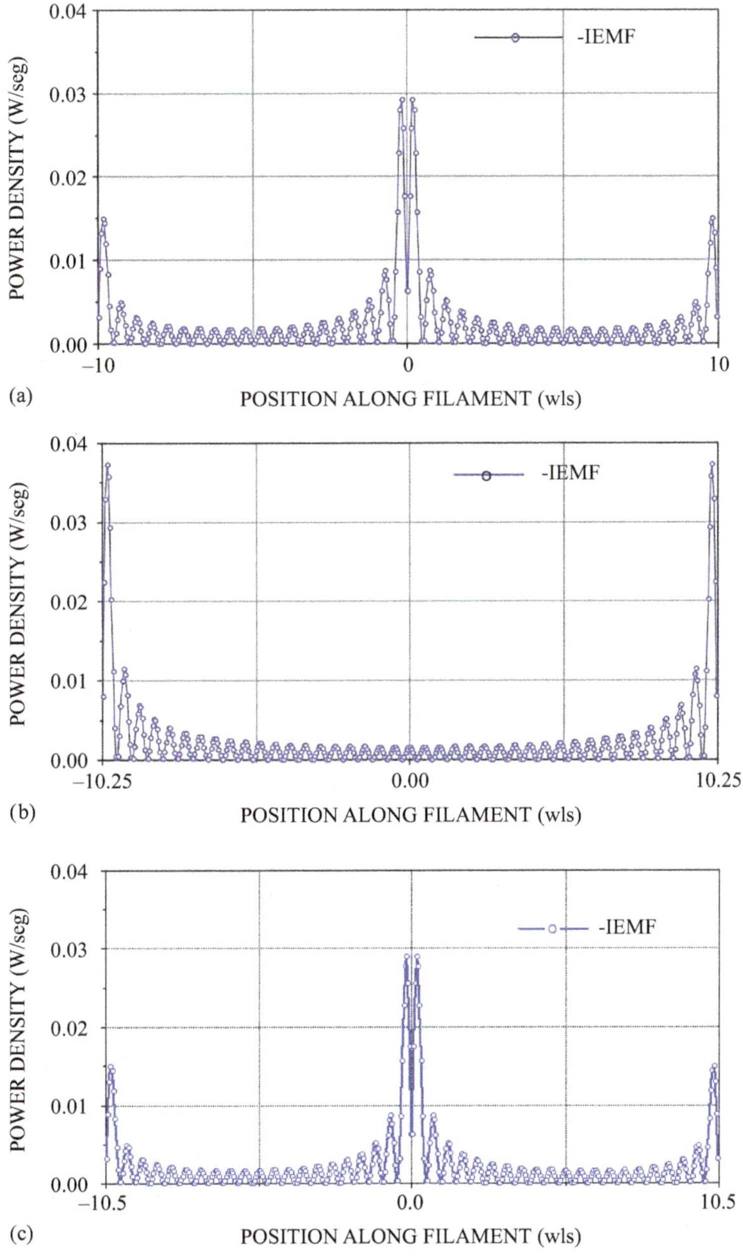

Figure 8.2 The -IEMF power densities for a 20 (a), 20.5 (b), and 21 wavelength (c) sinusoidal current filaments

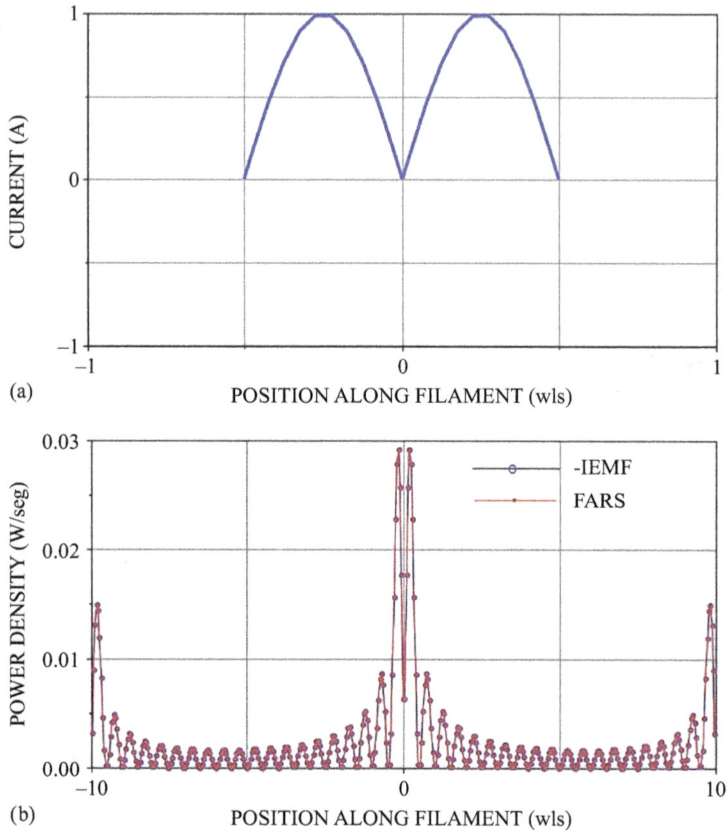

Figure 8.3 The source-region current (a) and the 1-W power density using FARS and the IEMF method for a 20-wavelelngth sinusoidal current filament (b)

20-wavelength case due to the increase in the negative current cusp at the center of the SCF.

The FARS and IEMF results of Figure 8.5 no longer exhibit the sharp variation around the center of the SCF shown in Figures 8.3 and 8.4. This is due to the fact that the current at the center of the 20.5-wavelength SCF lacks their current-slope discontinuities.

Continuing this sequence, FARS power density for a SCF of 20.75 wavelengths is presented in Figure 8.6. There is again a current-slope discontinuity at the center of the SCF that again results in a center peak in the power density. This peak, and those in Figures 8.3 and 8.4, is caused by a sign change in the current slop and an associated step change in the sign of the charge that results.

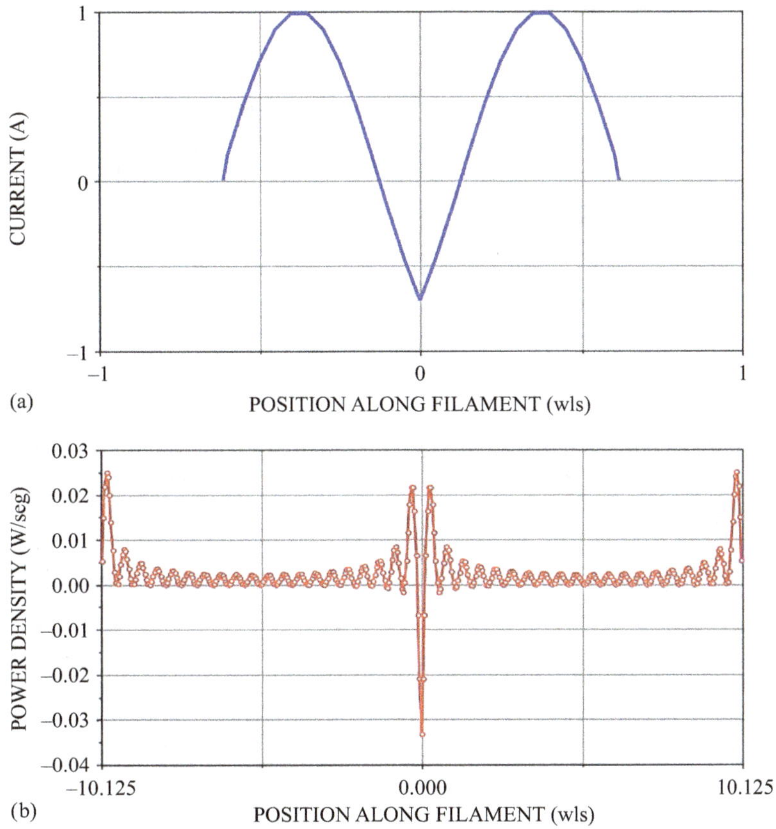

Figure 8.4 The source-region current (a) and the 1-W power density from using FARS for a 20.25-wavelelngth sinusoidal current filament (b)

The last result of this series in Figure 8.7 is for an SCF 21-wavelengths long. The current-slope discontinuity at the source is opposite to that for the 20-wavelength case of Figure 8.3, but the power densities at the ends and center of the two are nearly identical.

8.7 FDFARS results for straight wires

Straight-wire FARS results are included here for a variety of applications. In Figure 8.8, are plotted the variation of FARS in W/segment for center-fed dipole antennas of integer lengths varying from 1 to 5 wavelengths using 21–101 segments. Comparable results for dipoles varying from 0.5 to 4.5 wavelengths are shown in Figure 8.9 (using 11–91 segments). The effect of radiation loss is exhibited in both cases by the decreasing FARS values with increasing dipole length, being somewhat more pronounced for the

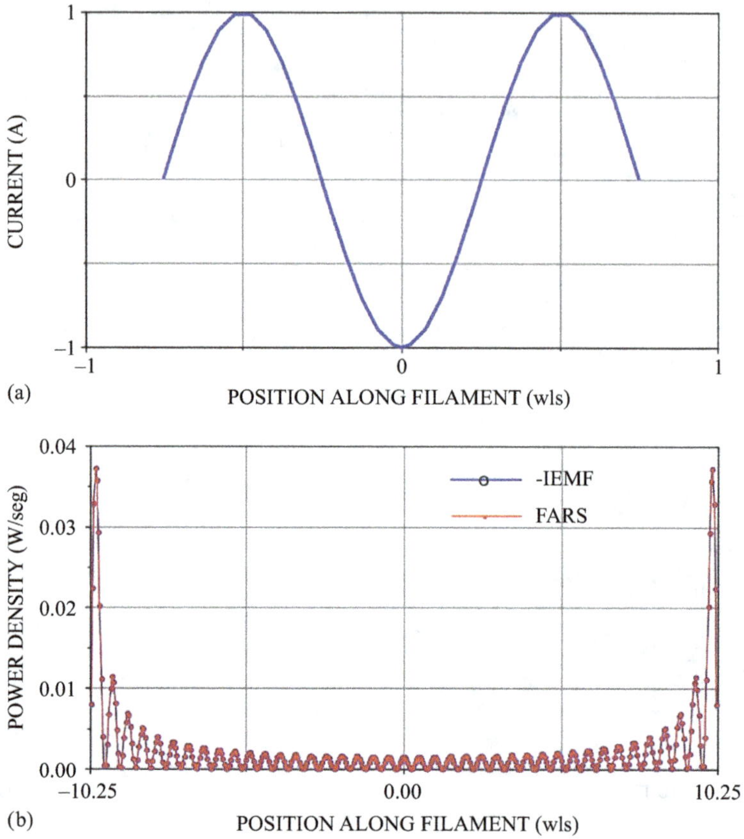

Figure 8.5 The source-region current (a) and the 1-W power density from using FARS and the IEMF method for a 20.5-wavelelngth sinusoidal current filament (b)

integer-wavelength case. The feedpoint radiation for the two cases is very different due to the current-slope discontinuity that occurs in the integer-wavelength dipole.

A comparison of a10-wavelength NEC dipole and a SCF with both having a 1-A maximum current is shown in Figure 8.10. Their respective power densities have very similar shapes but the SCF slightly exceeds that of the dipole due to attenuation of the dipole current with increasing distance from the feedpoint.

Results for scattering from a 20-wavelength straight wire are presented in Figures 8.11 and 8.12 respectively for broadside incidence and from 10-deg off axis on the right end of a 1-V/m plane wave. At broadside incidence, only 20 power-density maxima are produced while near-axial incidence results in 40, since in the latter case, both full-wave and half-wave modes are excited. The near-axial power

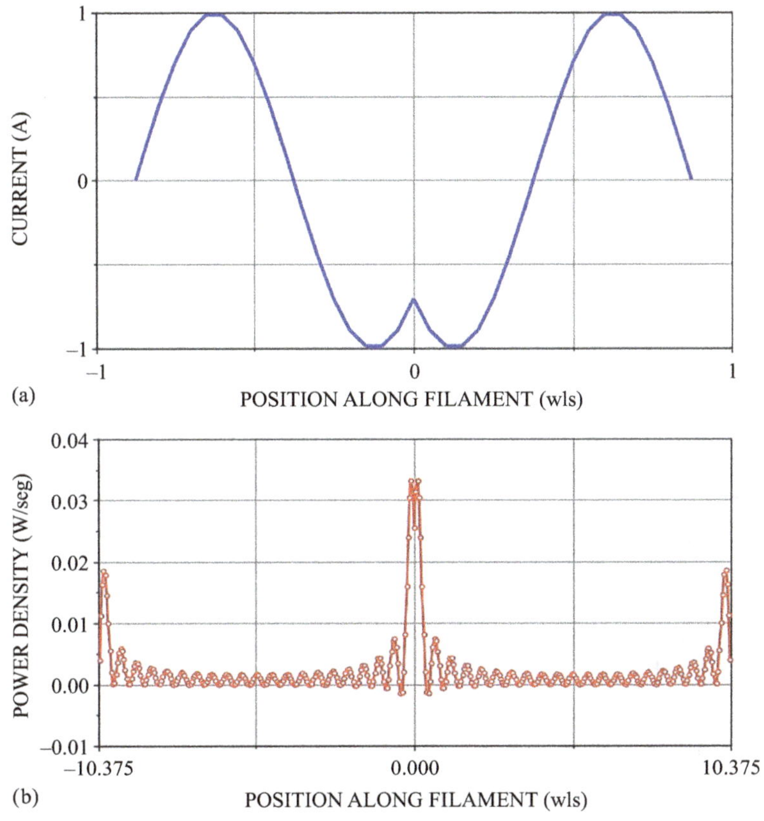

Figure 8.6 The source-region current (a) and the 1-W power density from using FARS for a 20.75-wavelelngth sinusoidal current filament (b)

distribution is dominated by maxima at both ends with that on the left end about three times larger due to a traveling-wave effect.

The effect of a step-in wire radius is demonstrated by the FARS results of Figure 8.13 where the total radiated power in each case is 1 W. The effect of the radius step is clear in the perturbation in the power density at ± 2.5 wavelengths where a near doubling of the radiation occurs.

A validation check of FARS for a 10-waveleength dipole antenna 10^{-20} wavelengths in radius having a peak current of 1-A is shown in Figure 8.14. The radial power flow 10^{-3} wavelengths from the dipole surface is compared with the FARS power density. The two curves overlap over most of the dipole with a smaller difference near its end and a larger difference at the center feedpoint. The peak in the feedpoint radial power flow is not radiated power as shown by FARS, but power that flows outward along the dipole arms. The respective total powers are 201.2 and 206.9 W for FARS and the radial flow.

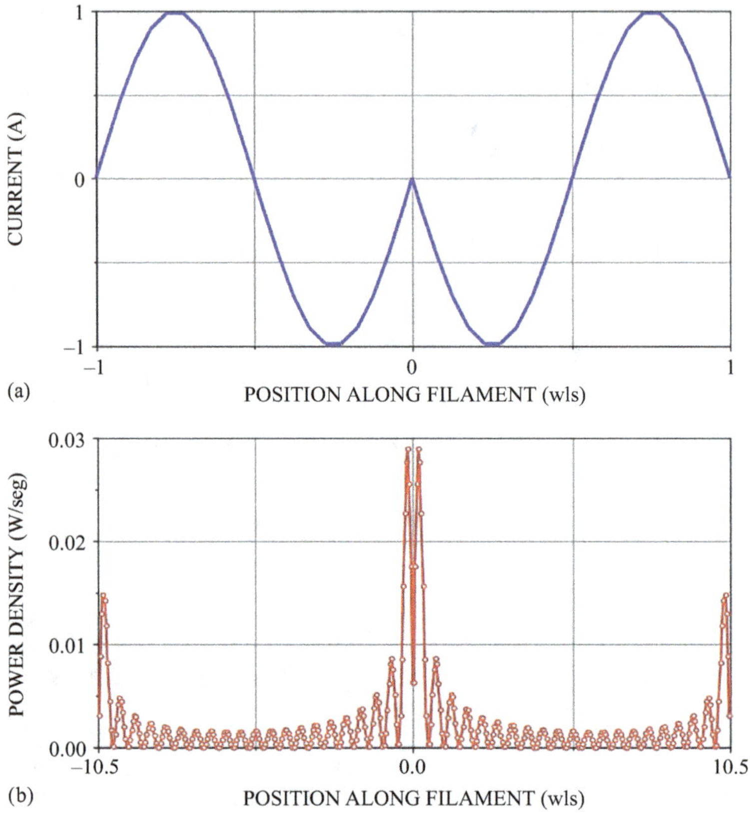

Figure 8.7 The source-region current (a) and the 1-W power density from using FARS for a 21-wavelelngth sinusoidal current filament (b)

8.8 FDFARS results for wire loops

Results for a circular-loop antenna and scatterer are presented in Figures 8.15 and 8.16. The antenna, the feedpoint of which is at 0, exhibits oscillations similar to those of the straight-wire dipole except that the lobes do not decay as much as for the straight wire. This is because the effect of loop curvature increases the radiation emitted from around it. For normal, or axial incidence, the scattered radiation forms a circular mode while edge incidence is more varied.

The power-density for radiation from a square-loop antenna in Figure 8.17 is seen to be more complex than for the circular loop antenna. In particular, at each 90-deg bend in the loop the power density is similar to the feedpoint FARS results of the dipoles in Figure 8.9. The effect of the charge direction-change acceleration at the bend is similar to the feedpoint behavior in that case.

The FARS power densities of a 10-wavelength circular-loop and dipole antenna are illustrated in Figure 8.18 where both have a 1-A maximum current.

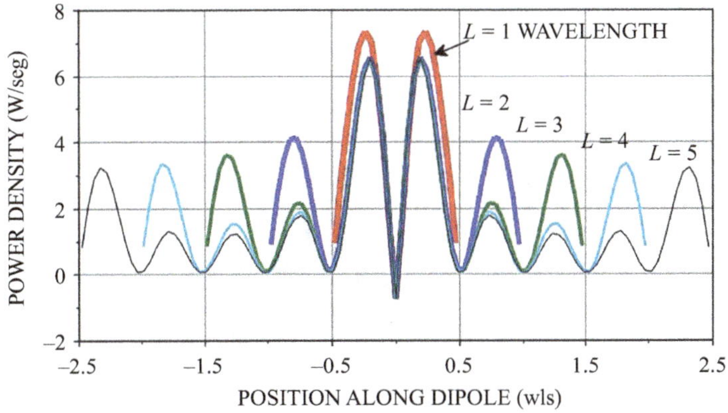

Figure 8.8 *FARS results for NEC dipoles varying from 1 to 5 wavelengths long for a maximum current of 1-A*

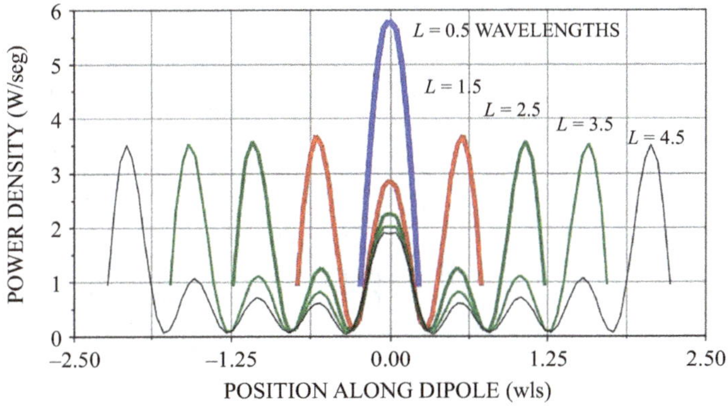

Figure 8.9 *FARS results for NEC dipoles varying from 0.5 to 4.5 wavelengths long for a maximum current of 1-A*

While the radiation from the vicinity of their feedpoints is somewhat similar for these two antennas, they differ substantially elsewhere. The influence of the loop's curvature is very obviously displayed relative to that of a straight wire. This is exhibited by the fact that the loop, having the same maximum current, radiates 357 W compared with 170.7 W radiated by the dipole.

8.9 FDFARS results for arrays

Antenna arrays occur in a variety of arrangements. A few are examined here using FDFARS to illustrate in a more detailed way their radiation characteristics.

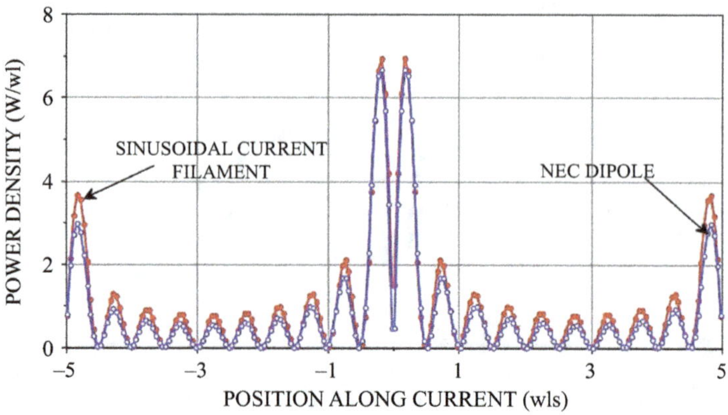

Figure 8.10 A comparison of FARS results for a 10-wavelength sinusoidal current filament and a NEC dipole with a 1-A maximum current

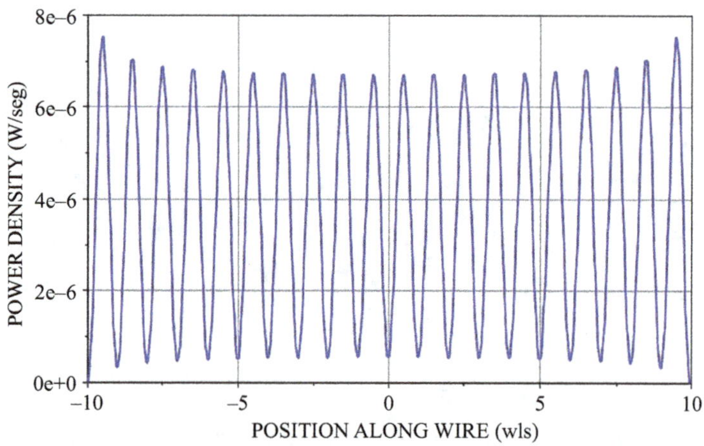

Figure 8.11 FARS power density for a 20-wavelength straight wire illuminated from broadside by a 1-V/m electric field parallel to the wire

The power densities of 9 half-wavelength, collinear dipoles spaced along the z-axis are shown in Figure 8.19a with a comparison of their NEC-computed input and their FARS powers per dipole in Figure 8.19b. The excitation voltages of the various dipoles are also indicated in part (a), having been chosen to provide significant differences between them. The FARS and NEC powers agree to within a few percent.

An array of 21 0.05-wavelength, collinear dipoles all excited by a 1-V source and located on the z-axis the ends of which are separated by 0.01 wavelengths was modeled using NEC to produce the FARS and IIEMF results in Figure 8.20. These

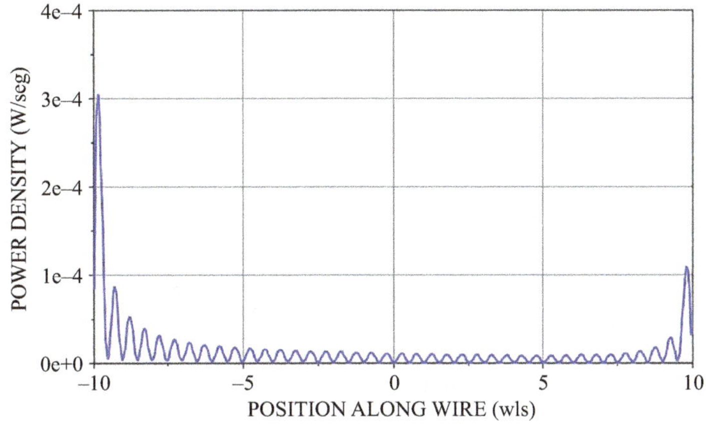

Figure 8.12 FARS power density for scattering by a 1-V/m plane wave incident at 10-deg off axial incidence from the right end of a 20-wavelength straight wire

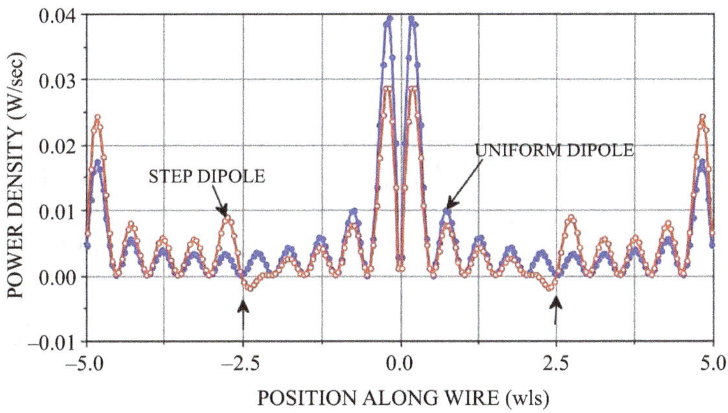

Figure 8.13 FARS results for 10-wavelength uniform and step-radius dipoles where the radius changes from 10^{-3} to 10^{-4} wavelengths at −2.5 wavelengths and the reverse at 2.5 wavelengths as indicated by the arrows

results exhibit agreement comparable to Figure 8.19 even though the "dipoles" are modeled using a single segment.

Results for a 12-element log-periodic array modeled using NEC and FDFARS are shown in Figure 8.21. The antennas on the left end, higher-frequency part of the array, are modeled using 5 segments. For the lower-frequency part of the array on the right end, 7 segments are used per antenna. The active region of the array is

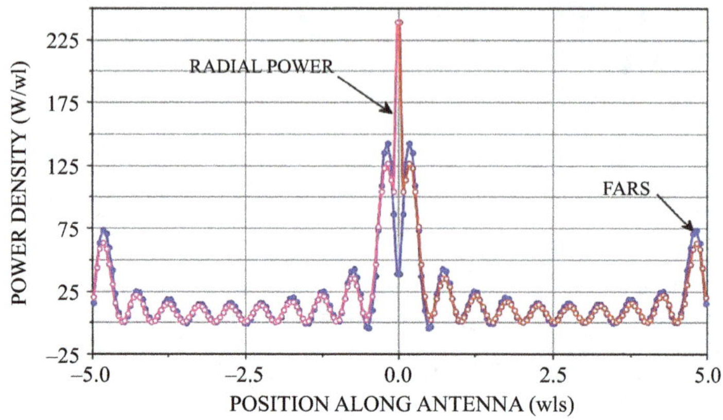

Figure 8.14 *A comparison of the FARS power density of a center-fed dipole antenna with the radial power flow at a distance of 10^{-3} wavelengths from the wire*

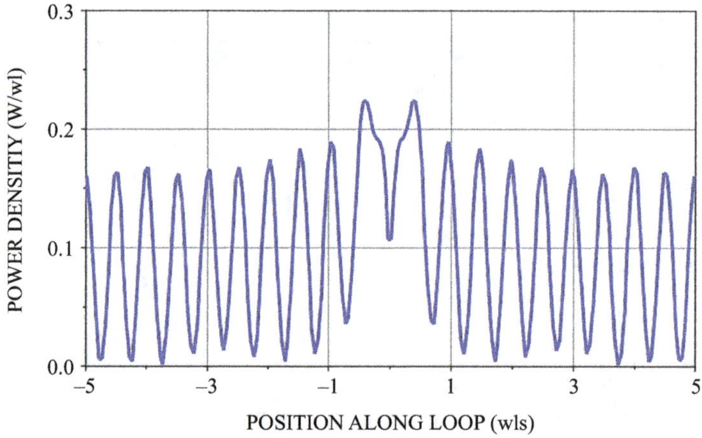

Figure 8.15 *The FARS power-flow density for a 10-wavelength circular loop antenna*

shared by elements 5 and 6, with the latter radiating the most power. Even though the power-density plot for element 6 is less than that for element 5, it has 2 more segments, which accounts for the greater total power.

The last of the array examples in Figure 8.22 displays the power-density FARS plots for a 5-element Yagi-Uda array where each antenna is modeled using 11 segments. The reflector is 0.504, the feeder is 0.48, and the directors are 0.456 wavelengths long. The FARS and NEC input powers for this case are 0.0284 and 0.0278 watts respectively for an excitation voltage of 1-V. With the reflector

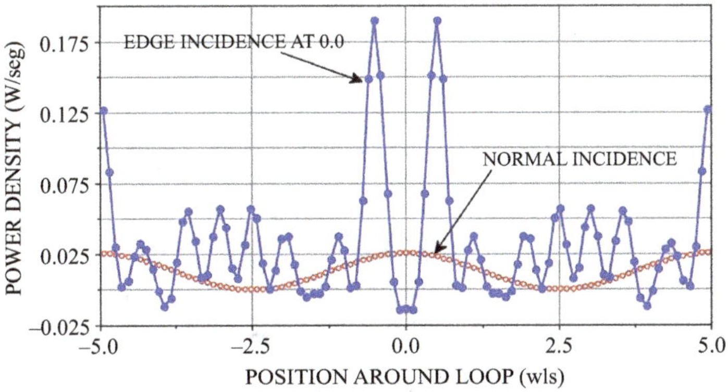

Figure 8.16 The FARS power-flow densities for a plane wave scattered from a 10-wavelength circular loop at normal incidence and from edge-on, using 10 samples per wavelength

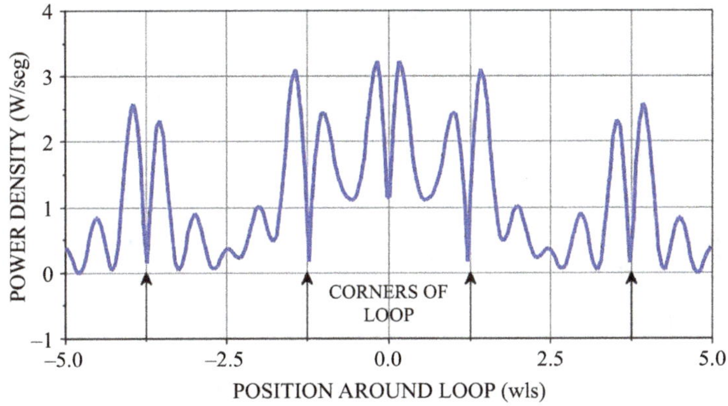

Figure 8.17 The FARS power-density of a 10-wavelength square-loop antenna

located at 0, the parallel feeder and directors are located at 0.1, 0.2, 0.4 and 0.6 wavelengths. The negative power of the reflector has the effect of reducing the backward radiation from this antenna array.

8.10 Other geometries

Some further results are included here to demonstrate FDFARS for different geometries. In Figure 8.23a is shown the power density for a 10-wavelength dipole excited by a 2-wire transmission line. In Figure 8.23b, the power density of a 10-wavelength center-fed dipole normalized to the peak of Figure 8.23a is superimposed to demonstrate the difference between the two kinds of excitations. The

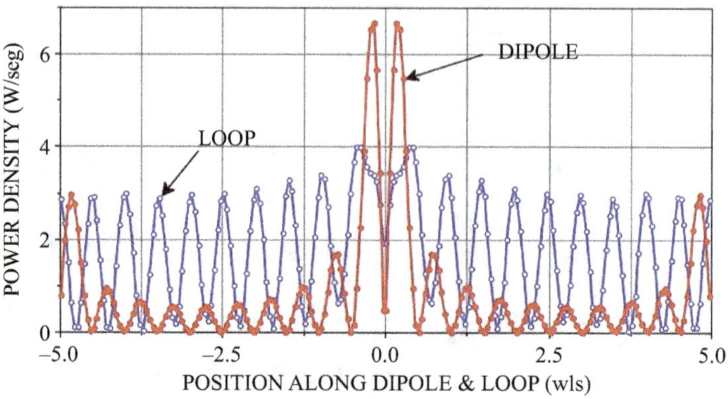

Figure 8.18 A comparison of the power densities of the 10-wavelength loop and dipole having current maxima of 1-A

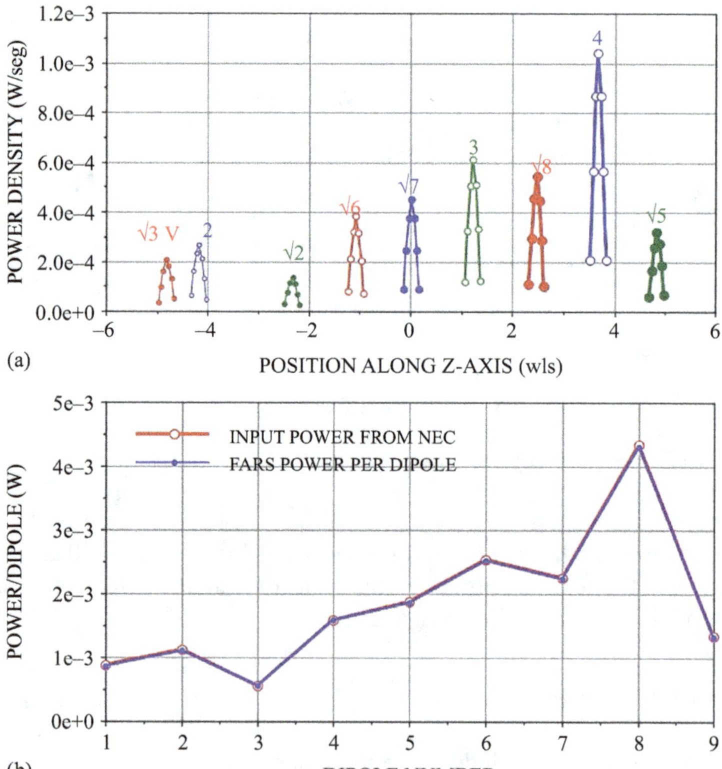

Figure 8.19 FARS power-density results (a) with the excitation voltages listed above each dipole plot and the total power per dipole from FARS and the NEC input power using the IEMF method (b).

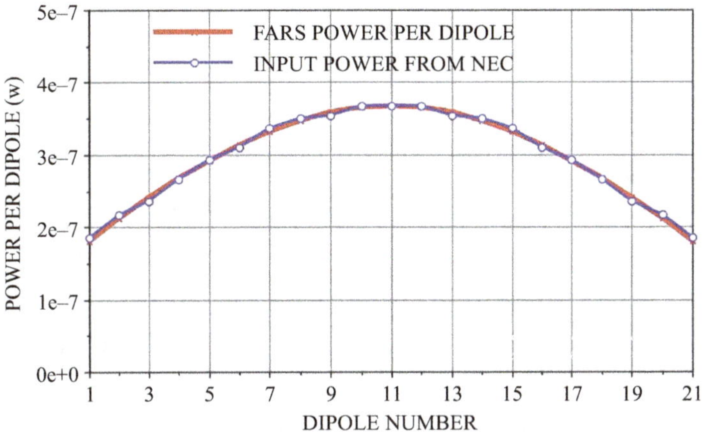

Figure 8.20 FARS and NEC IEMF input powers for a 21-element collinear array

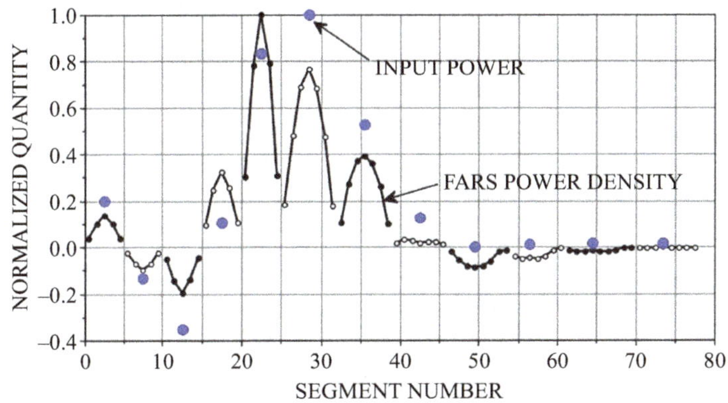

Figure 8.21 The FDFARS power densities and NEC input powers for a 12-element log-periodic array

center-fed dipole radiated-power density is similar, but not identical to that of the transmission-line feed, demonstrating their different effects.

The effects of introducing bends 2 wavelengths from each end of a 10-wavelength dipole are shown in Figures 8.24 and 8.25. FARS power densities are presented for only the right half of the 2-bend wire, to display more detail, in Figure 8.24 for a no-bend case as a reference to a maximum bend of 90 deg. The power densities near the bend can be seen to increase as the bend becomes larger. This fraction of the FARS power originating within 0.5 wavelengths of the bend is plotted in Figure 8.25 to illustrate this behavior more quantitatively. From a straight wire to a bend of 45 deg, the effect is only about 0.02 in the power fraction, but from 45 to 90 deg is more than 0.2.

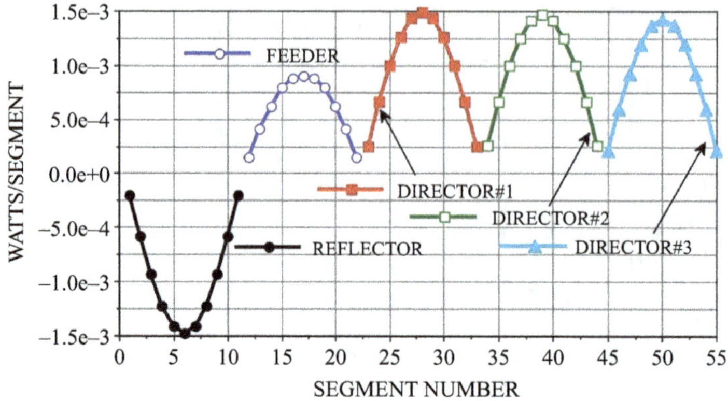

Figure 8.22 The FARS power densities for a 5-element Yagi-Uda array

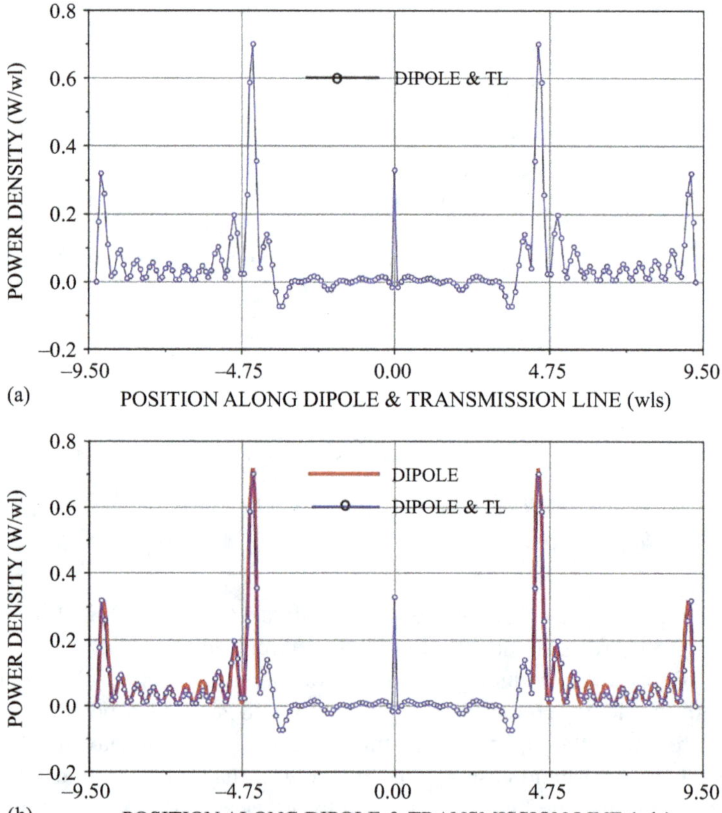

Figure 8.23 The FARS power densities of a 10-wavelength dipole connected to a two-wire transmission line (a) and for the two halves of a center-fed 10-wavelength dipole superimposed for comparison (b)

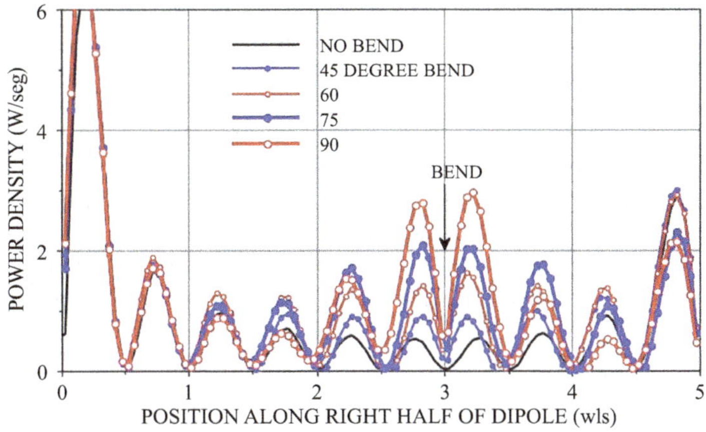

Figure 8.24 FARS power-density results on one-half of a 10-wavelength dipole having symmetric bends three wavelengths from each end and having a 1-A current magnitude

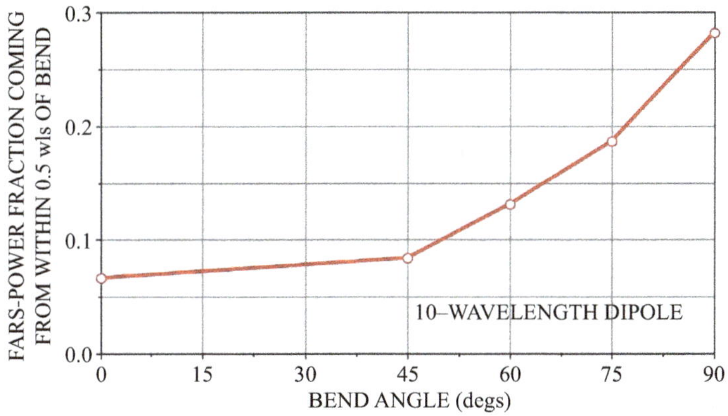

Figure 8.25 The FARS power-density contribution to the radiation in the vicinity of the bend as a function of the bend angle

In Figures 8.26 and 8.27, the shielding effect of a wire cage on an enclosed wire located on the axis of the cage is presented for two different situations. In both cases, the radius of the cage wires is chosen according to the equal-area rule such that were they unrolled they would form a closed cylinder. For the case of Figure 8.26, a 0.5-wavelength dipole is located centrally in a 4-wavelength-long cage. As a reference, the radiated power for the same dipole without the cage is computed using the NEC input power and the FARS power, PFARS are 4.25×10^{-3} W. On adding the cage wires the dipole power becomes -2.041×10^{-5} W while the cage FARS power is 2.077×10^{-5} W, for a net power of 3.68×10^{-7} W, more than 40 dB down from the isolated dipole. This demonstrates the effect of the equal-area's shielding effect as well as the numerical field cancellation that occurs when a radiating source is contained within a PEC enclosure.

The equal-area cage result shown in Figure 8.27 is for the situation where a 10-wavelength, center-fed dipole is centrally located on the axis of the cage where the isolated dipole is also included as a reference. The NEC input power and the FARS radiated power agree for the cage-enclosed dipole to within 0.2% with the cage wires partially screening radiation from that part of the longer dipole that is enclosed.

A scattering example for the cage problem is presented in Figure 8.28 where FARS results are shown for a 1-wavelength-long dipole located on the axis of a

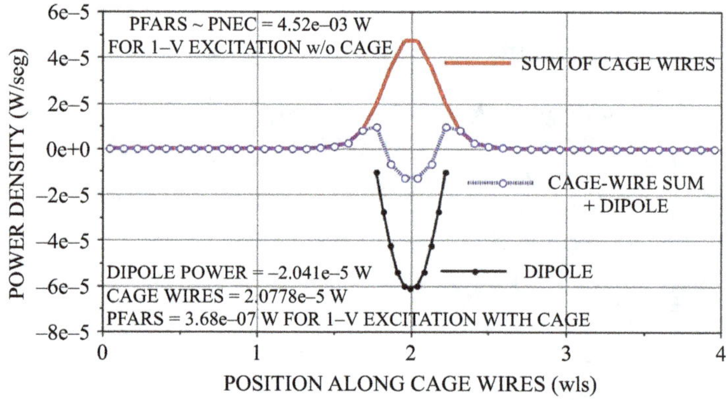

Figure 8.26 The FARS power densities for an isolated 0.5-wavelength dipole and when it is enclosed by a 4-wavelength wire cage the wires of which when unrolled would form a closed surface about it

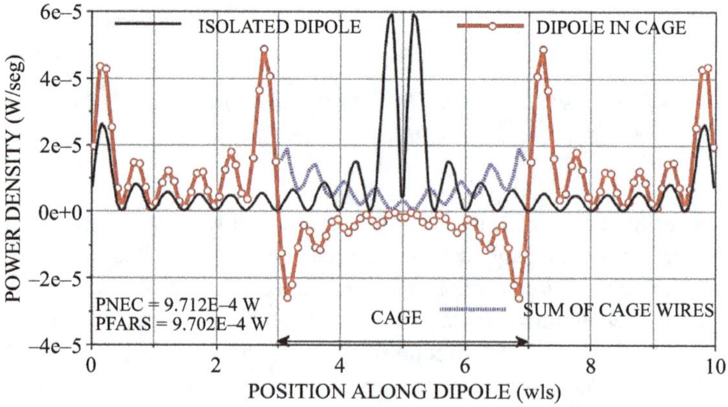

Figure 8.27 Results for a 10-wavelength dipole enclosed by a cage of wires the radii of which satisfy the equal-area rule

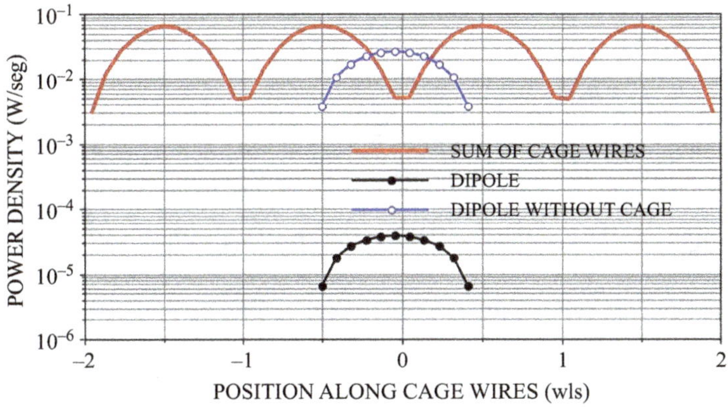

Figure 8.28 FARS results for plane-wave scattering from a 10-wire cage with an enclosed 1-wavelength dipole located on its axis

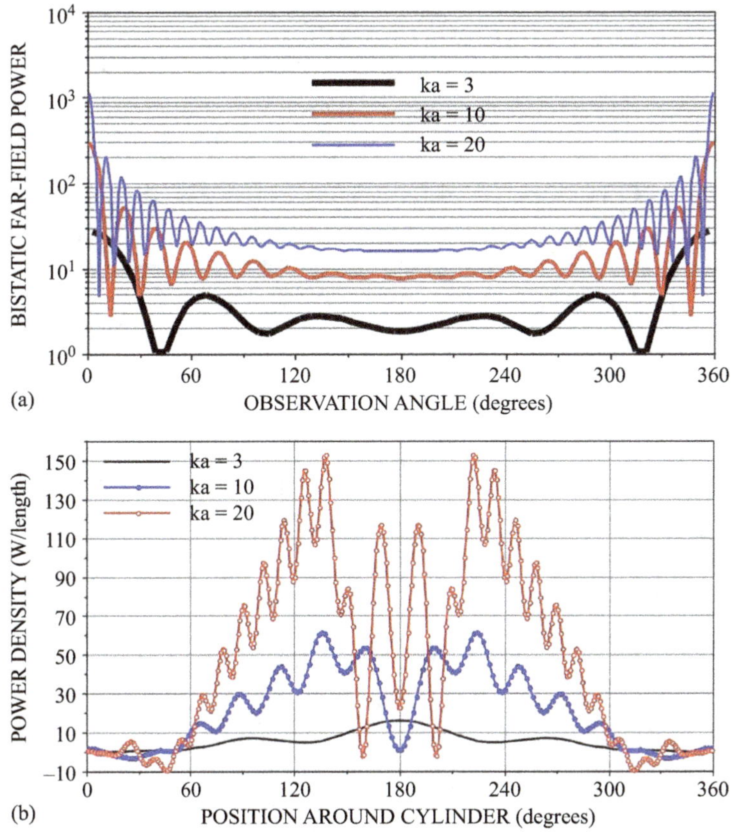

Figure 8.29 The bistatic scattered power pattern (a) and the FARS radiated power density (b) for the TE polarization

4-wavelength-long cage when illuminated by a normally incident 1-V/wl plane. Also included is the power density for the dipole absent the cage. The shielding effect of the equal-area cage is about 30 dB.

The last examples of FDFARS applications are for scattering of a normally incident plane wave from an infinite, PEC cylinder. The case of TE incidence is shown in Figure 8.29 and TM incidence in Figure 8.30 with the waves incident from 180 deg. Both the bistatic scattered power and FARS power density for the TE case are highly oscillatory due to excitation of a circulatory creeping wave around the cylinder.

Corresponding results for the TM case in Figure 8.30 reveal a much less oscillatory bistatic power pattern. The FARS radiated power density is similar to that for the TE case while having an amplitude that is about 40% less. For both cases, radiation from the extreme shadow side of the cylinder is near zero.

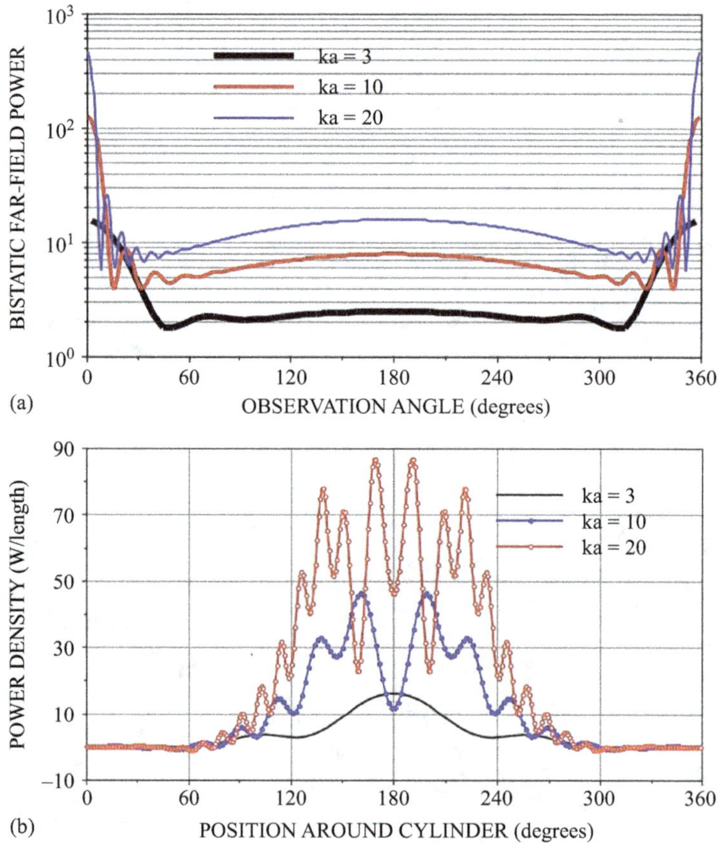

Figure 8.30 The bistatic scattered power pattern (a) and the FARS radiated power density (b) for the TM polarization

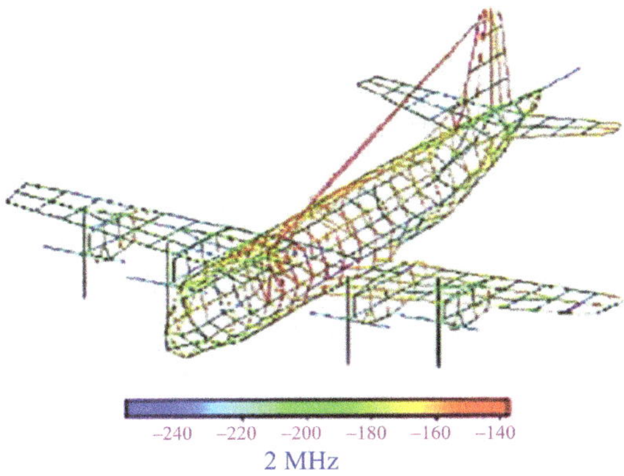

Figure 8.31 FARS results for a moderately large aircraft with wire antenna extending from the fuselage to the tail [8]

8.11 A more complex geometry

An example of applying FDFARS to a wire-grid model of an aircraft is included in Figure 8.31 [8]. The long-wire antenna attached between the forward part of the fuselage and the top of the tail is excited at a frequency of 2 MHz. The antenna radiates along its length with some additional radiation being emitted from the top of the fuselage that provides a fairly effective ground plane.

8.12 Summary

This chapter is devoted to the frequency-domain version of FARS (FDFARS), including some background about its development and demonstrating its application to a variety of radiating geometries. Among the specific topics and results included are:

1. Demonstration that radiation from a sinusoidal current filament occurs all along its length as determined by the induced-electromotive-force (IEMF) method.
2. Confirmation of the SCF IEMF results by the initial application of FARS.
3. Showing how the FARS results are affected by the current variation in the center region of the antenna.
4. The distribution of FARS power for variable dipole lengths.
5. Comparison of the length-wise radiation distribution of the SCF with a NEC dipole.
6. The radiation distribution for scattering from a long wire at normal and oblique incidence.
7. The effect of a step radius on the distribution of radiation from a straight wire.

8. The close correlation between radial power flow near a straight-wire antenna and the FARS results.
9. The FARS power density of circular-loop antennas decays more rapidly than on a straight wire due to the loop curvature.
10. A square-loop antenna exhibits doublet-shaped FARS radiation at the bends similar to the feedpoint power density on a straight wire having a current-slope discontinuity.
11. Axial-incidence scattering from a circular loop produces a single circular mode in the radiated power while edge-incidence results in multiple lobes.
12. The FARS radiated powers for several antenna arrays agree within a few percent of the NEC input powers.
13. FARS results provide interesting insights about the radiation distributions of other geometries such as a transmission-line excited dipole, a bent-wire dipole, and dipoles enclosed in wire cages.
14. Equal-area cages are shown to screen the radiated powers of enclosed dipole antennas and the received power for an incident plane wave.
15. Radiation distributions for TE and TM normally incident on a PEC infinite, circular cylinder reveal the excitation of traveling waves around the circumference.

References

[1] E. K. Miller, PCs for AP and other EM reflections, *IEEE AP-S Magazine*, 41 (2), pp. 82–86, 1999; Vol. 41, No. 3, pp. 83–88, 1999.

[2] C. Balanis, *Antenna Theory, Analysis and Design*, Harper & Row, New York, NY, 1982.

[3] E. K. Miller, Comparison of the radiation properties of a sinusoidal current filament and a PEC dipole of near-zero radius, *IEEE Antennas and Propagation Society Magazine*, 48(4), pp. 37–47, 2006.

[4] E. K. Miller and G. J. Burke, A multi-perspective examination of the physics of electromagnetic radiation, *Applied Computational Electromagnetics Society Journal*, 16(3), pp. 190–201, 2001.

[5] J. D. Jackson, How an antenna launches its input power into radiation: the pattern of the Poynting vector at and near an antenna, *American Journal of Physics*, 74(4), pp. 280–288, 2006.

[6] S. Castillo, The interpretation of body current distributions in designing low-RCS scatterers, *1997 North American Radio Science Meeting, Montreal, Canada*, p. 736.

[7] J. F. Shaeffer, B. A. Cooper, K. W. Hom, R. C. Baucke, and N. A. Talcott, A review of bistatic k-space imaging for electromagnetic prediction codes for scattering and antennas, *IEEE AP-S Magazine*, 39(5), pp. 2121–29, 1997.

[8] Private communication from Professor Christopher Trueman, Concordia University, Montreal, Quebec, Canada, 2000.

Chapter 9
Time-domain FARS and frequency-domain FARS compared

Abstract

Time-domain far field analysis of radiation sources (TDFARS) and frequency-domain far field analysis of radiation sources (FDFARS) have been shown in Chapters 7 and 8 to determine the quantitative distribution of radiated power and radiated energy on a variety of wire objects. The FDFARS results reveal their frequency dependence through a series of lobes related to the wavelength. For antennas, there are nominally 2 lobes per wavelength with larger maxima at the feedpoint and ends. For scattering the lobe structure can vary from 1 to 2 lobes per wavelength. Similar lobe structure is found for other wire objects. The TDFARS results, on the other hand, are not lobed but vary smoothly over an object with maxima at the open ends of a wire and at the feedpoint as well as opposite the feedpoint of a circular or square loop. The absence of lobes is because the time-domain current and charge are not standing waves.

The comparisons included here are normalized to a 1-W total power or 1-J total energy on a per-segment basis for a nominal spatial sampling density of 20 per m. Discussion is limited here to wires of several wavelengths so that physical features that may not be as pronounced for shorter objects are observable.

Keywords: Energy density; FARS; FDFARS; Power density; Radiated energy; Radiated power; Spatial power distribution; TDFARS; Time-domain

9.1 Comparison of time-domain and frequency-domain far-field analysis of radiation sources

A spatial integration over an object at each time stop is used to obtain the time-domain power density TDFARS [1]. A further integration over time until the spatial distribution stabilizes then provides the energy density. The incremental far-field power-flow density $p_m^{(TDF)}(\mathbf{r}, t)$ is given by

$$p_m^{(TDF)}(\mathbf{r}, t) = Lim(r \to \infty)\left[\frac{r^2}{\eta_o}\mathbf{e}_m(\mathbf{r}, t)\bullet\mathbf{E}(\mathbf{r}, t)\right] \tag{9.1}$$

with η_o being the free space wave impedance. By integrating over 4π steradian observation angles, the total TDFARS linear power density, i.e. the power radiated from segment m, $P_m^{(TDF)}(t)$ at time t, with the superscript (TDF) indicating TDFARS, is given by

$$P_m^{(TDF)}(t) = \int_0^{2\pi} p_m^{(TDF)}(\mathbf{r}, t)\sin\,\theta d\theta d\varphi. \tag{9.2}$$

The per-segment linear energy density, $W_m^{(TDF)}(t)$, up to time t can then be written as

$$W_m^{(TDF)}(t) = \int_{-\infty}^{t} P_m^{(TDF)}(t')dt'. \tag{9.3}$$

The FDFARS results are obtained in a similar fashion. This incremental *FARS* power, $p_m^{(FDF)}(\theta,\varphi)$ for segment s_m in observation direction θ,φ as

$$p_m^{(FDF)}(\theta,\varphi) = \lim r \to \infty\left(r^2 \frac{1}{2\eta}\mathrm{Re}[\mathbf{e}_m(\theta,\varphi)\cdot\mathbf{E}^*(\theta,\varphi)]\right) \tag{9.4}$$

where $\mathbf{e}_m(\theta,\varphi)$ is the electric field due to source m and η_o is the medium impedance. The total power contributed by each segment, $P_m^{(FDF)}$, comes from integrating the incremental result $p_m^{(FDF)}(\theta,\varphi)$ over all θ,φ to obtain

$$P_m^{(FDF)} = \int_0^{\pi}\int_0^{2\pi} p_m^{(FDF)}(\theta,\varphi)\sin\,\theta d\theta d\varphi \tag{9.5}$$

9.2 Results for straight wires

The initial result presented here in Figure 9.1 is a comparison of FDFARS [2] and TDFARS for a 10-m straight wire. The FDFARS results are for 300 MHz so that the wavelength is 1 m. A time-harmonic excitation of TDFARS was also computed for a

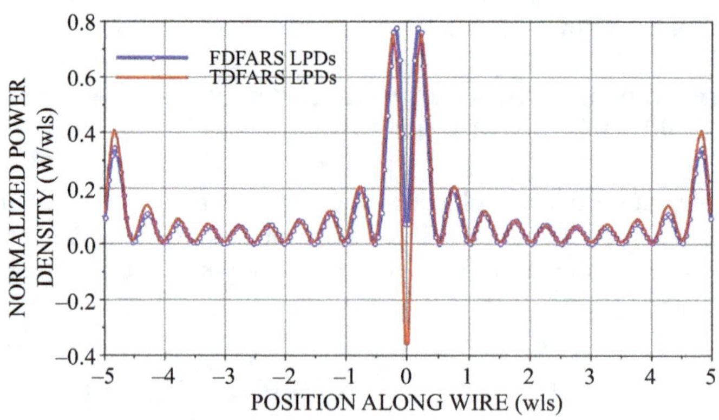

Figure 9.1 A comparison of TDFARS and FDTARS for a 10-wavelength, center-fed dipole 10-m antenna

frequency of 300 MHz. The numerical models used here and in the following have 20 segments per meter or wavelength. This example was run to cross-validate the two approaches. These results agree to within a few percent except at the center feedpoint and wire ends, an acceptable outcome considering such substantial differences in their numerical treatments.

A comparison of time-harmonic excitation using FDFARS and impulsive excitation using TDFARS for the same 10-m dipole is shown in Figure 9.2. While both results have maxima near the ends and a discontinuity at the center feedpoint, these features of FDFARS are much more prominent than those of TDFARS.

Scattering results for a 10-m wire illuminated from broadside by a plane wave are presented in Figure 9.3. The FDFARS for the antenna of Figure 9.2 exhibit 2 lobes/wavelength, while those for the scattering case of Figure 9.3 have half as many lobes. The TDFARS in Figure 9.3 fall of linearly from a center maximum while the FDFARS maxima are nearly uniform along the wire with a slight upward trend towards the ends.

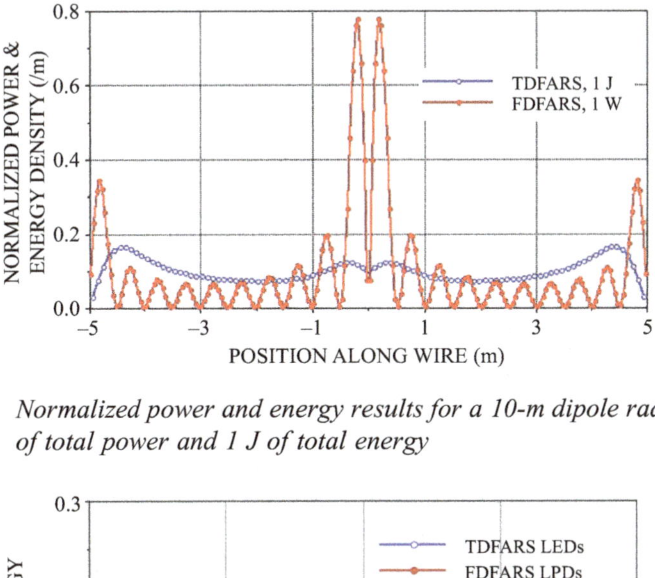

Figure 9.2 Normalized power and energy results for a 10-m dipole radiating 1 W of total power and 1 J of total energy

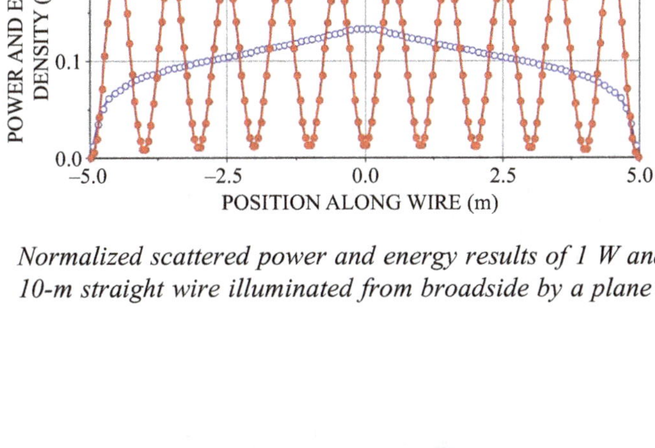

Figure 9.3 Normalized scattered power and energy results of 1 W and 1 J for a 10-m straight wire illuminated from broadside by a plane wave

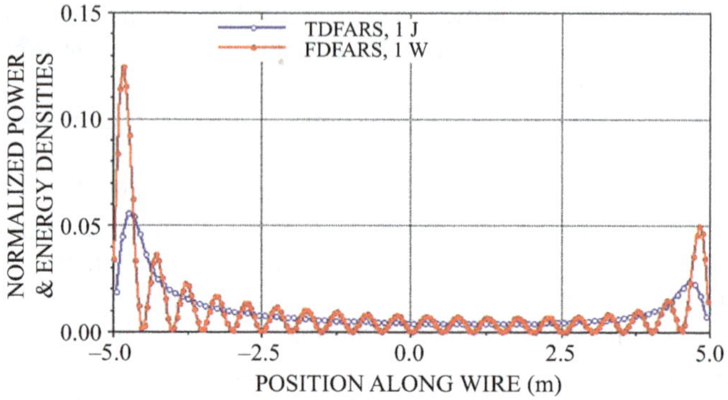

Figure 9.4 Normalized scattered power and energy results of 1 W and 1 J for a 10-m straight wire illuminated by a plane wave at 10 deg off axial incidence from its right end

Scattering results from a 10-m straight wire for a plane wave incident at 10-degrees off the right end are presented in Figure 9.4. In contrast with the normal-incidence results of Figure 9.3, the FDFARS for near-axial incidence have 2 lobes/wavelength rather than 1. Both the TDFARS and the FDFARS energy and power plots have far-end maxima that are more than twice those of the near-end results. This result is consistent with results obtained using the geometrical theory of diffraction where the wire ends are found to dominate the scattering in this situation.

9.3 Results for wire loops

FARS results for a circular wire loop are presented in Figure 9.5 where the feedpoint for both antennas is at 0 on the plots. As previously shown for straight wires, the FDFARS power density for a 10-m loop exhibits about 2 lobes per wavelength, while TDFARS for the 10-m loop varies smoothly away from the feedpoint. However, opposite the feedpoint a maximum occurs in the radiation that is larger than that at the feedpoint. This reason for this effect has been discussed in Section 7.6 of Chapter 7. There, it is shown how radiation maxima coincide due to counter-propagating current maxima and charge minima.

Comparable FARS results are shown in Figure 9.6 for square loops. The feedpoint is at 0 with the vertices of the square loops located at ± 1.25 and ± 3.75 m (marked by the vertical lines). The FDFARS again have about 2 lobes per wavelength (m) although very different in their amplitude variation from the circular loop. The TDFARS energy has sharp cusps at the bends of the square loop and has a maximum opposite the feedpoint similar to that of the circular loop. Note that the corner cusps, or doublets, resemble the FARS behavior where the loops are excited. This is due to the fact that charge-direction change around the corner results in acceleration-caused radiation like that caused by a localized voltage source.

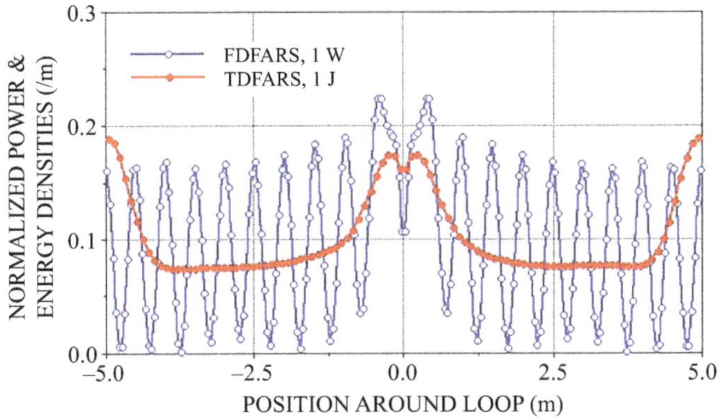

Figure 9.5 *Normalized power and energy densities of 10-m circular-loop antennas*

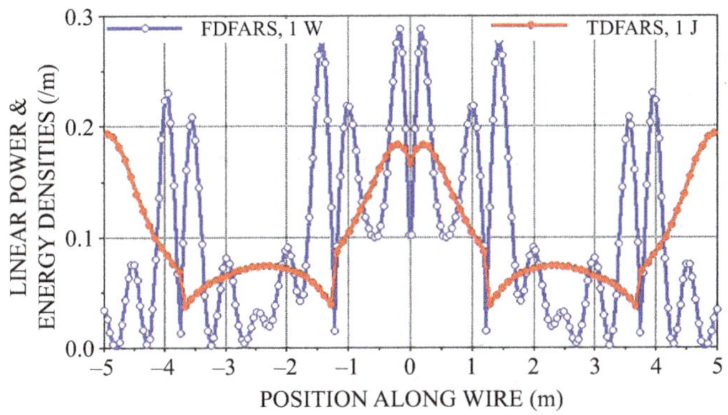

Figure 9.6 *Normalized power and energy densities of 10-m square-loop antennas*

9.4 The transmission-line excited dipole antenna

A comparison of applying FARS to a transmission-line-fed dipole antenna is shown in Figure 9.7. The time-domain energy distribution on the dipole now has a higher maximum at the dipole feedpoint from the transmission line rather than at its ends as in the case of Figure 9.2 for a dipole excited at its center by a voltage source. The frequency-domain power distribution for the transmission-line feed is, on the other hand, similar to that seen in Figure 9.2 with higher radiation peaks at the transmission-line connection to the dipole compared with its ends. Another distinct difference between the TDFARS and FDFARS results is the transmission line where the former is essentially flat while the latter exhibits some wave-dependent effects.

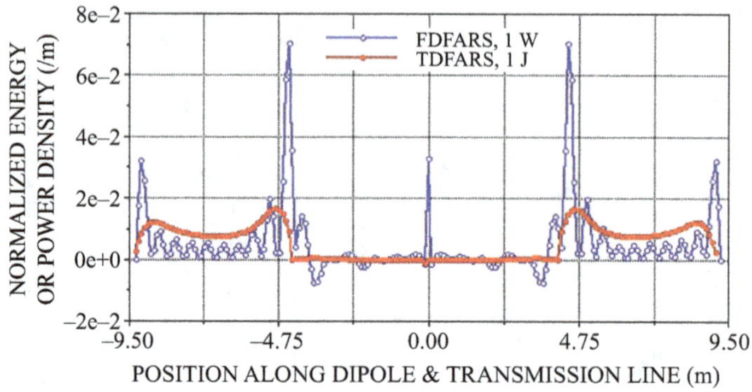

Figure 9.7 Normalized power and energy densities for a 10-m dipole excited by a two-wire transmission line

9.5 A dipole enclosed by a wire cage

Results are presented in Figures 9.8 and 9.9 for a dipole antenna centered on the axis of a cage of longer parallel wires. The cage wires satisfy the "equal-area rule" such that were they unrolled, their total area would form a continuous surface around the dipole to simulate the effect of a solid enclosure.

The numerical models for these two examples employ slightly different parameters while having the same goal of testing how well a wire cage will "shield" the radiation from a dipole antenna. Circumferential wires are not needed for this example because it is only the parallel electric fields of the dipole that form its radiating components.

Plotting the power and energy densities of the center dipole as positive quantities results in the corresponding densities of the cage wires to be negative. This is the equivalent situation to that of an antenna being enclosed by a PEC continuous surface or envelope. The equivalence principle requires that the fields of the antenna be exactly cancelled by the sources it induces on the enclosing envelope to produce zero fields outside the envelope. Thus, if the antenna fields are taken as being positive, those of the envelope sources will be equal and negative. This is not to say that negative power is radiated, but that the total external field must be zero.

Thus, in both the FDFARS and TDFARS examples, positive FARS values are associated with the center dipole while negative FARS values are similarly associated with the cage wires. On adding the dipole and cage wires FARS values the net result is small but nonzero, due to the total radiation not being zero for either case. The reason could be attributed to numerical error, but both the NEC and TWTD models are well validated. Possibly, the equal-area rule may not be entirely accurate in this situation. Another more likely explanation is that since the cage ends are not enclosed some field leakage occurs out the ends. This possibility seems more probable in the time-domain model as shown in the cage-wire plot.

The power density from FDFARS for a wire cage of 10 4-m wires with a centered, 0.5-m dipole for a 1-m wavelength is presented in Figure 9.8. The sum of

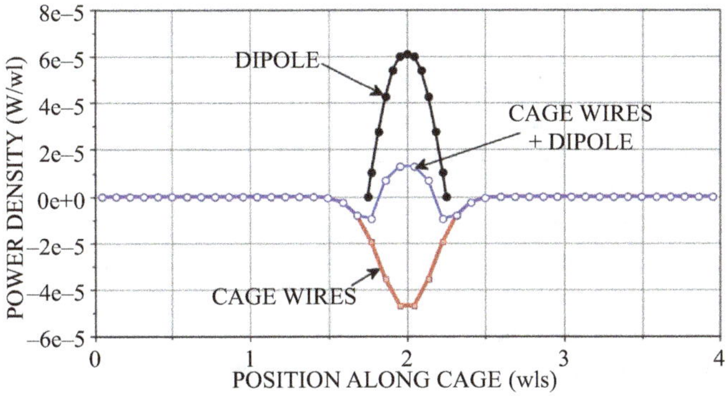

Figure 9.8 The FDFARS results for a half wavelength, 11-segment dipole located inside a cage of 10 parallel, 4 wavelength wires of 44 segments

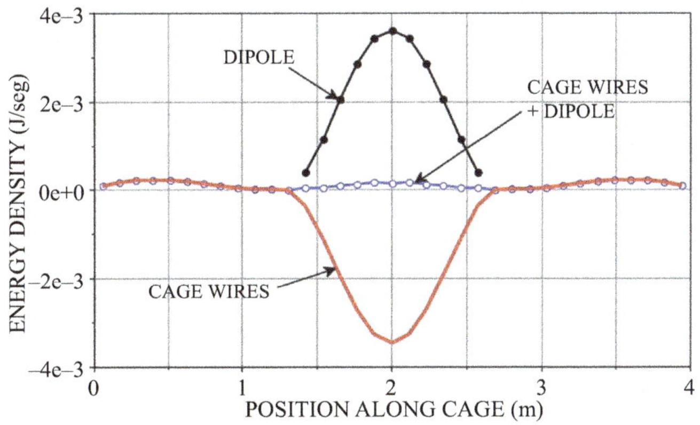

Figure 9.9 The TDFARS results for an 11-segment dipole located inside a cage of 8 parallel, 35-segment wires

the cage and dipole linear power densities is about 40 dB below that for the unshielded dipole. Their opposite-signed values occur because the net far field is reduced from that which either one alone would produce.

Corresponding results from TDFARS for a wire cage are shown in Figure 9.9. There are 8.4-m long cage wires of 35 segments enclosing an 11-segment dipole about 1.2 m long. More leakage occurs here than for the frequency-domain case, as the dipole is not as well enclosed, but the results are otherwise similar.

9.6 Summary

Comparison of FDFARS and TDFARS results for the wire objects discussed here shows that the most distinctive difference between them is that the frequency-domain FARS exhibit one to two lobes per wavelength, a feature absent in the time-domain results. This because of the standing-wave current distributions normally found in the frequency domain. These lobes occur coincident with the current maxima caused where counter-propagating waves meet. A similarly caused maximum occurs on a loop opposite the feedpoint in the time domain.

The FARS results for both domains are were found in Chapters 7 and 8 to sometimes have negative values over parts of an object. Results presented for a dipole located in a cage of parallel wires demonstrated this phenomenon as a logical outcome for situations where some parts of an object reduce the radiated field as is the case of the dipole located in a wire cage. Negative FARS values do not represent the radiation of negative power or energy, but simply the reduction of the overall electromagnetic field.

References

[1] E. K. Miller, Time-domain far-field analysis of radiation sources, *IEEE Antennas and Propagation Society Magazine*, Vol. 53, No. 5, pp. 82–97, 2011.

[2] E. K. Miller, PCs for AP and other EM reflections, *IEEE AP-S Magazine*, Vol. 41, No. 2, pp. 82–86, 1999; Vol. 41, No. 3, pp. 83–88, 1999.

Chapter 10

The radiation properties of some specified currents

Abstract

This chapter focuses on radiation from various specified current distributions, with an emphasis on the classic sinusoidal current filament (SCF). The SCF provided a basis for early attempts to predict the pattern and radiation characteristics of an elementary physical antenna, the wire dipole. This was because closed-form, analytical expressions had been developed for its near and far fields and radiated power as a function of its length. The SCF made a good test case for FDFARS because the effective input power required to maintain the SCF can be determined from the Induced ElectroMotive Force (IEMF) method. As was shown in Chapter 8, the IEMF results for the SCF were instrumental in development of FDFARS.

The SCF is examined further in this chapter. One item discussed is establishing why the total power it radiates oscillates between maximum and minimum values as its increasing length, L, passes through even and odd-number of half wavelengths. Radiated-power maxima occur when its length is an even number of half wavelengths while minima occur for an odd number of half wavelengths. The reason why the average radiated power increases as the Log(kL) is shown to be due to radiation that takes place between the center feedpoint and its ends.

The distributions of radiated power derived from FDFARS and the IEMF method are presented for several other specified currents to provide further validation of both methods while also demonstrating the wide applicability of these procedures.

Keywords: FDFARS; IEMF method; Incremental IEMF method; Sinusoidal current filament (SCF); Specified current filament; Uniform current filament (UCF)

10.1 Introduction

The sinusoidal current filament (SCF), (10.1) has been of interest in electromagnetics for a long time. One reason for this is the fact that the SCF is a good approximation to the actual current on a wire dipole antenna. This made it a useful tool for approximating the electromagnetic performance of antenna types of great practical interest in the pre-computer era. Another valuable property of the SCF was that it is near (10.2) and far fields (10.3) and the length dependence of its radiated power (10.4) were worked out in closed form in the early 20th century [1–5] as summarized in the NEC documentation [6–8].

$$I(z) = I_o \sin\left[k\left(\frac{L}{2} - |z|\right)\right], \frac{-L}{2} \le z \le \frac{L}{2}. \tag{10.1}$$

$$E_\rho^{(S)} = \frac{-j\eta_o I_o}{4\pi k\rho}\frac{e^{-jkr}}{r}\left\{k(z-z')\cos(kz') + \left[1 - \frac{(z-z')^2}{r^2}(1+jkr)\sin(kz')\right]\right\}_{z'=z_1}^{z'=z_2} \tag{10.2a}$$

$$E_z^{(S)} = \frac{j\eta_o I_o}{4\pi k\rho}\frac{e^{-jkr}}{r}\left\{k\cos(kz') - (1+jkr)\frac{(z-z')^2}{r^2}[\sin(kz')]\right\}_{z'=z_1}^{z'=z_2} \tag{10.2b}$$

$$E_\theta^{(S,RAD)}(\theta) = j\eta_o\frac{I_o e^{-jkr}}{2\pi r}\left[\frac{\cos\left(\frac{kL}{2}\cos\theta\right) - \cos\left(\frac{kL}{2}\right)}{\sin\theta}\right] \tag{10.3a}$$

with $r = \left[\rho^2 + (z-z')^2\right]^{1/2}$.

The superscript S in (10.2) and (10.3) denotes these as the fields of the sinusoidal current. For reference, the corresponding fields of the cosinusoidal current are given by

$$E_\rho^{(C)} = \frac{-j\eta_o I_o}{4\pi k\rho}\frac{e^{-jkr}}{r}\left\{-k(z-z')\sin(kz') + \left[1 - \frac{(z-z')^2}{r^2}(1+jkr)\cos(kz')\right]\right\}_{z'=z_1}^{z'=z_2}$$

$$E_z^{(C)} = \frac{j\eta_o I_o}{4\pi k\rho}\frac{e^{-jkr}}{r}\left\{-k\sin(kz') - (1+jkr)\frac{(z-z')^2}{r^2}[\cos(kz')]\right\}_{z'=z_1}^{z'=z_2}.$$

Note that (10.3a) can also be written as

$$E_\theta^{(S,RAD)}(\vartheta) = j\eta_o\frac{I_o e^{-jkr}}{2\pi r}\frac{1}{\sin(\theta)}\left[e^{(jkL/2)\cos(\theta)} + e^{-(jkL/2)\cos(\theta)} - 2\cos(kL/2)\right] \tag{10.3b}$$

which may superficially appear to be the equivalent of three-point sources of radiation, two of unit amplitude at $\pm L/2$ and another at the center of variable amplitude $2\cos(kL/2)$. If this were actually the case, it would be logical to conclude that the SCF must not radiate from along its length. But it is not clear from the radiated-power expression (10.4a) that this is the case since it contains

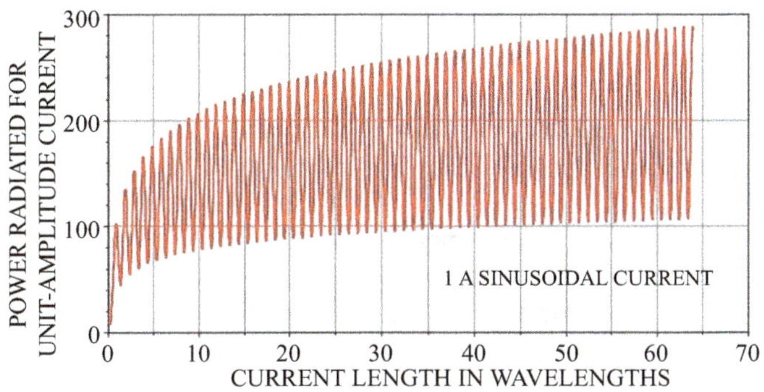

Figure 10.1 The power radiated by a SCF of 1-A amplitude as a function of its length in wavelengths

non-oscillatory, logarithmic terms that involve its length. Indeed, when the SCF radiated power of (10.4) is plotted as a function of L, there is a systematic increase in the average radiated power as shown in Figure 10.1:

$$P^{(RAD)} = K\Big\{C + \ln(kL) - C_i(kL) + \frac{1}{2}\sin(kL)[S_i(2kL) - 2S_i(kL) + \frac{1}{2}\cos(kL)] + \frac{1}{2}\cos(kL)\left[C + \ln\left(\frac{kL}{2}\right) + C_i(2kL) - 2C_i(kL)\right]\Big\} \tag{10.4a}$$

$$K = \eta_o \frac{|I_o|^2}{4\pi} \tag{10.4b}$$

where

$$C = 0.5772 \text{ (Euler's constant)},$$

with

$$C_i(x) = -\int_x^\infty \frac{\cos y}{y} dy = \int_\infty^x \frac{\cos y}{y} dy \tag{10.4c}$$

and

$$S_i(x) = \int_0^x \frac{\sin y}{y} dy. \tag{10.4d}$$

The minima in the radiated power of Figure 10.1 occur when the variable center term is zero, i.e., when the SCF is an odd number of half-wavelengths long. When this occurs the current slope is zero at the center of the current filament as is the charge. For a SCF that is an even number of half-wavelengths long, the current slope has a maximum discontinuity at the center and an associated charge maximum, to produce a maximum in the radiated power.

Plots of the minima and maxima samples of the radiated power from Figure 10.1 are included in Figure 10.2 along with best-fit curves of these results.

Each of these curves varies as the log of the filament length, confirming that the SCF radiates from along its length as well as from its ends and center in accordance with (10.4).

NEC was used to model a variable-length, center fed dipole having a radius of 10^{-3} wavelengths. The dipole length was systematically increased by adding equal-length segments to each of its ends to cover a length range from about 0.25 to 10 wavelengths. In order to make a meaningful comparison with the unit-amplitude SCF, the NEC radiated power was then scaled to the value that would be produced were the current peak nearest the feed point to be one amp. This computation produced the results shown in Figure 10.3.

It can be seen that the NEC results also exhibit a log(kL) trend, and indeed the minimum values lie on a best-fit curve given by 38.9 + 38.7 × log(L/wl), compared with the corresponding curve of Figure 10.2 which varies as 39.2 + 37.6 × log(L/wl).

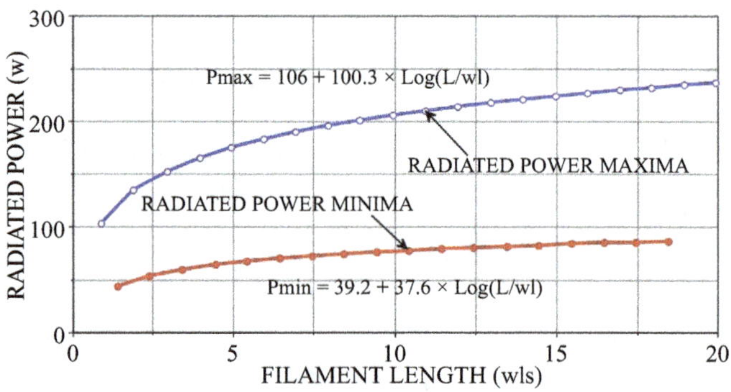

Figure 10.2 The maxima and minima radiated powers of Figure 10.1 are found to vary as the log of the SCF length as illustrated by these best-fit curves

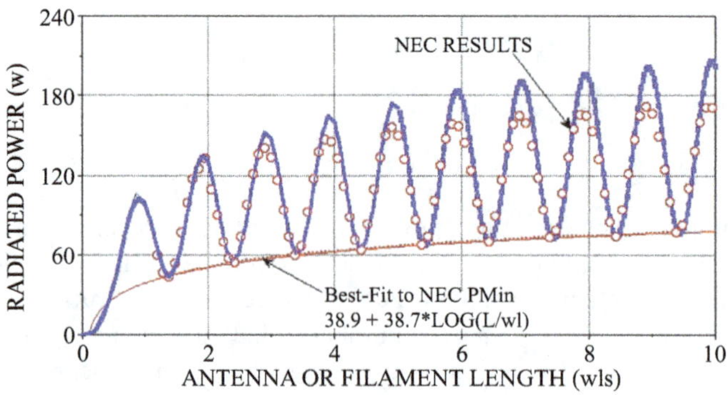

Figure 10.3 A comparison of 1-A NEC results with the SCF radiated power with a best fit to the minima levels as a function of length in wavelengths

Also of interest is the fact that the NEC radiated power essentially tracks that of the SCF except for the maxima, where the NEC values are systematically less. A possible explanation for this latter difference might be the fact that the NEC currents decay slightly away from the source region whereas the SCF does not and thus can be expected to radiate more power. For comparison, the power that would be radiated by an infinite cylindrical antenna of the same radius having a unit-amplitude current at the feed point, which should represent an asymptote for the NEC results, is about 330 W.

10.2 Using the IEMF method to analyze radiation from the SCF

The radiation properties of the SCF have been discussed elsewhere [9] in connection with the original development of FDFARS. Although shown by a numerical evaluation using the IEMF (induced-EMF) method and confirmed by the theoretical expression for the radiated power, why the P^{rad} grows in proportion to $\log(kL)$ may not be obvious. The reason for this phenomenon is examined in more detail here. The power-density plots that follow are presented either in units of power/wavelength or power/segment. A sampling rate of 20 numerical samples, or segments, is used per wavelength unless otherwise indicated for these computations. Such a sampling rate is needed to adequately resolve the spatial distribution of the quantities of interest. The IEMF method is implemented by computing the tangential electric field (or E-field) of the total SCF at the center of each of the numerical segments used in the numerical model. The distributed IEMF power density is obtained using

$$p^{(IEMF)}(z) = \frac{1}{2}\mathrm{Re}\left[E_{inc}^{(TAN)}(z)I^*(z)\right] \tag{10.5a}$$

where $I(z)$ is the specified current and $E_{inc}^{(TAN)}(z)$ is evaluated at a radial distance of 10^{-3} wavelengths from the current. The IEMF power is negative in the following examples, but for comparison purposes with the positive FARS results it is shown as –IEMF. The segmental and total IEMF powers are given by integrating (10.5a)

$$P^{(IEMF)} = \int_L p^{(IEMF)}(z)dz \approx \sum_1^N p^{(IEMF)}(z_m)\Delta_m \tag{10.5b}$$

In Figures 10.4–10.6 are plotted results for a 5-wavelength SCF with each providing further details of this source. The tangential electric field along the SCF at a radial distance of 10^{-3} wavelengths is first presented in Figure 10.4. The field varies smoothly along the filament with maxima near its ends and center. For comparison, the current itself is added in Figure 10.5 where the center peak is found to be associated with the charge maxima caused by the current-slope discontinuity at the filament's center and at the ends where another charge maxima occur where the current goes to zero.

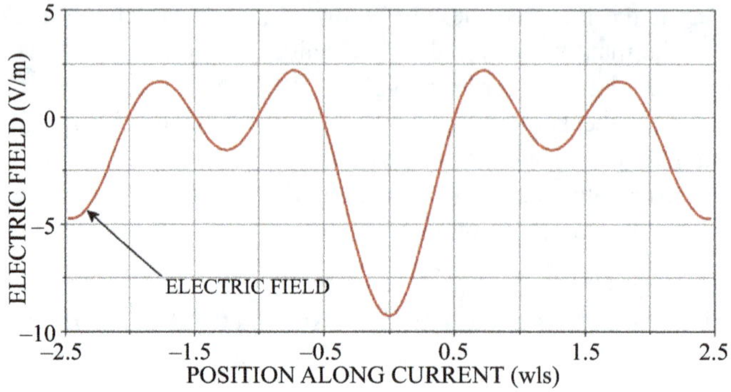

Figure 10.4 The tangential electric field at a radial distance of 10^{-3} wavelengths for a 5-wavelength SCF

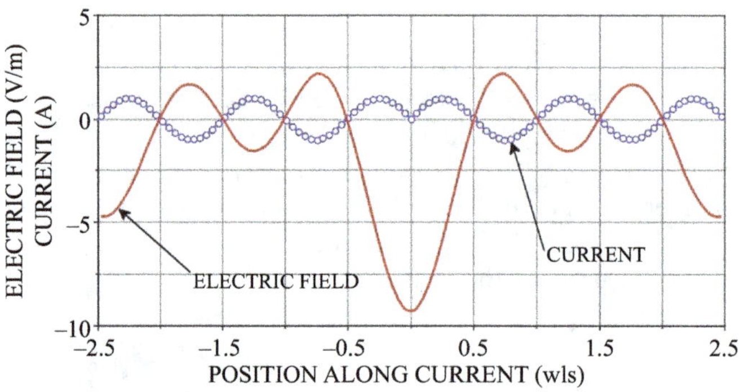

Figure 10.5 The 5-wavelength SCF and its tangential electric field at a radial distance of 10^{-3} wavelengths

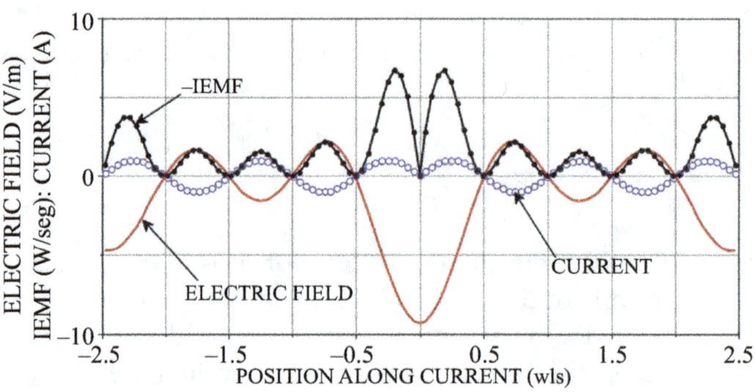

Figure 10.6 The 5-wavelength SCF, its tangential electric field at a radial distance of 10^{-3} wavelengths, and the IEMF power distribution along its length

The negative of the IEMF power density for this problem from (10.5a) is plotted in Figure 10.6.

The FDFARS result for the 5-wavelength SCF is plotted with the −IEMF result in Figure 10.7a. This comparison provides a mutual validation for FARS and the IEMF method for a standing-wave current such as a SCF. The effect of changing the length of a SCF to an odd number of half wavelengths is illustrated by the FARS results of Figure 10.7b for a SCF of 5.5 wavelengths. Here, the centered radiation peaks at the center of the 5-wave-length filament have disappeared because there is no current-slope discontinuity at the center of the odd-half-wavelength filament.

The FARS result for a 10-wavelength uniform SCF is plotted in Figure 10.8a, for comparison with a tapered SCF in (10.8b). In the latter case the SCF increases linearly from 0 to 1 on its left half and then decreases linearly back to 0 on its right half. The radiation peaks that occur at the ends of the uniform SCF in Figure 10.8a are gone in (10.8b) with the radiation concentrated towards the center of the current filament. The far-field pattern of the tapered SCF Figure 10.8c lacks most of the lobes that are present in the pattern of the uniform SCF.

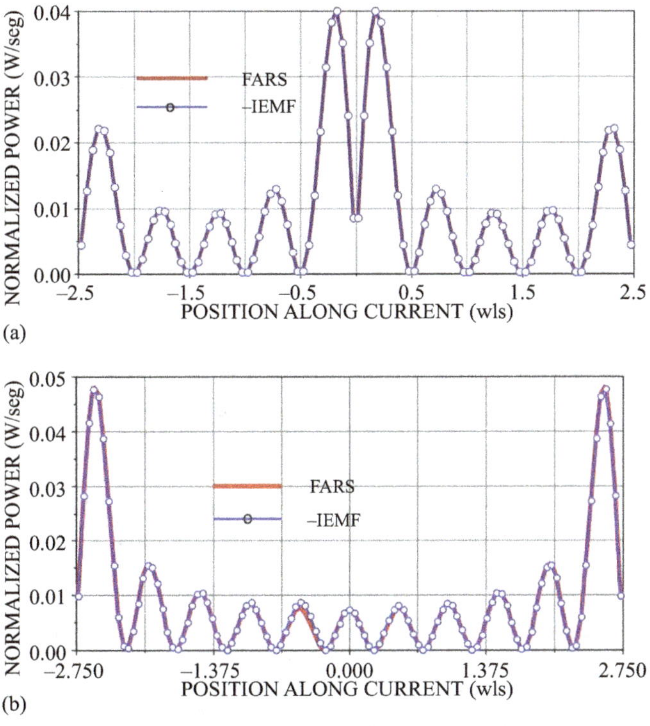

Figure 10.7 A comparison of the FARS and −IEMF 1-W total power distributions for the 5-wavelength SCF (a) and a 5.5 wavelength SCF (b)

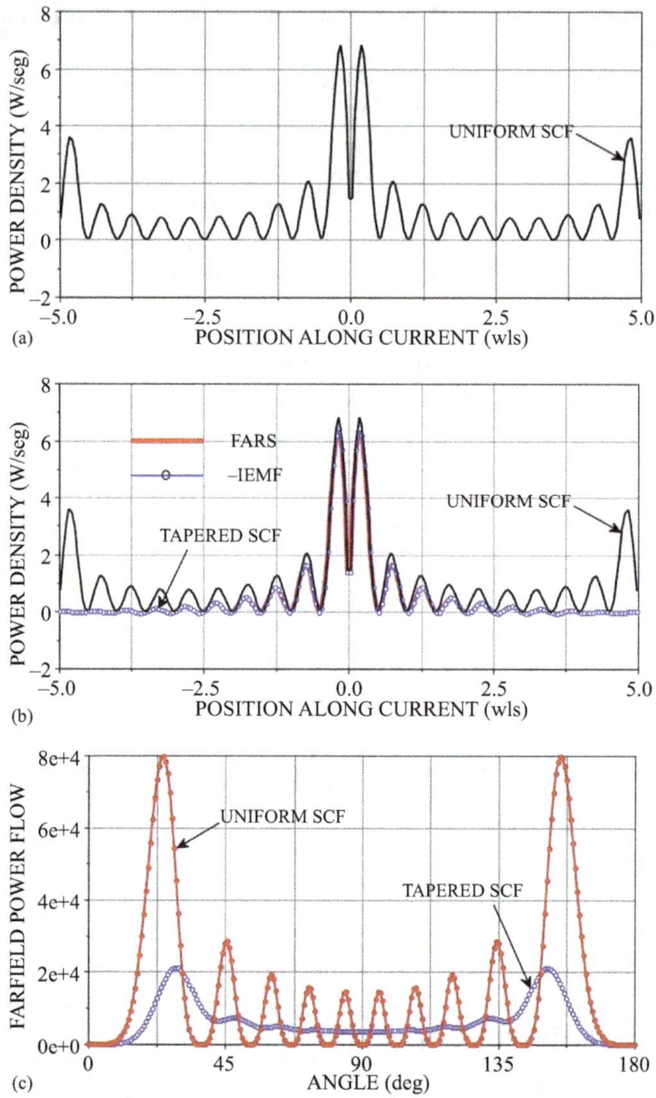

Figure 10.8 The FARS results for a 10-wavelength uniform SCF (a), the FARS and –IEMF power distributions for a tapered SCF (b) and the farfield power flow for both (c)

The influence of the $1/\sin(\theta)$ term on the radiated-power distribution of the 10.5-wavelength SCF is also illustrated by the results of Figure 10.9. That term is what causes the growth in radiated power with an increasing length of the SCF. Removing it changes the SCF distributed radiation to that of three-point sources with the strength of the center source becoming zero for the odd-half-wavelength case.

Additional examples of these effects are further illustrated by the results of Figures 10.10–10.13. Figures 10.11 and 10.13 show the FARS and −IEMF results for SCFs of 20 and 20.5 wavelengths long in w/segment for a total radiated power

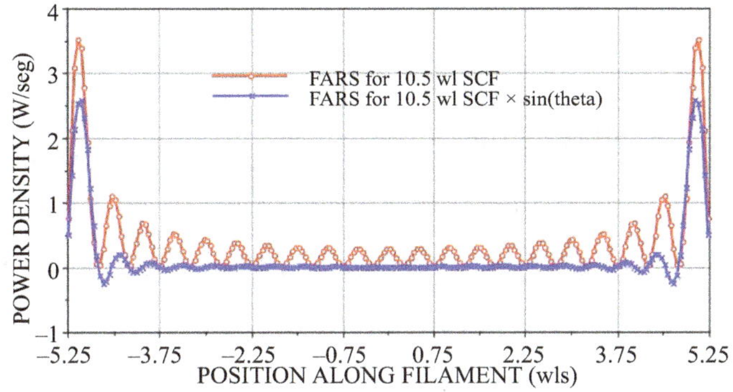

Figure 10.9 FARS results for SCFs of 10.5 wavelengths with and without the 1/sin(θ) term

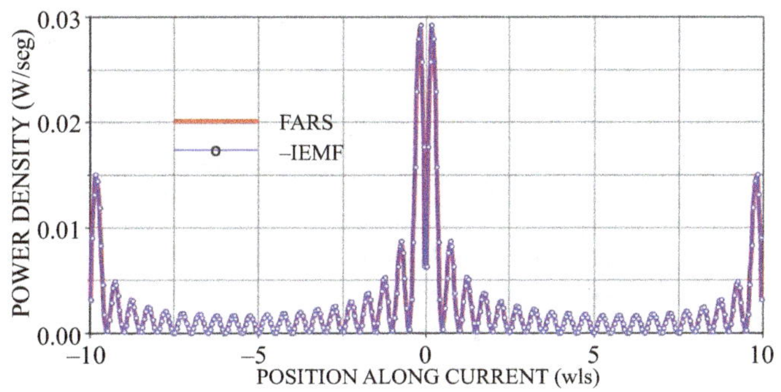

Figure 10.10 FARS and −IEMF results for a 20-wavelength SCF

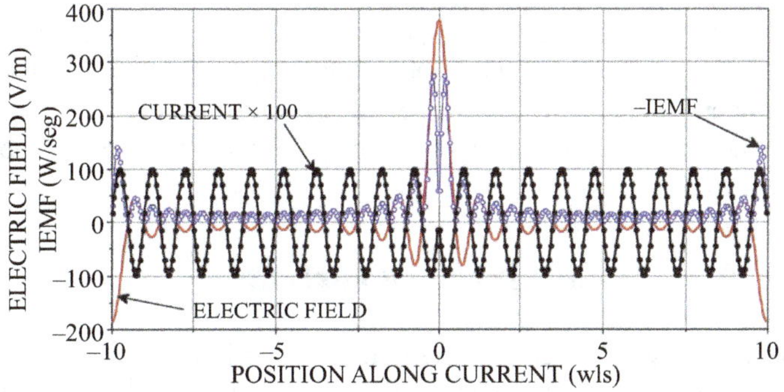

Figure 10.11 The tangential electric field, 100 × the current and −IEMF for a 20-wavelength SCF

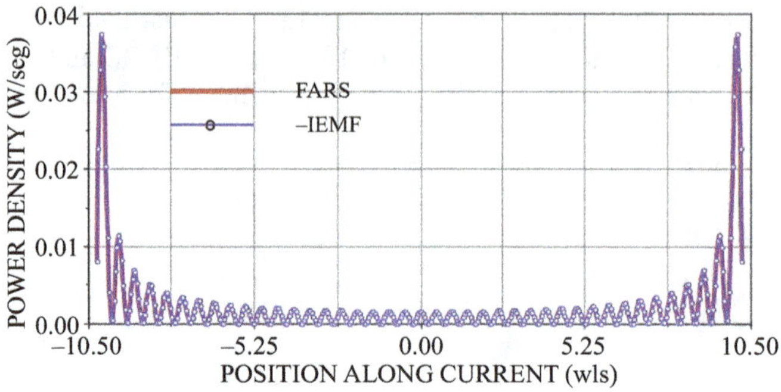

Figure 10.12 FARS and −IEMF results for a 20.5-wavelength SCF

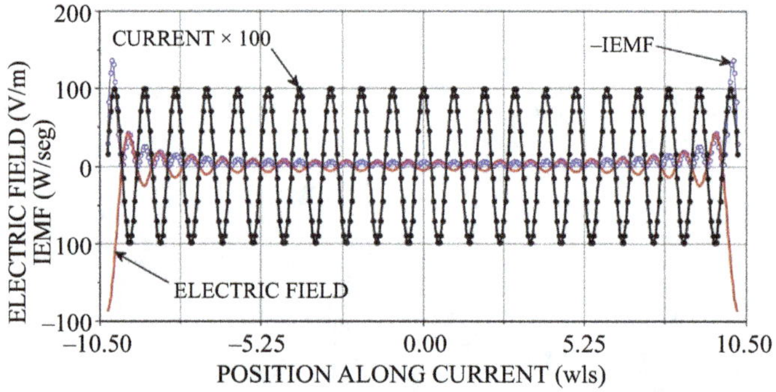

Figure 10.13 The tangential electric field, 100 × the current and –IEMF for a 20.5-wavelength SCF

of 1-W. In Figures 10.11 and 10.13, the tangential electric fields are included together with the current × 100 to show it on this scale with the –IEMF. Note that the –IEMF plot is used to emphasize the case where it is shown with the electric field and its alternating sign relative to the current distribution.

Figure 10.14 displays the right-hand halves of the FARs results for SCFs 20 and 20.5 wavelengths long to emphasize the point that the end radiation peak is essentially independent of the length of a unit-amplitude current. This plot also clearly demonstrates the minima and maxima patterns of the radiated power seen in Figures 10.1–10.3.

10.3 Results for other specified currents

Results for a variety of other specified current distributions are presented in this section to demonstrate the broader applicability of the FARS and IEMF approaches. The radiation-power densities for a 10-wavelength, 1-A uniform current obtained from FARS and the IEMF method is presented in Figure 10.15 where the

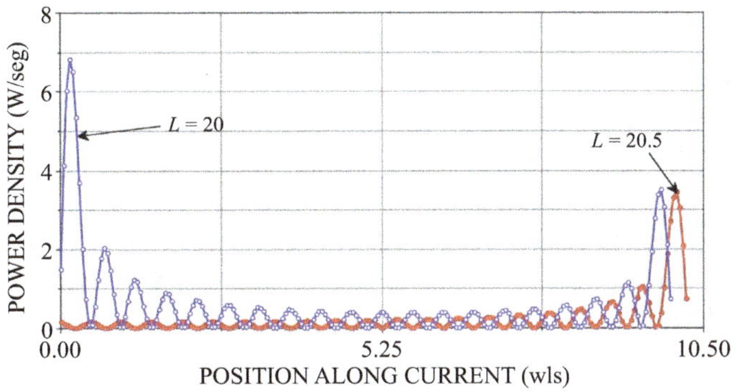

Figure 10.14 FARS results for the right-hand half of 20 and 20.5-wavelength SCFs

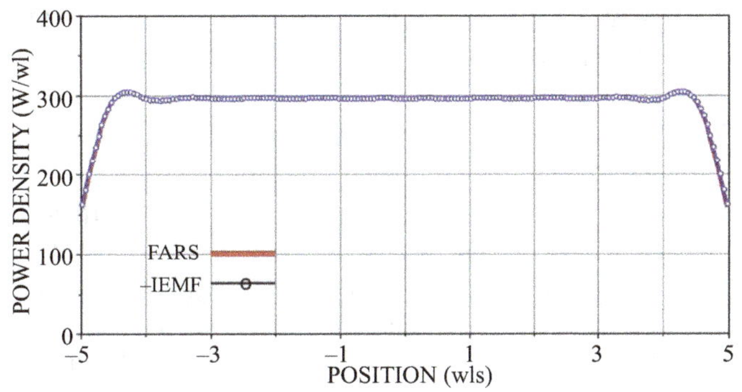

Figure 10.15 The FARS and –IEMF radiation-power density for a 10-wavelength 1-A uniform current filament

total power is 2,981 W. The radiation distribution along the current filament is essentially uniform over the central 8 or so wavelengths. Somewhat similarly, a 2-wavelength triangular current having a peak amplitude of 1-A produces a radiated-power distribution that is a near replica of that current as shown in Figure 10.16 with a radiated power of about 324 W.

The next specified current explored here is the traveling-wave, as given by,

$$I(z) = I_o e^{-jkz}, 0 \leq z \leq L \tag{10.6}$$

where L is its length and I_o is its amplitude. The radiated electric field of the TWC at distance r is thus

$$E_\theta(\theta) = j\eta \frac{kLI_o e^{-jkr}}{4\pi r} e^{-j(kL/2)(K-\cos\theta)} \sin\theta \frac{\sin[(kL/2)(\cos\theta - K)]}{(kL/2)(\cos\theta - K)}, \tag{10.7}$$

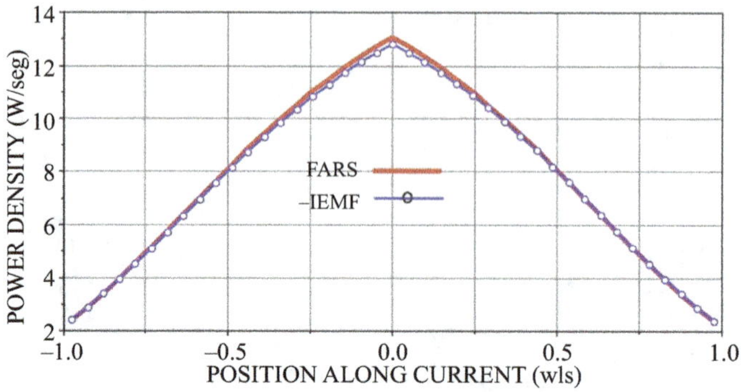

Figure 10.16 The FARS and −IEMF radiation-power density for a 2-wavelength 1-A peak triangle current filament

where $K = 1$ when the wave number $k = k_o$, as is assumed here [5]. The radiated power of the TWC then follows as

$$P^{(RAD)} = \frac{\eta}{4\pi} I_o^2 \left[1.415 + \ln\left(\frac{kL}{\pi}\right) - C_i(2kL) + \frac{\sin(2kL)}{2kL} \right]. \tag{10.8}$$

The FARS and –IEMF results for TWCs of 2.5, 5, and 10 wavelengths and with FARS only for 20-wavelength case are plotted in Figure 10.17. Radiation maxima are concentrated at both ends of the current with an interior minimum that decreases as the filament length is increased. The 20-wavelength TWC and 20.5 wavelength SCF FARS results are compared in Figure 10.18, using the odd-half wavelength case for the latter to eliminate its center peaks. The major difference between the two currents is that the standing-wave structure of SCF produces a lobed structure in FARS as contrasted with a traveling wave.

By contrast with the uniform and triangular currents of Figures 10.15 and 10.16, the FARS and −IEMF results in Figure 10.19 for a current varying linearly from 1 to 0 have no resemblance to the current variation. Still, as for all the previous cases presented thus far in this chapter, the FARS and −IEMF results exhibit hardly any discernable quantitative difference.

In the next two graphs, results are presented for current distributions that are variations of the basic SCF. Figure 10.20 displays FARS and –IEMF results for three-wavelength 1-A SCF and 1 − SCF. The power density of the 1 − SCF has fewer lobes but radiates 1,266 W compared with 147 W for the SCF itself. In Figure 10.21, are results for 2 wavelength currents, a SCF and 1 ± SCF current filaments whose lobes are substantially shifted from each other. The FARS and –IEMF results for the 1 – SCF are 741 and 723 W, respectively, a difference of about 3%.

The concluding example of this series demonstrates the result of squaring the SCF in Figure 10.22. The lobe structure of the squared SCF is qualitatively similar but of increased amplitude. The radiated powers are 210 W for the squared SCF and 147 W for the SCF.

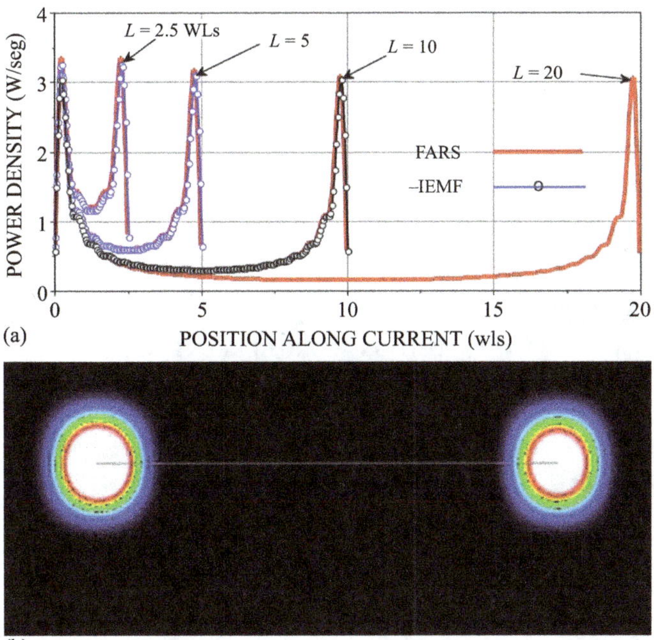

(b)

Figure 10.17 (a) The FARS and −IEMF radiation-power densities for traveling-wave currents of variable lengths and (b) the broadside bistatic k-space image of a TWC [11]

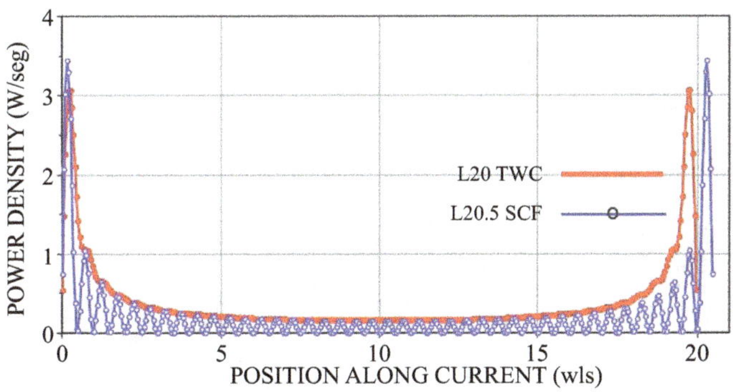

Figure 10.18 A comparison of FARS results for a SCF 20.5 wavelengths long and a TWC 20 wavelengths long

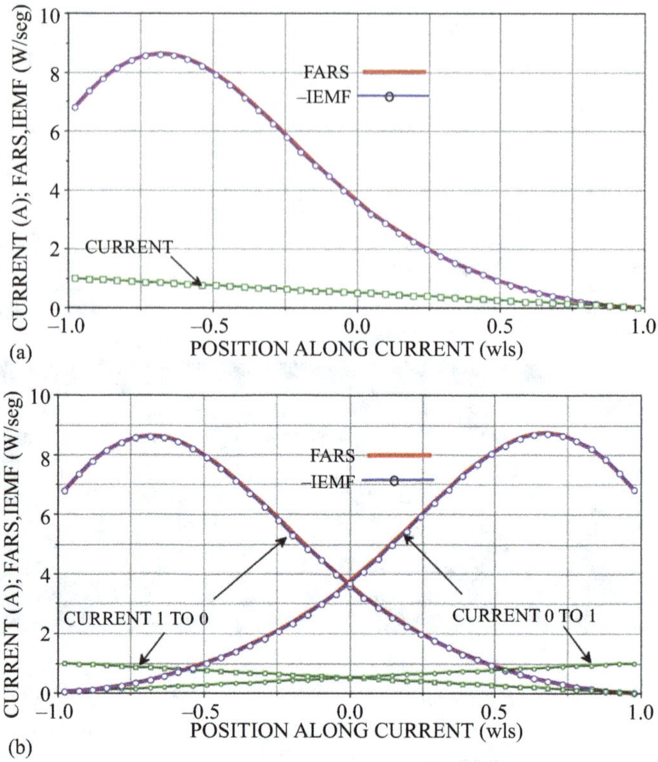

Figure 10.19 The FARS and –IEMF radiation-power density for a current filament varying linearly from left to right over a range of 1–0 (a) and the complementary case of a variation of 0–1 (b)

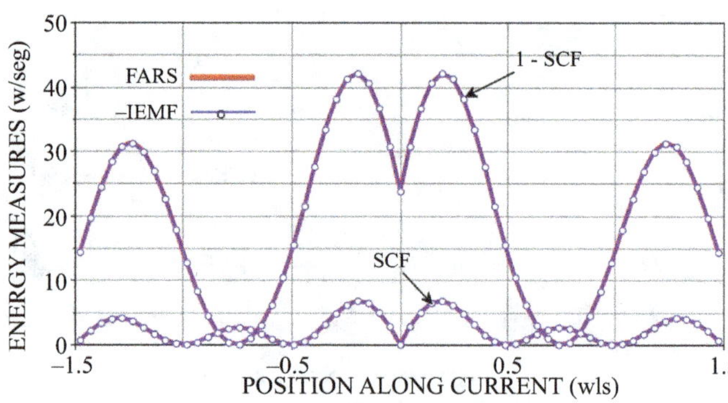

Figure 10.20 A comparison of the FARS and –IEMF results for a three-wavelength SCF and 1 – SCF

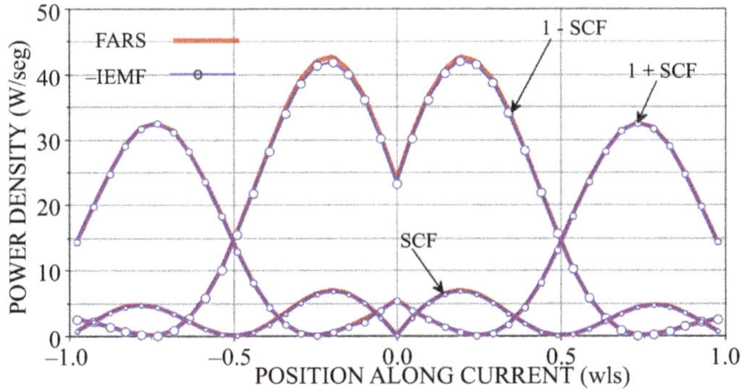

Figure 10.21 *A comparison of FARS and –IEMF results for a two-wavelength currents, a SCF, 1 + SCF and 1 – SCF*

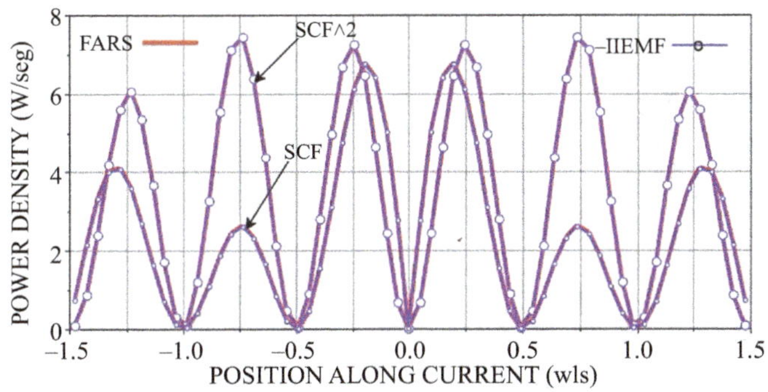

Figure 10.22 *A comparison of FARS and –IEMF results for the SCF and the square of the SCF*

10.4 The Schelkunoff–Feldman distributed radiation resistance

In a remarkable article published in 1942 [10], Schelkunoff and Feldman introduced the idea of a distributed radiation resistance for a SCF. An extension of their work is discussed in more detail in Chapter 12. The results below are based on their original article, included here to complete this discussion of specified currents.

Their derivation of the SCF distributed radiation resistance was limited to a current an odd number of half wavelengths long, and is given by

$$R(z) = \frac{60h^2}{h^2 - z^2}; -h \leq z \leq +h. \tag{10.9}$$

a plot of which for a 9.5-wavelength SCF is included in Figure 10.23.

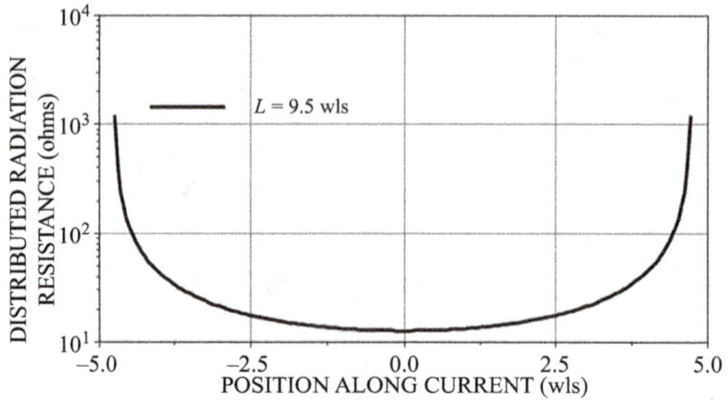

Figure 10.23 The Schelkunoff–Feldman distributed radiation resistance of a 9.5-wavelength SCF

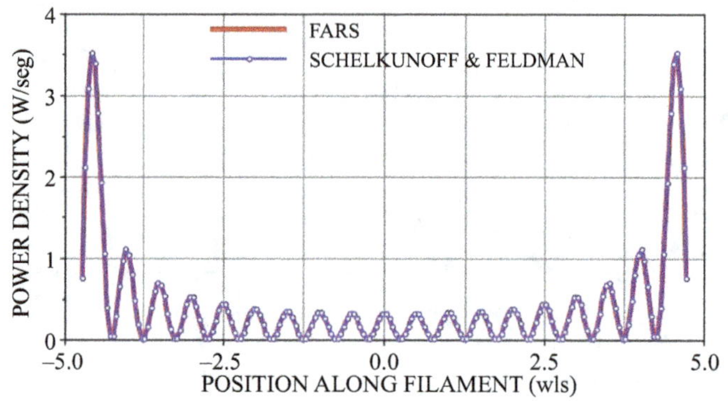

Figure 10.24 Comparison of the FARS distributed radiation and that obtained using the Schelkunoff–Feldman distributed radiation resistance for a 9.5 wavelength SCF

If the distributed radiation resistance is a valid concept, then it should provide an alternate way of determining the distributed radiated power from a SCF. This is confirmed by the results of Figure 10.24 that are obtained using

$$P_{rad}(s) = \frac{1}{2} I^2(s) R_{rad}(s) \tag{10.10}$$

where they are found to be graphically indistinguishable from the FARS results for a 9.5-wavelength SCF. This comparison provides another validation example for FARS as well. An extension of the Schelkunoff–Feldman radiation resistance to an arbitrary-length current (discussed in Chapter 12), is demonstrated in Figure 10.24, for a 10-wavelength SCF.

10.5 Summary

This chapter has explored the application of FARS and the IEMF method to a variety of specified current filaments. The results presented serve as a mutual validation of these quite-different ways of finding where far-field radiation originates from such sources. These results show as well that radiation generally occurs from over the entire extended source besides more obvious places like current-slope discontinuities.

The SCF was first examined in detail since it provides a useful approximation of the current distribution on a physical dipole. The length dependence of the power radiated by a unit-amplitude SCF oscillates between two extrema that occur at half-wavelength intervals. For a SCF of length L, maxima occur when $L/2$ is an even number of half wavelengths due to the maxima in the current-slope discontinuity at the center of the SCF. The minima occur when $L/2$ is an odd number of half-wavelengths as this slope discontinuity vanishes. Both extrema increase as log(kL) due to the radiation that occurs between the center and ends of the SCF. The current, tangential electric field, IEMF and FARS power were plotted to demonstrate their spatial relationships and to show that the input power from the IEMF was balanced by the radiated power determined from FARS. Also demonstrated is the fact that the end radiation from the SCF is essentially independent of the filament length.

Several other variations of the SCF were also investigated to determine how their distributed radiation would vary. It was generally found that the distributed radiation was spatially similar to the magnitude of the current for the SCF variations as was also the case for uniform or triangular current distributions. But a traveling-wave and a linearly varying current had quite different spatial radiation distributions.

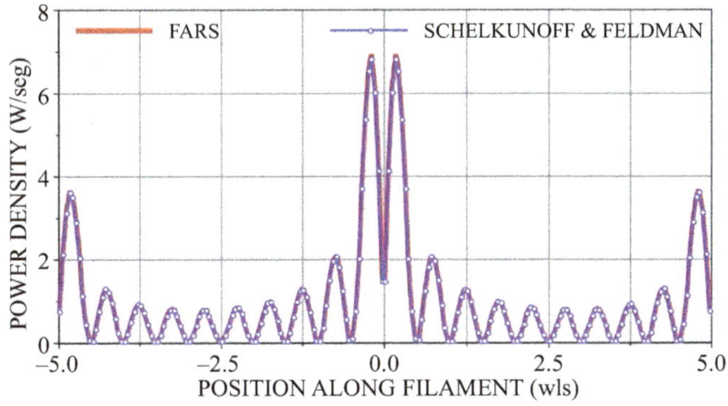

Figure 10.25 Comparison of the FARS distributed radiation and that obtained using an extension of the Schelkunoff–Feldman distributed radiation resistance for a 10.0 wavelength SCF

The concept of a distributed radiation resistance, $R_{rad}(z)$, introduced by Schelkunoff and Feldman [10] was investigated. Results derived for the distributed radiation from a SCF using $\frac{1}{2}I^2(z)R_{rad}(z)$ were found to agree with FARS results to within a few percent, to validate the S–F approach.

References

[1] L. Brillouin, Sur l'Origin de la Resistance de Rayonnement, *Radioelectricite*, 3, 141–152, 1922.

[2] A. A. Pistokors, The radiation resistance of beam antennas, *Proc. IRE*, 17, 562–579, 1929.

[3] R. Bechmann, On the calculation of radiation resistance of antennas and antenna combinations, *Proc. IRE*, 19, 1971–1980, 1931.

[4] J. A. Stratton, *Electromagnetic Theory*, McGraw-Hill Book Company, New York, NY, 1941, p. 455–457.

[5] C. A. Balanis, *Antenna Theory*, Harper & Row, New York, NY, 1982, pp. 118–124.

[6] G. J. Burke, E. K. Miller, and A. J. Poggio, The Numerical Electromagnetics Code (NEC)—a brief history, in: *2004 IEEE AP-S International Symposium and USNC/URSI National Radio Science*, Monterey, CA.

[7] G. J. Burke, *Numerical Electromagnetics Code—NEC-4: Method of Moments, Part I, User's Manual (NEC-4.1), UCRL-MA-109338 Pt. I*, January 1992.

[8] G. J. Burke, *Numerical Electromagnetics Code—NEC-4: Method of Moments, Part II, Program Description—Theory, UCRL-MA-109338 Pt. II*, January 1992.

[9] E. K. Miller, PCs for AP June 1996 and October 1996, *IEEE Antennas and Propagation Society Magazine*,

[10] S. A. Schelkunoff and C. B. Feldman, On radiation from antennas, *Proc. IRE*, November 1942, pp. 511–516.

[11] J. F. Shaeffer, B. A. Cooper, K. W. Hom, R. C. Baucke, and N. A. Talcott, A review of bistatic k-space imaging for electromagnetic prediction codes for scattering and antennas, *IEEE AP-S Magazine*, 39(5), 2121–2129, 1997.

Chapter 11

The incremental FARS (IFARS) and incremental IEMF (IIEMF) methods

Abstract

Incremental versions of far-field analysis of radiation sources (FARS) and the induced electromotive force (IEMF) method are implemented in this chapter in connection with moment-method models for two reasons. One is to provide a more geometrically detailed way of determining the power-density distribution of radiation from over the surface of an object. A second is to make this possible without the need to integrate the power flow over the entire far-field sphere. The generalizations of these two methods are denoted here as incremental FARS (IFARS) and incremental IEMF (IIEMF). These incremental procedures are developed and their application demonstrated using various NEC models. Both involve determining the pairwise interaction between current samples on a PEC object. Mutual validation of both techniques is confirmed by the agreement shown between their numerical results in the following.

Keywords: Antenna radiation; FARS; FARS and IEMF compared; IEMF method generalized; Incremental FARS; Incremental IEMF method; Inward and outward power flow; Radiation power density; Scattered radiation

11.1 Some background

Far-field analysis of radiation sources in the time domain (TDFARS) and in the frequency domain (FDFARS) were introduced previously in Chapters 7 and 8, respectively. The induced electromotive force (IEMF) method was compared with and found to confirm FDFARS in Chapter 10 for various specified current distributions.

The rationale for an extended version of the IEMF method in an incremental fashion is outlined here in Section 11.2. This is followed in Section 11.3 by development of the incremental version of IEMF, or incremental IEMF (IIEMF). The incremental version of FDFARS or IFARS is described in Section 11.4. A variety of numerical results to demonstrate and further validate far-field analysis of radiation sources (FARS) and mutually validate IFARS and the IIEMF method are included in Section 11.5.

11.2 Generalizing the IEMF method

The utility of FARS and the IEMF method for analyzing where and how much radiation originates from various wire objects has been demonstrated in previous chapters. The incremental extensions that are developed and employed here enhance their applicability beyond what has already been presented. While the use of the IEMF method for PEC objects has in practice been primarily limited to antenna-source models, it will be shown in the following that it has a utility beyond this limited application.

One of the most used applications of the IEMF method has been to determine the input admittance (or impedance) of a wire antenna, a dipole for example. Consider a dipole of length L lying on the z-axis of a Cartesian coordinate system that is excited at its center, $-\Delta/2 \leq z \leq \Delta/2$, by an applied electric field E_{exc} across a wire segment of length Δ, in the context of the method of moments. The resulting source voltage $V_{exc} = -E_{exc}\ \Delta$ produces a current $I(z)$ that induces an opposing electric field $E_{ind} = -E_{ex}$ over that segment and 0 elsewhere to satisfy the boundary condition that $E_{tan} = 0$ on the wire's surface. The dipole's input admittance Y is given by

$$Y = \frac{I(0)}{E_{exc}\Delta} = \frac{I(0)}{V(0)}. \tag{11.1}$$

The power flow normal to the surface of a PEC wire is identically 0 everywhere on it because of the $E_{tan} = 0$ boundary condition. The power flow normal to the wire must therefore be zero or must consist of equal inward- and outward-directed components, a condition that also holds over the source segment. The applied electric field there produces a nonzero inward power flow at its surface that is balanced by an outward power flow onto the antenna. This source power subsequently propagates in both directions towards the ends of the dipole, radiating and diminishing in amplitude with distance. This situation is partially illustrated conceptually in Figure 11.1 for a section of curved wire.

These inward- and outward-directed power flows produced by the current on a 100-segment, 5-wavelength NEC dipole excited at its 2 center segments are plotted in Figure 11.2. These NEC-computed normal power flows are equal and opposite numerically, agreeing to within a percent or so, except in the source region where the induced current produces only an outward power flow. Its outward power flow is balanced by the inward power coming from the excitation, which is not shown on this plot.

Demonstrated in Figure 11.2 is the fact that any exciting field applied to a PEC wire will produce a current flow that induces an equal and opposite field to cancel the excitation. It is reasonable to ask whether this phenomenon also applies to the mutual interaction fields of that current? Every one of the N moment-method-model segments produces a self-field that cancels the collective, incremental interaction fields due to the other N–1 segments. Could the effects of these interaction fields be to have the same effect as an exciting field on the power balance along the antenna?

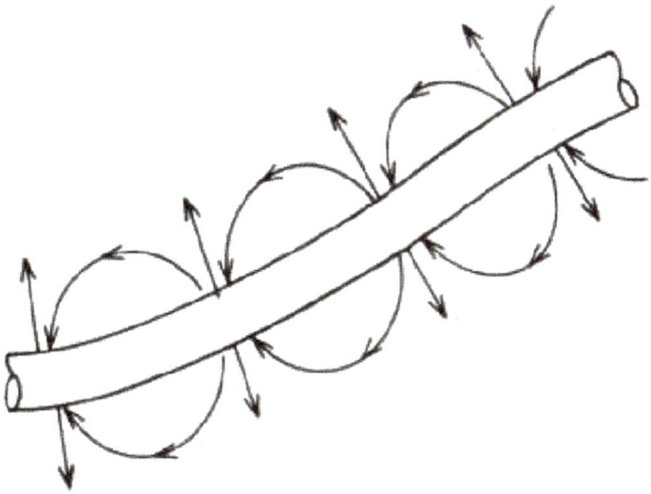

Figure 11.1 A conceptual depiction of near-field interaction and radiation power from a PEC wire [1]

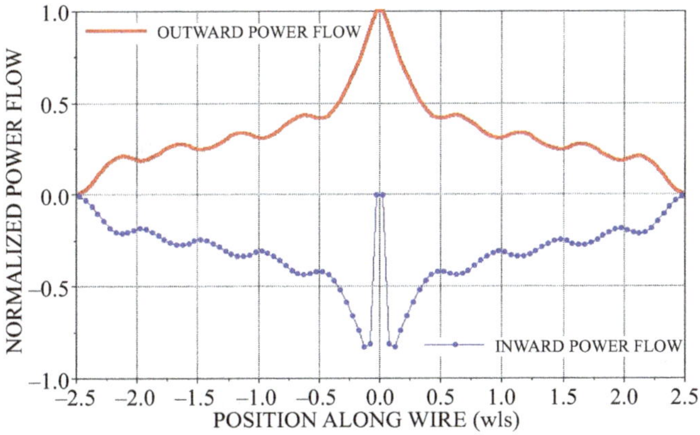

Figure 11.2 The normalized inward and outward power flows due to the current induced along a 5-wavelength, dipole antenna excited at its 2 center segments obtained using the incremental version of the IEMF method as described below

If so, then could it be feasible to explore whether incremental versions of both FARS and IEMF (denoted here as IFARS and IIEMF) could provide the total radiated power per segment? This might provide a useful extension of the IEMF method for application to a PEC object over regions having no exciting fields that would otherwise seem to be inappropriate.

11.3 FARS and its incremental version

11.3.1 The basic frequency-domain FARS

The FARS approach summarized here is based on the bistatic far fields of a source distribution at a single frequency [2–4]. A time-domain version was developed by Miller and Burke [5] and is discussed in Chapters 7 and 9 of this book. Shaeffer *et al.* [6] describe a similarly motivated imaging approach based on the bistatic far fields. But rather than imaging *per se*, FARS seeks instead to determine at a given far field observation point, the quantitative contribution made to the total power there by each incremental source, *m*, on the boundary of a PEC object, a quantity designated as $p_m^{(FARS)}(\vartheta,\varphi)$. A subsequent integration over the far-field sphere then yields the total radiated power from each source $P_m^{(FARS)}$. The power-flow density, $p_m^{(FARS)}(\vartheta,\varphi)$ for segment *m* in observation direction θ,φ at distance *r*, is defined as [2]

$$\begin{aligned} p_m^{(FARS)}(\vartheta,\varphi) &= \frac{1}{2\pi r^2}\mathrm{Re}\left[\mathbf{E}(\theta,\varphi)\bullet e_m^*(\theta,\varphi)\right] \\ &= \frac{1}{2\pi r^2}\mathrm{Re}\left[\mathbf{E}_\vartheta(\theta,\varphi)e_{m,\vartheta}^*(\theta,\varphi)+\mathbf{E}_\varphi(\theta,\varphi)e_{m,\varphi}^*(\theta,\varphi)\right], m=1,\ldots,M \end{aligned} \tag{11.2a}$$

where

$$E_\vartheta(\theta,\varphi)=\sum_{m=1}^{M}e_{m,\vartheta}(\theta,\varphi) \tag{11.2b}$$

$$E_\varphi(\theta,\varphi)=\sum_{m=1}^{M}e_{m,\varphi}(\theta,\varphi) \tag{11.2c}$$

are the total far electric fields with the theta and phi far electric-field components due to the source on segment *m*, $e_{m,\theta}(\theta,\varphi)$ and $e_{m,\varphi}(\theta,\varphi)$. Also, η is the medium impedance and *M* is the total number of source segments. The treatment here is limited to current filaments or wire objects or simplicity. On integrating over the far field sphere the FARS, or radiated, power for segment *m*, $P_m^{(FARS)}$, is obtained from (11-2a) as

$$\begin{aligned} P_m^{(FARS)} &= r^2\left[\int_0^{2\pi}\int_0^{\pi}p_m^{(FARS)}(\theta,\varphi)\sin\,\theta d\theta d\varphi\right] \\ &= \frac{1}{2\eta}\mathrm{Re}\left(\int_0^{2\pi}\int_0^{\pi}\left[E_\theta(\theta,\varphi)e_{m,\theta}^*(\theta,\varphi)+E_\varphi(\theta,\varphi)e_{m,\varphi}^*(\theta,\varphi)\right]\right)\sin\,\theta d\theta d\varphi, m=1,\ldots,M. \end{aligned} \tag{11.3}$$

Expanding the two fields in the square-bracketed term on the right-hand side of (11.3) obtains

$$[\ldots]\Rightarrow\mathrm{Re}\left(e_1e_m^*+e_2e_m^*+\ldots+e_me_m^*+\ldots+e_{M-1}e_m^*+e_Me_m^*\right) \tag{11.4}$$

and (11.4) representing the form of both angle components and where, for simplicity, the angle subscript is suppressed. The far-field power-flow density for each incremental E-field, e_m in (11.4) consists of a self-term, $|e_m|^2$, and M-1 mutual interaction terms of the form $e_p e_m^*$, $m \neq n$.

11.3.2 Incremental FARS

The original version of FARS in (11.2) above involves successively forming the conjugate dot product of the incremental far field of each individual segment with the total far field. Each incremental source explicitly appears in the computation only once while otherwise being included implicitly as part of the total field. Alternatively, (11.2a) can be written instead as

$$p_m^{(FARS)} = \frac{1}{2\eta r^2}\mathrm{Re}\left[e_m^2 + \sum_{n=1}^{M,n\neq m} \mathbf{e}_n \cdot \mathbf{e}_m^*\right], m = 1,\ldots, M \tag{11.2d}$$

for each θ and ϕ field component, where $\boldsymbol{e}_m$ and $\boldsymbol{e}_n$ represent the vector far field of segment m and n respectively. Each incremental field will appear explicitly a total of $2N$ times of the total N^2 pairwise terms in (11.2a), consisting of two identical self-terms and $2N-2$ interaction terms. The two interaction, or incremental, power-flow densities, $p_{mn}^{(IFARS)}$ and $p_{nm}^{(IFARS)}$, can be written for the two sources, m and n in (11.2d) as

$$p_{mn}^{(IFARS)} = \frac{1}{2\eta r^2}\mathrm{Re}\left[\mathbf{e}_m \cdot \mathbf{e}_n^*\right] \text{ and } p_{nm}^{(IFARS)} = \frac{1}{2\eta r^2}\mathrm{Re}\left[\mathbf{e}_n \cdot \mathbf{e}_m^*\right] \tag{11.5a,11.5b}$$

and designated here as the Incremental FARS (IFARS) results. (Incremental quantities, whether from IFARS or IIEMF, are identifiable by having two subscripts.) From (11.5), it follows that

$$\mathrm{Re}\left[\mathbf{e}_m \cdot \mathbf{e}_n^*\right] = \mathrm{Re}\left[\mathbf{e}_n \cdot \mathbf{e}_m^*\right] \tag{11.6a}$$

$$\text{then } p_{mn}^{(IFARS)} = p_{nm}^{(IFARS)}, \tag{11.6b}$$

Note that the two separate IFARS terms for the far-field interaction powers of any pair of sources are equal. This equality was implicitly incorporated in the original version of FARS by assigning the first term of (11.4) to source n and the second term to source m through the dot product of the total field with the conjugate of source n and source m incremental fields, respectively.

An equivalent but more symmetric form for the incremental power flows in (11.5a) and (11.5b) could instead be written as a consequence of (11.6b)

$$p_{mn}^{(IFARS)} = \frac{1}{2\eta r^2}\mathrm{Re}\left[\frac{1}{2}\left(\mathbf{e}_m \cdot \mathbf{e}_n^* + \mathbf{e}_n \cdot \mathbf{e}_m^*\right)\right] = p_{nm}^{(IFARS)} \tag{11.7a}$$

leading to equal interaction powers from integration over the far-field sphere, given by $P_{mn}^{(IFARS)}$ and $P_{nm}^{(IFARS)}$, respectively.

Note that the neither the $p_{mn}^{(IFARS)}$ values nor are the individual segment results, $P_m^{(FARS)}$ guaranteed to be positive, but the total radiated power must be non-negative.

The intermediate results represented by $p_{mn}^{(IFARS)}$ and $P_{m}^{(FARS)}$ show how the segment-pair interaction far fields and the individual segments affect the total far field.

The plot in Figure 11.3 of the incremental FARS, or IFARS, results for a 21-segment, unit-amplitude, 1-wavelength SCF shows the pairwise interactions over the entire current for segments 3, 5, 10, and 17. Results for segments 5 and 17, located at the same distance from either end of the SCF, are reciprocal. As a quantitative example, the reciprocal points of these curves are numerically equal with the 10th point of the segmet-5 curve agreeing with the 5th point of the segment-10 curve. Also note that the total FARS power for each of the segments in Figure 11.3 is provided by an integral of its respective plot, a result that applies all other segments as well.

For comparison, a corresponding result for a 1-wavelength NEC dipole excited by 1-V is plotted in Figure 11.4 for the same segments as for the SCF in

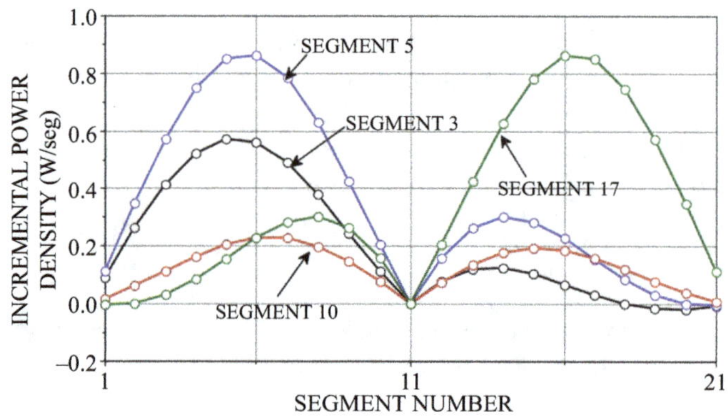

Figure 11.3 Incremental FARS power densities for several segments of a unit-amplitude, 1-wavelength sinusoidal current filament

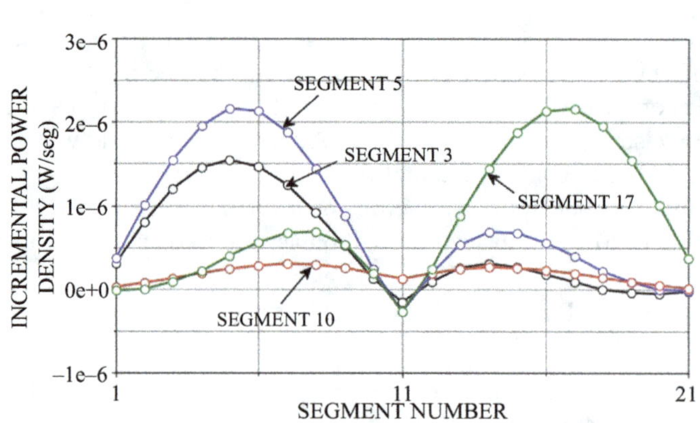

Figure 11.4 Incremental FARS power densities for several segments of a 1-wavelength NEC dipole center-excited by 1-V

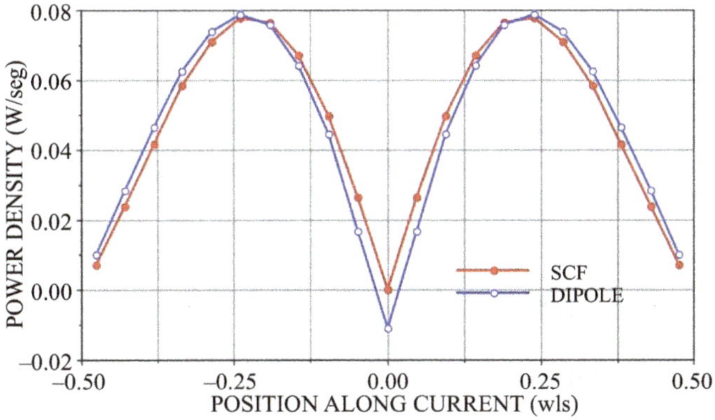

Figure 11.5 FARS results for the 1-wavelength SCF and NEC dipole normalized to a radiated power of 1-W

Figure 11.3. The curves for the SCF and dipole are of similar shape for the most part, with the dipole values lower because of its lower current.

The total FARS results for the SCF and NEC dipole of Figures 11.3 and 11.4 normalized to 1-W radiated power are compared of in Figure 11.5. Their power densities are slightly different and not identical as might be expected from their IFARS results, most likely being due to the difference in their effective excitations.

11.4 The incremental IEMF method

The moment-method treatment, when applied to an electric-field integral equation (EFIE) for a PEC, leads to a matrix relationship of the form

$$\sum_{n=1}^{N} Z_{mn} I_n = V_m^{(exc)}, m = 1, \ldots, N \tag{11.7}$$

with I_n being the sampled current. The matrix Z is symmetric about its diagonal, i.e., $Z_{mn} = Z_{nm}$ for a Galerkin procedure. For the electric-field integral equation $V_m^{(exc)}$ is the voltage across segment m when using an appropriate testing function. Piecewise uniform testing is one choice that has been used whose integral of the exciting electric field is the voltage over segment m.

The impedance matrix Z is normally used to find the current response for a specified excitation by inverting or otherwise solving (11.7). An alternative is to use (11.7) to find the field distribution caused by a known current, as is done here. The IIEMF method uses the incremental, induced voltage $V_{mn}^{(ind)}$ on segment m due to the current on segment n and vice-versa, operating on a pair-wise basis over the entire structure as given by

$$V_{mn}^{(ind)} = Z_{mn} I_n \tag{11.8a}$$

and

$$V_{nm}^{(ind)} = Z_{nm}I_m. \tag{11.8b}$$

Note that the first subscript on V here indicates the segment on which an induced voltage is caused by the incremental or segment current denoted by the second subscript. The near-field IIEMF power due to the interaction voltage on segment m caused by the current on segment n is given by

$$p_{mn}^{(IIEMF)} = \frac{1}{2}\text{Re}\left[V_{mn}^{(ind)}I_m^*\right] = \frac{1}{2}\text{Re}\left[Z_{mn}I_nI_m^*\right] \tag{11.9}$$

and similarly the IIEMF power on segment n caused by the current on segment m is

$$p_{nm}^{(IIEMF)} = \frac{1}{2}\text{Re}\left[V_{nm}^{(ind)}I_n^*\right] = \frac{1}{2}\text{Re}\left[Z_{nm}I_mI_n^*\right], \tag{11.10}$$

The total IEMF power on segments m and n is then given by

$$P_m^{(IEMF)} = \sum_{n=1}^{N} p_{mn}^{(IIEMF)} \tag{11.11a}$$

and

$$P_n^{(IEMF)} = \sum_{m=1}^{N} p_{nm}^{(IIEMF)}. \tag{11.11b}$$

Because of the PEC boundary condition on the tangential electric field, the self-field of a given segment must cancel the total interaction fields caused by the remaining N–1 segments and any excitation field. Taking the radiated power as a

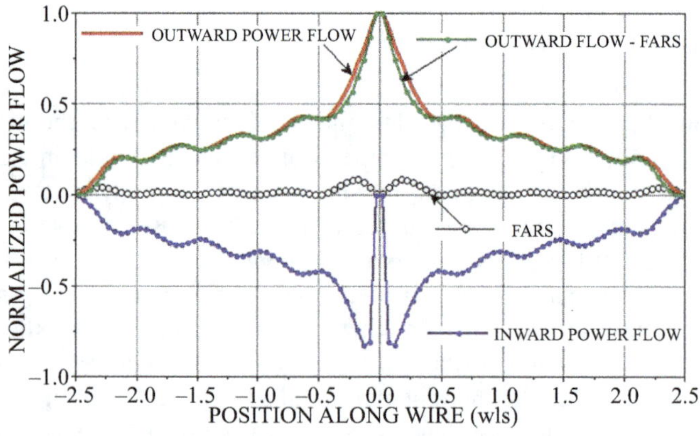

Figure 11.6 The inward and outward power-flow densities along a 5-wavelength, center-excited dipole together with the FARS radiated power to show the fraction of the outward flow due to radiation

positive quantity, the net remaining power flows must be negative to produce a net normal power flow of zero on a PEC object. For comparison purposes with FARS results, however, the IEMF-derived powers in the following are plotted as positive quantities. The difference between the total positive outward power flow and the radiated power is due to interactive, near-field power flow. The power-flow plots in Figure 11.6 illustrate that a relatively small fraction of the outward power flow is due to radiation where the radiated power distribution from FARS is included. The basic idea behind the IIEMF approach is that the net interaction electric field over a given segment induces the equivalent response as does an applied source field, effectively providing an input power to it.

11.5 Numerical results

Results for prescribed real currents such as the SCF and UCF of the radiated-power density obtained using the IEMF method are demonstrated in Chapter 10. Presented in Figure 11.7 are the IFARS results shown previously in Figure 11.3 for comparison with IIEMF results for a 1-wavelength SCF. These different results are graphically indistinguishable providing mutual validation for each. When the power density is plotted on a segment-to-segment basis, it is referred to as an incremental power, as in Figure 11.7. Also note that since all of the IEMF power densities are found using the IIEMF method, this is how they are referred to in the numerical results included here.

A similar comparison was initially developed of the IFARS and IIEMF results for the 1-wavelength NEC dipole as plotted in Figure 11.8 where the mutual-interaction powers $p_{3m}^{(IIEMF)}$ and $p_{m3}^{(IIEMF)}$, $m = 1$–21 for the 21-segment NEC dipole are shown. But the results of Figure 11.8 are neither equal nor reciprocal as are their IFARS counterparts shown in Figure 11.4. However, when the average of these two curves is computed, i.e. their sum divided by 2, also shown in Figure 11.7, the

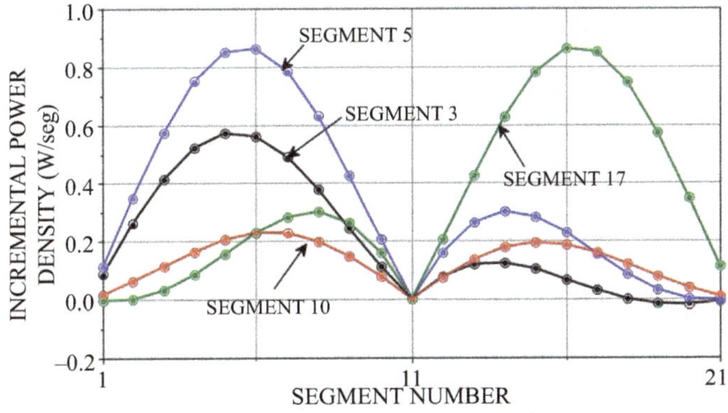

Figure 11.7 A comparison of IFARS from Figure 11.3 (the open circles) and IIEMF power (the solid circles) densities for a 1-wavelength SCF

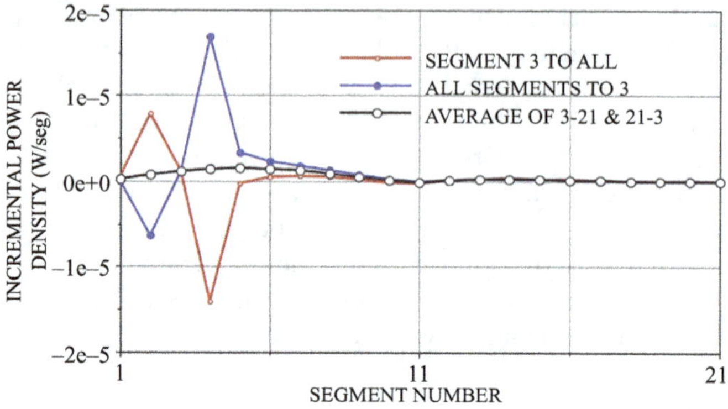

Figure 11.8 The $p_{3m}^{(IIEMF)}$ and $p_{m3}^{(IIEMF)}$ results and their average values for a 1-wavelength NEC dipole modeled using 21 segments

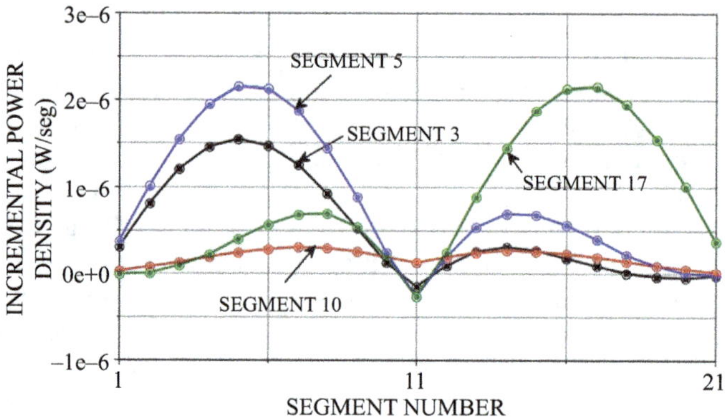

Figure 11.9 The IIEMF results (the solid circles) equivalent of the IFARS results of Figure 11.4 (the open circles) for the 1-wavelength NEC dipole modeled using 21 segments excited by 1 V

result is essentially identical to those of IFARS. A comparison of IFARS and IIEMF results for the 1-wavelength dipole is shown in Figure 11.9 for the same segment pairs as used for the SCF in Figure 11.7.

The difference between the SCF and dipole pairwise IIEMF results is not clear, but seems likely to be due to the fact that while the SCF is pure real the NEC current is a complex quantity. A complex traveling-wave current also exhibits a similar discrepancy. But, as for the NEC dipole, averaging $p_{nm}^{(IIEMF)}$ and $p_{mn}^{(IIEMF)}$ of the traveling-wave current yields IIEMF values and IFARS that agree to within a percent or so. The fact that the IFARS and IIEMF-derived, averaged results agree this well provides some validation for what might otherwise appear to be an *ad hoc* procedure.

Thus, the incremental power density for segments m and n and total IEMF power density for segment m are determined in the following using, respectively

$$p_m^{(av)} \approx \frac{1}{2}\left(p_{nm}^{(IIEMF)} + p_{mn}^{(IIEMF)}\right) = \frac{1}{2}\left[\frac{1}{2}\mathrm{Re}\left[Z_{mn}I_nI_m^*\right] + \frac{1}{2}\mathrm{Re}\left[Z_{nm}I_mI_n^*\right]\right] = p_n^{(av)} \tag{11.12}$$

and

$$P_m^{(IIEMF)} = \frac{1}{2}\sum_{n=1}^{N}\left(p_{nm}^{(IIEMF)} + p_{mn}^{(IIEMF)}\right) \tag{11.13}$$

A half-wavelength NEC dipole having 11 segments and excited at segment 3 was modeled as shown in Figure 11.10 in order to test the averaging approach for a non-symmetric problem. The FARS and IIEMF power densities again are found to agree to within a percent or better.

A scattering example is presented in Figure 11.11 for a normally incident plane wave of 20 V/wavelength using NEC for the same half-wavelength straight wire. An agreement between the FARS and IIEMF results is obtained similar to those of the previous examples. A total scattered power of 0.24 W is obtained from both, which also matches the collected power P_{coll} from

$$P_{coll} = \frac{1}{2}\mathrm{Re}\int_{-L/2}^{L/2}\left[V(z)I^*(z)\right]dz. \tag{11.14}$$

Examples of scattering results from FARS and IIEMF for a 10-wavelength, 201-segment straight wire for normal incidence are shown in Figure 11.12 and for 10-deg off axial incidence from the right in Figure 11.13. As before, these results are also basically graphically indistinguishable, providing further validation of both the FARS and the IIEMF methods. The near-axial incidence case of Figure 11.13

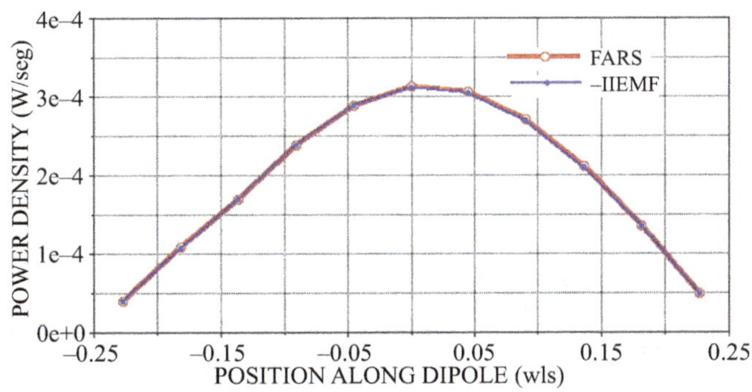

Figure 11.10 Power densities derived from FARS and the IIEMF method for a half-wavelength NEC dipole modeled using 11 segments and excited by a 1-V source at segment 3 near the left end

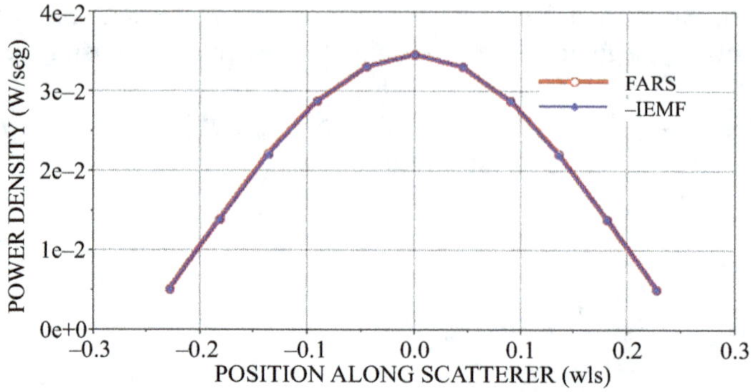

Figure 11.11 *The scattered power density for a normally incident plane wave on a half-wavelength straight wire from FARS and the IIEMF method*

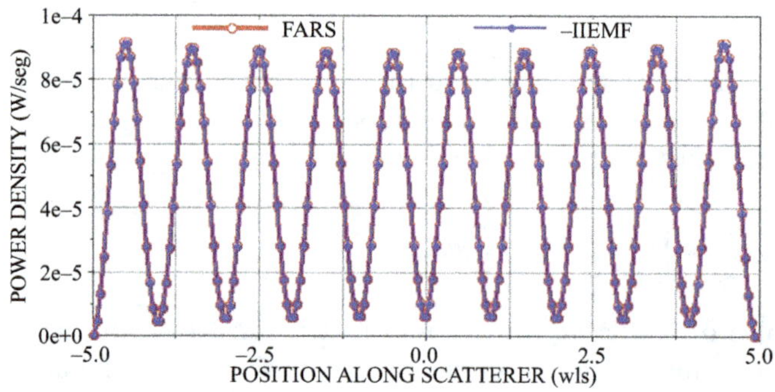

Figure 11.12 *The scattered power density for a 10-wavelength straight wire caused by a normally incident plane wave*

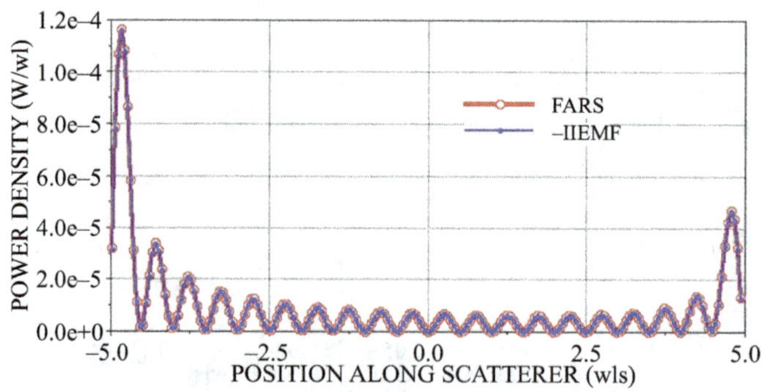

Figure 11.13 *The scattered power density for a 10-wavelength straight wire caused by a plane wave incident at an angle of 10 deg from the right end*

exhibits two principle radiation sources at the ends of the wire, with the far-end peak nearly three times that of the near end. This is evidently caused by the traveling-wave effect of the excitation on the current. This case is much more asymmetric than the antenna example of Figure 11.10 further supporting using (11.12) to compute the IIEMF power.

Repeating the FARS and IIEMF computation for 10-wavelength, 201-segment wires excited at two segments to produce more complex radiated-power distributions are shown in Figures 11.14a and 11.15a. The excitations are at segments 64 and 137 located about ±3.25 wavelengths from either end. The two sources are +1-V in Figure 11.14 and, in Figure 11.15, they are ±1-V, respectively. The normalized FARS result and the current magnitude are shown in part (b) of each figure.

The radiation in Figure 11.14a is mostly confined to the region between the two feedpoints while negative FARS peaks occur at the respective feedpoints. Peaks in the power density line up with those in the current magnitudes except near the feedpoint. Why this occurs is discussed in Chapter 7 in connection with Figures 7.25 and 7.25. The current peaks and associated charge minima coincide with FARS peaks because of how propagation radiation is generated. The input

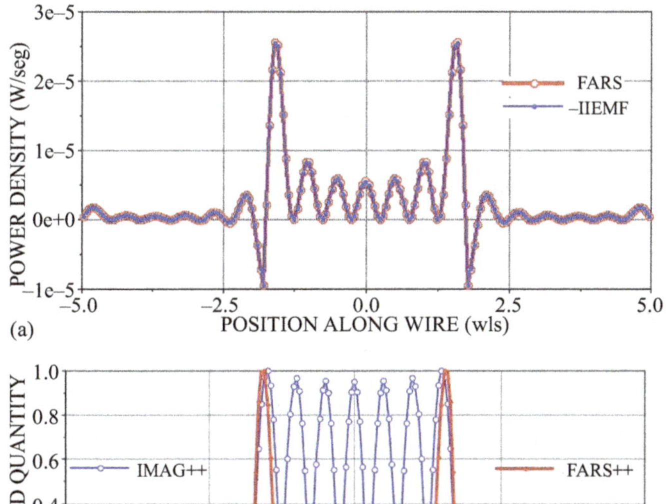

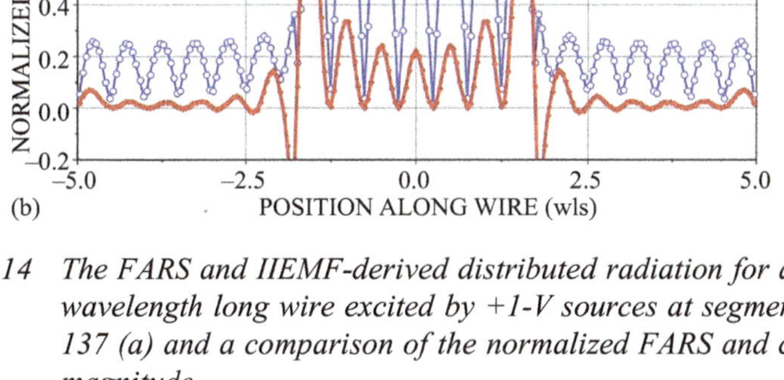

Figure 11.14 The FARS and IIEMF-derived distributed radiation for a 10-wavelength long wire excited by +1-V sources at segments 65 and 137 (a) and a comparison of the normalized FARS and current magnitude

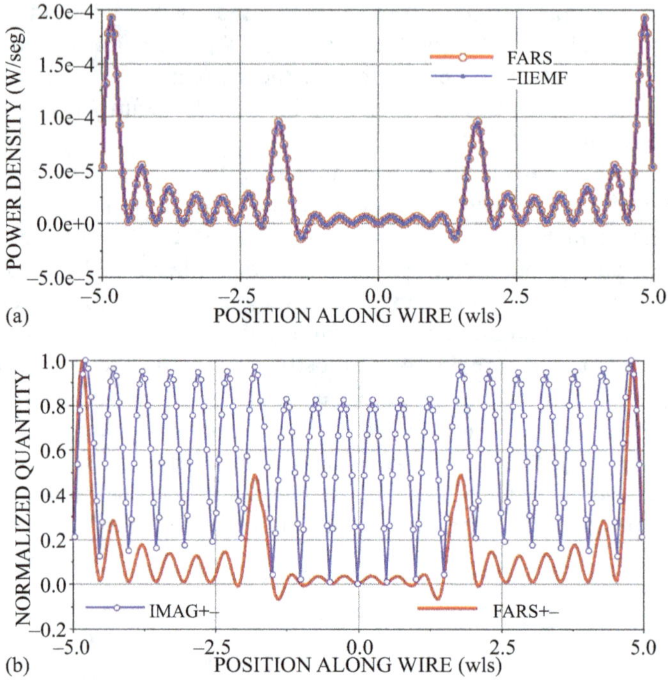

Figure 11.15 The FARS and IIEMF-derived distributed radiation for a 10-wavelength long wire excited by ±1-V sources at segments 65 and 137 (a) and a comparison of the normalized FARS and current magnitude

power computed by NEC and the FARS and IIEMF powers for this case are $4.82\text{x}10^{-4}$ W, 4.83×10^{-4} W and 4.79×10^{-4} W, respectively, an agreement comparable with other results included here.

On using excitations of opposite sign, as shown by the results of Figure 11.15, the distributed radiation becomes concentrated between the feedpoints and ends of the wire with relatively little coming from the central part of the wire. As before, the maxima of the current magnitude and the distributed radiation are aligned except close to the feedpoint. These results again demonstrate the dependence of the acceleration-caused radiation on the current and charge. For the $\pm$ case the respective NEC, FARS and IIEMF powers are 5.06×10^{-3} W, 5.03×10^{-3} W and 4.98×10^{-3} W.

Incremental power densities similar to Figure 11.9 but for a longer 5-wavelength, 101-segment NEC dipole are plotted in Figure 11.16 where an agreement of better than 1% is again obtained between the IFARS and IIIEMF results. A last comparison of IFARS and IIEMF computations is shown in Figure 11.17 for 5 center-excited NEC dipoles varying from 0.5 to 4.5 wavelengths long.

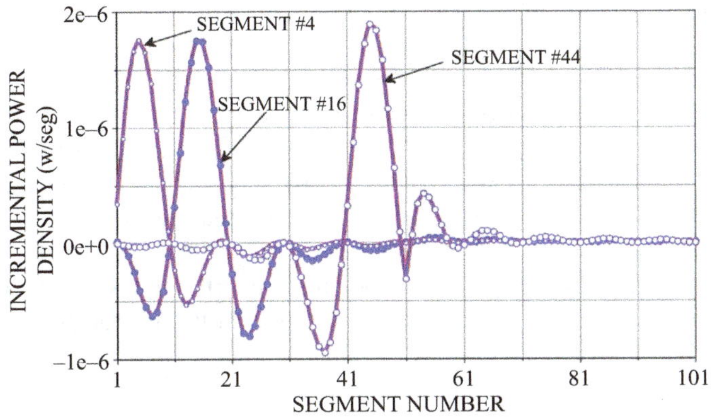

Figure 11.16 *Incremental power densities for a 5-wavelength, 101-segment NEC dipole for segments number* m = *4, 16, and 44 and* n *scanning over the 101 segments of the dipole with IFARS shown as solid lines and the IIEMF results as data points*

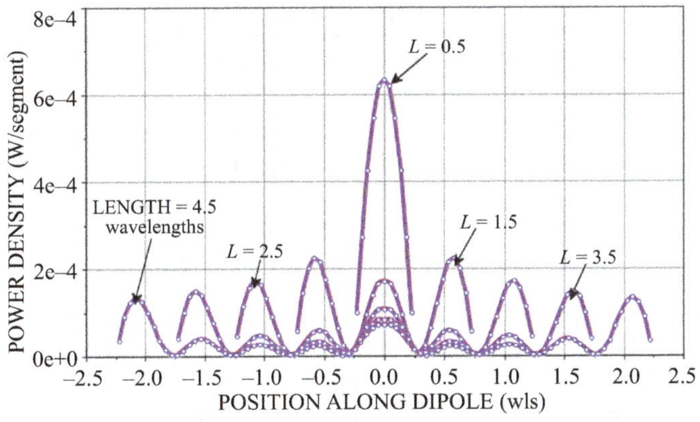

Figure 11.17 *Radiated linear power densities for NEC dipoles an odd number of half wavelengths long, obtained from FARS (solid lines) and IIEMF (data points)*

11.6 Discussion

The implementation and use of the IEMF method as an incremental process in the way presented in this chapter might appear unconventional and even questionable considering its validity as presented relies entirely on numerical results. The IEMF method is usually employed to approximate the input impedance of an antenna from an assumed current distribution, of which numerous and varied examples will be found in an Internet search. The discovery that the interactive powers between different segments of a PEC wire are not reciprocal for complex-valued currents using the IIEMF method was unexpected, a result that does not occur for real-valued currents. In spite of that issue, the numerical results obtained from averaging the pairwise powers of two segments agree to within one percent or better with the IFARS and FARS results for a given problem.

The results included in this chapter show that the IIEMF method is a viable alternative to FARS for determining the quantitative distribution of the power radiated by a PEC object. It is worth emphasizing that the good agreement between FARS and the IIEMF method comes from very different information. FARS is based only on the far field while the IIEMF instead is based on a near-field quantity, the current on an object. Also, observe that obtaining the IIEMF results requires only the original impedance matrix and the currents computed from it for their computation.

11.7 Summary

This chapter has introduced the concept of incremental, far-field, paired current segments in a moment-method context as incremental far-field analysis of radiation sources (IFARS). This involves exploiting the fact that the conjugate dot product of the total far electric field of a source distribution can be usefully analyzed in terms of the products of fields of incremental source pairs. Also introduced was development of a complementary approach, the IIEMF method. The IIEMF method has been shown to offer an independent way of quantitatively determining the distribution of radiated power over the source of interest. This is in contrast with the standard IEMF method that yields zero power on a PEC wherever the excitation is zero and thus cannot show where radiation originates.

References

[1] E. K. Miller and J. A. Landt (1980), "Direct time-domain techniques for transient radiation and scattering from wires", Invited Paper in *Proceedings of the IEEE*, Vol. 68, pp. 1396–1423.

[2] E. K. Miller (1999a, 1999b), "PCs for AP and other EM reflections," *EEE AP-S Magazine*, Vol. 41, No. 2, April, pp. 82–86; Vol. 41, No. 3, June, pp. 83–88.

[3] E. K. Miller (2000), "Further investigation using far-field analysis of radiation sources (FARS)," in: *IEEE International Symposium on Antennas and Propagation*, Salt Lake City, UT, Hilton Hotel, July 16–21, pp. 2058–61.

[4] E. K. Miller and G. J. Burke (2001), "A multi-perspective examination of the physics of electromagnetic radiation," *Applied Computational Electromagnetics Society Journal*, Vol. 16, No. 3, pp. 190–201.

[5] E. K. Miller and G. J. Burke (2000), "Time-domain far-field analysis of radiation sources," in: *IEEE International Symposium on Antennas and Propagation*, Salt Lake City, UT, Hilton Hotel, July 16–21, pp. 1546–49.

[6] J. F. Shaeffer, B. A. Cooper, K. W. Hom, R. C. Baucke, and N. A. Talcott (1997), "A review of bistatic k-space imaging for electromagnetic prediction codes for scattering and antennas," *IEEE AP-S Magazine*, Vol. 39, No. 5, pp. 2121–29.

Chapter 12

The Schelkunoff–Feldman radiation resistance

Abstract

This chapter revisits the concept of a distributed radiation resistance developed by Schelkunoff and Feldman. Their article addressed the possibility of accounting for radiation from a wire antenna via a distributed resistance added to whatever dissipative resistance an imperfectly conducting wire might have.

In order to retain its complete development of this novel concept, their entire article is included here. This is followed by an extension of their development. While they limited the results of their analytical treatment to currents an even number of half wavelengths long their treatment is shown numerically here to be valid for an arbitrary length. This is demonstrated by using their distributed radiation resistance to determine the quantitative distribution of radiation from along the current. Results thus obtained are found to agree with the frequency-domain FARS and IEMF results discussed in previous chapters, providing validation for both approaches. Also, numerical computations from FARS are shown to provide an alternate way to evaluate the distributed radiation resistance.

Keywords: Distributed Radiation Resistance; FARS; Induced EMF Method; Radiation Resistance; Sinusoidal Current.

On Radiation from Antennas*

S. A. SCHELKUNOFF, ASSOCIATE, I.R.E., AND C. B. FELDMAN, ASSOCIATE, I.R.E.

Summary-This paper presents some theoretical remarks and experimental data relating to applications of the transmission-line theory to antennas. It is emphasized that the voltage, the current, and the charge are affected by radiation in different ways, a fact which should be considered in any adaptation of linear equations to antennas.

It is shown experimentally and theoretically that in an antenna of length equal to an integral number of half wavelengths, which is energized at a current antinode, the effect of radiation on the current and the charge (but not on the voltage) can roughly be represented by adding to the resistance of the wires another fairly simple term.

*S. A. Schelkunoff and C. B. Feldman, "On Radiation from Antennas," *Proceedings of the I. R. E.*, November, 1942, pp. 511–516. When the IEEE was organized it absorbed the IRE but according to IEEE Explore, IEEE no longer owns the copyright to this article.

I. GENERAL DISCUSSION

For many years it has been known that the current distribution in an antenna approximates that obtained by regarding the antenna as an open-circuited transmission line with certain series inductance and shunt capacitance per unit length. In other words, it has been known that effects of radiation on the current distribution are relatively small. During the last decade much thought has been given to improving upon this approximate picture by an inclusion of radiation effects, while remaining within the framework of the transmission-line theory. A theoretical question is whether or not such an improvement is possible within the limits of the line theory. While experimentally minded engineers, not being particularly worried by such questions, have achieved a measure of success, theoreticians have been beset with doubts. These doubts are only natural, since the only safe mathematical approach to the problem is through Maxwell's field equations, and these apparently make the antenna problem quite different from line problems.

In the line theory we write

$$\frac{dV}{dx} = -(R + i\omega L)I, \frac{dI}{dx} = -(G + i\omega C)V, \tag{1}$$

where V and I are the voltage across the line and the current in it; the line parameters *R*, *L*, *G*, and *C* are, respectively, the series resistance, the series inductance, the shunt conductance, and the shunt capacitance, all per unit length. In applying these equations to a pair of parallel wires (Fig. 1), we say that R is the resistance of the wires per unit length and that it represents the power dissipated in heat. If the medium surrounding the wires is nondissipative, we say that *G* is equal to 0. We define *C* as the ratio of the electric charge per unit length to the voltage across the line. The quantity "$-i\omega LI$" is the counter electromotive force of induction in "series" with the wires; it represents the reaction of the changing magnetic flux between the wires on the line voltage. This is the picture if radiation is ignored; and if we ignore radiation, there is nothing to prevent us from applying the picture to diverging wires (Figs. 2 and 3). If there is any difference between the three structures in Figures 1, 2, and 3, it is in the magnitude of radiation effects.

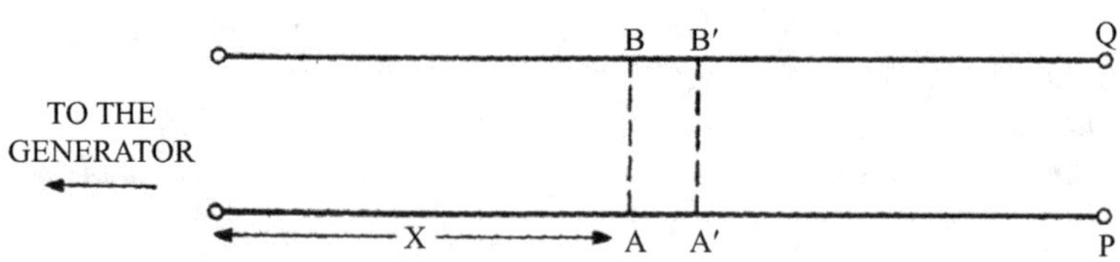

Fig. 1

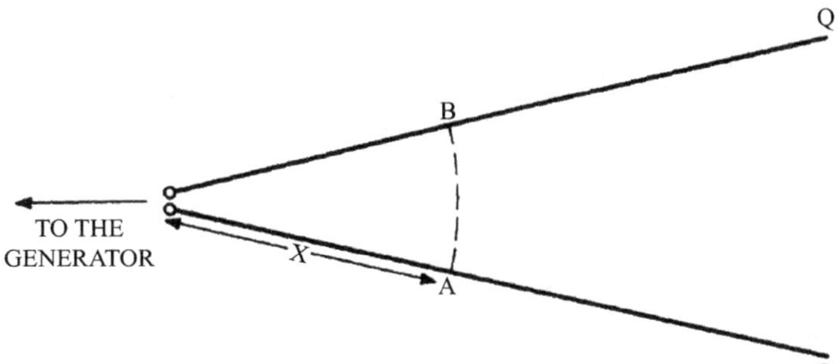

Fig. 2

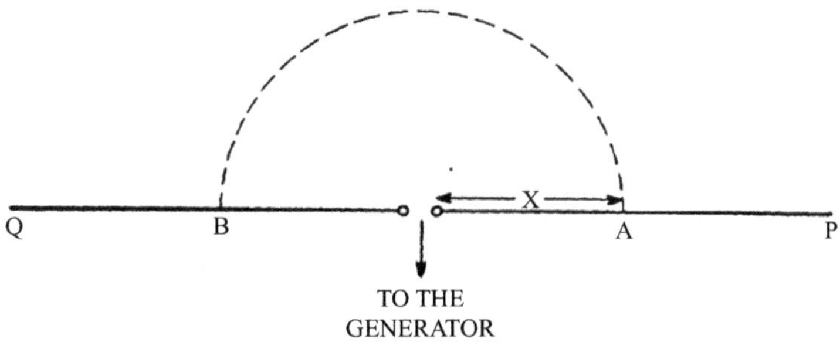

Fig. 3

One of the effects of dissipation is to cause attenuation of the voltage and current waves. Since radiation represents additional loss of power, it has been thought that it should cause further attenuation. In order not to confuse the issue, let us assume, for the time being, that the wires are perfectly conducting and that the surrounding medium is nondissipative; then, if R and G are different from 0, they must represent the effect of radiation.

One line of thought runs somewhat as follows. Ignoring radiation and solving (1) we find the first approximation to the current distribution $I(x)$ and charge distribution $q(x)$. Using this current distribution, we compute the retarded electric potential U and the retarded magnetic vector potential Π

$$U(x) = \frac{1}{4\pi\varepsilon}\int \frac{q(x')e^{-\beta r}}{r}dx',$$

$$\Pi(x) = \frac{1}{4\pi}\int \frac{I(x')e^{-\beta r}}{r}dx', \tag{2}$$

where the integration is extended over the wires; $\beta = 2\pi/\lambda = \omega/c$; and r is the distance between the current element $I(x')dx'$ and a typical point in space

$$r = \sqrt{(x-x')^2 + \rho^2}, \rho^2 = y^2 + z^2 \tag{3}$$

Strictly speaking, the integrals should be taken over the surface of the wires; but if the wires are thin, good approximations are given by (2).

From U and Π we compute the field with the aid of the following equations

$$\mathbf{E} = i\omega\mu\Pi - \nabla U, \mathbf{H} = \nabla \times \Pi \tag{4}$$

Next, we obtain the total radiated power by forming the Poynting vector *EXH* and integrating it over a surface enclosing the antenna.

At this point we usually leave the solid ground and make the following *a priori* physical assumptions: (1) Equations (1) *can* take care of radiation, (2) the effect of radiation is expressed solely by a series resistance R_{rad}, and (3) R_{rad} is uniform. Granting tentatively these assumptions, we can compute R_{rad} so that the integral of $R_{\text{rad}}\, I^2$ along the antenna is equal to the radiated power. We can now use (1) for recomputing the current distribution. P. 0. Pedersen [1] used this procedure and obtained a current distribution which checked rather well the measured current distribution.

Nevertheless, not all is well with this procedure. Let us examine the three *a priori* assumptions. There are no logical objections to the first assumption, if we permit R, L, G, and C to vary in any manner with x since V and I are definite functions of x, we can always use (1) as *definitions* of the line parameters. Naturally, these equations are useless for computing V and I unless we can find R, L, G, and C by independent means; but utility, or lack of it, does not affect the assumption. However, the second assumption, G_{rad} equals 0, is incorrect. It can be shown that V is not affected by radiation and that the ratio of dI/dx to V is a complex number.

This difficulty can be obviated if we write our transmission equations in terms of electric charge $q(x)$ and electric current $I(x)$ rather than the voltage-current form (1). In the first place, we have the equation of conservation of electric charge

$$\frac{dI}{dx} = -i\omega q, \tag{5}$$

which takes place at the second equation in (1). Next, we define V arbitrarily so that $C = q/V$ is independent of x and, finally, we write the first equation of (1) as follows

$$\frac{dq}{dx} = -(R_{rad}C + i\omega LC)I, \tag{6}$$

Of course, V as defined above is fictitious; but then it does not appear in our ultimate equations (5) and (6) and we need not be concerned with it. Besides, the measurements are carried out on q and I.

In an earlier paper by one of the present authors [2], equation (6) was obtained directly from Maxwell's equations. There it was made clear that RC and LC were complicated functions of x. Comparing the voltage-current transmission equations (1) to the current-charge equations (5) and (6), we observe that the latter are simpler in the sense that the "line" parameters enter into only one equation. Once we are certain that the form of equations (5) and (6) is permissible, we step on a more certain ground with regard to Pedersen's method of obtaining a second approximation to the current distribution in antennas.

It appears that while radiation does not affect the voltage distribution in antennas, it affects the current and the charge in more or less the same manner that dissipation does. Pedersen has assumed *ab initio* that R_{rad} is constant. In the third part of this paper we derive from Maxwell's equations the following approximate expression

$$R_{rad} = \frac{60L}{L^2 - x^2} \tag{7}$$

for the case when the length $2L$ of the antenna (of the type shown in Fig. 3) is an integral number of half wavelengths and the generator is at a current antinode. This function is represented by the solid curve in Fig. 4, where the experimental points, obtained under various conditions, are also marked.

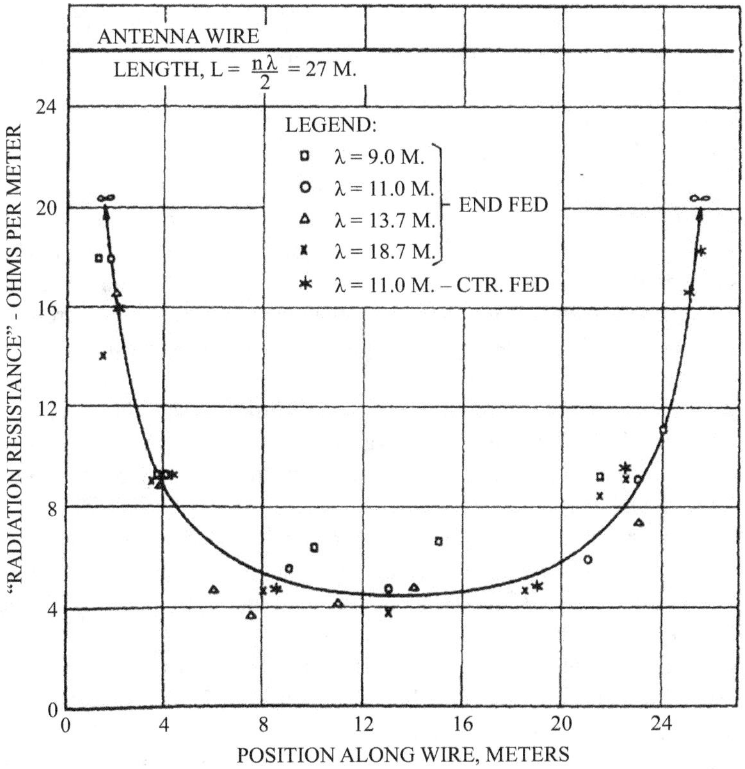

Fig. 4

For other antenna lengths, the formula becomes more complicated and indicates greatly increased radiation resistance near the current nodes. Before proceeding with an account of our experiments and the mathematical derivation of (7), we shall, following another line of thought, give a descriptive resume of a recent antenna theory [2] which is particularly useful for computing input impedances of antennas. Let us go back to (1) and write them in a form suitable to the nondissipative case

$$\frac{dV}{dx} = -i\omega LI, \frac{dI}{dx} = -i\omega CV. \tag{8}$$

These equations apply exactly to an infinitely long pair of conical conductors diverging from a common apex as shown in Figs. 2 and 3. In the limit, they apply to an infinitely long pair of parallel cylindrical wires. In any of the above cases L and C are constant. These equations will also apply with a high degree of accuracy to diverging cylindrical wires if the latter are thin; in this case, however, L and C will vary with the distance x.

The transverse voltage V is defined as the line integral of the electric intensity between the wires, taken along a path lying completely in an equiphase surface, that is, in a plane perpendicular to the parallel wires and in a sphere concentric with the apex of diverging wires.

Now let the wires terminate at P and Q. It turns out that if we subtract a certain "complementary" or "local" current $I_c(x)$ from the actual total current $I_t(x)$, the remainder $I(x) = I_t(x) - I_c(x)$ and the transverse voltage $V(x)$ will satisfy (8). The complementary current is 0 at the apex of the diverging wires; in the case of parallel wires, it diminishes as the distance from the terminals PQ increases. For the case of parallel wires this fact was noted by Carson [3]. Thus, the input impedance of diverging wires and long parallel pairs will be given as the ratio of the applied voltage to the "principal current" $I(0)$, the latter being equal to the total current $I_t(0)$.

The total current vanishes at P; hence, the principal current there becomes equal to $-I_c(P)$. The relation between the total voltage and the principal current at PQ is that which would be obtained if the complementary current were disregarded and in its stead a terminal impedance equal to $V(P)/-I(P)$ were assumed. This terminal impedance, which may be called the "radiation impedance," is a complex quantity, the real part of which represents the radiated power. For large values of the characteristic impedance, the radiation impedance is substantially proportional to the square of the characteristic impedance.

In the elementary theory, the complementary current and the complementary charge are ignored. In effect, this makes the terminal radiation impedance infinite and the transmission line becomes electrically "open." In the complete theory, the voltage remains sinusoidal, although the end of the line is no longer a voltage antinode; of the electric current and electric charge only the major parts are sinusoidal. The effect of the discontinuity of the line at PQ on these major parts is represented completely by the radiation impedance.

The complete theory approaches the elementary theory as either the characteristic impedance or the wavelength becomes infinite. In accordance with what has just been said, we are permitted to think that a wave emerging from a generator is guided by the wires until it reaches the ends where the transmission system suddenly changes [4]; there, reflection takes place. The amount of reflection is not the same at all points of the wavefront so that besides the reflected wave which is guided back to the generator, there appears a "complementary" field which compensates for the uneven reflection.

The uneven reflection may be pictured as follows. Consider the wavefront passing through the ends of the wires and apply Huygens' principle or, rather, it is more explicit form known as the equivalence principle [5–8]. The secondary sources, near the wire but on its opposite sides, are 180 degrees out of phase. Being so near each other, their effect on the field beyond the antenna region is very small and they emit very little energy. But the energy arriving from the generator is denser near the wires than elsewhere; hence, the greatest reflection will occur over the area of the wavefront in the neighborhood of the wires.

The quantitative counterpart of the above qualitative antenna theory will be found in the paper [2] already referred to. The theory is sufficient for obtaining engineering information about performance of many different types of antennas.

We have just discussed several ways in which the line equations can be used in conjunction with the field equations in dealing with radiation problems. At present, the method which has been discussed last appears to be the most promising from the practical point of view and it possesses the theoretical advantage of being a child of the field theory. The earlier method is of interest largely because it shows that the old controversy as to whether radiation takes place continuously or "snaps off" the ends of the antenna is rather meaningless. One can look at it either way.

II. EXPERIMENTAL PART

The experiment concerns verification of (7). The method consists of exciting an elevated antenna wire with a transmitting oscillator, and measuring the shape of the nodes (minima) of current and charge. The space with wires embedded in it is replaced by space without wires.

The total attenuation suffered by a wave traveling to the open end and back to the node in question can be deduced from the nodal shape, assuming a constant characteristic impedance. Measurements made upon several nodes between the generator and the open end yield a curve of total attenuation versus position on the wire whose slope can be interpreted as the attenuation exponent for that position. In order to obtain the resistance from the attenuation exponent it is necessary to know the characteristic impedance. One method of measuring the characteristic impedance is to measure attenuation for two antennas whose ohmic resistances are different. Such a resistance variation permits the radiation effect to be subtracted out and the characteristic impedance to be calculated from the remaining attenuation. The theory underlying this method of measurement is as follows.

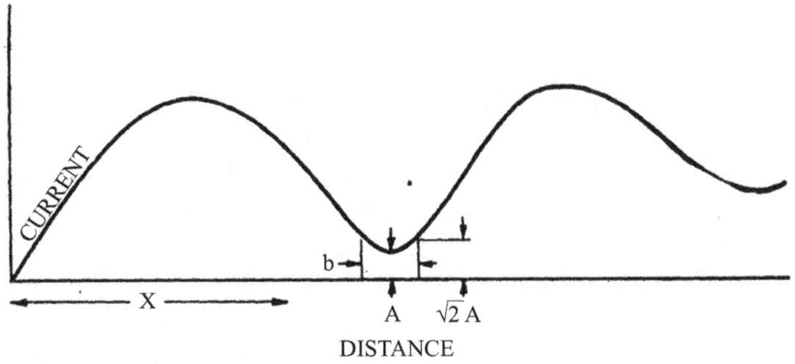

Fig. 5

Assuming that the resistance $R(x)$ of a transmission line is small compared with its characteristic impedance K, we can write an approximate expression for the current at distance x from one end of the wire (Fig. 5) in the form

$$I(x) = \sinh(A + i\beta x) \equiv A \cos \beta x + i \sin \beta x, \tag{9}$$

where $A(x)$ is the total attenuation to point x

$$A(x) = \frac{\int_0^x R(x)dx}{2K}. \tag{10}$$

From (10) we have

$$R(x) = 2K\frac{dA}{dx}. \tag{11}$$

The values of A corresponding to the current minima and charge minima can be related to the shape of the latter. Thus the minima of $I(x)$ are near points for which $\sin \beta x = 0$, so that minimum amplitudes are given by $A(x)$. The point for which the amplitudes are to $\sqrt{2}A$ occur where the real and the imaginary parts of (9) are equal; that is, (Fig. 5),

$$\sin\frac{\beta b}{2} = A\cos\frac{\beta b}{2}, or \frac{\beta b}{2} = A, \quad \cdot A = \frac{\pi b}{\lambda}. \tag{12}$$

Setting

$$A_1(x) = \frac{\int_0^x R_{rad}(x)dx}{2K} \quad \text{for zero wire resistance}$$

$$A_2(x) = \frac{\int_0^x R_{rad}(x)dx + rx}{2K} \quad \text{for resistance wire} \tag{13}$$

where r is the resistance per unit length, we obtain,

$$2K = \frac{rx}{A_2(x) - A_1(x)},$$

$$R_{rad}(x) = \frac{rx}{A_2(x) - A_1(x)}\frac{dA_1(x)}{dx}. \tag{14}$$

The test antenna was a No. 14 B & S gauge wire 27 meters long stretched horizontally under the roof of a large barn. Copper and Nichrome wires were used. A wooden platform erected 7 feet beneath the wire enabled an operator to carry apparatus with which to measure current and charge distribution. The current measuring equipment employed a small loop highly balanced against response to potential gradient, while charge-measuring equipment employed a receiver with weak capacitive coupling to the antenna. Measurements were made at several wavelengths from 9 to 18 meters and with the oscillator coupled to the wire in various ways near one end and at the middle. A plot of the total attenuation measured at 11 meters wavelength appears in Fig. 6. The smooth curves, whose slopes are to be determined graphically, were drawn in among the scattered experimental points to conform with the knowledge that the difference should increase linearly. The slope of this linear curve determines the characteristic impedance, knowing the difference in high-frequency resistance. The slope of the copper-wire curve (the high-frequency resistance is negligible) determines the radiation resistance at any point, knowing the characteristic impedance. The Nichrome-wire resistance was 3.56 ohms per meter at 11 meters wavelength, while the copper-wire resistance was 0.27 ohm per meter. Using these data the characteristic impedance is found to be 470 ohms which agrees well with calculation and other measurements. Radiation-resistance values so determined are plotted in Fig. 4 together with a curve calculated from (7).

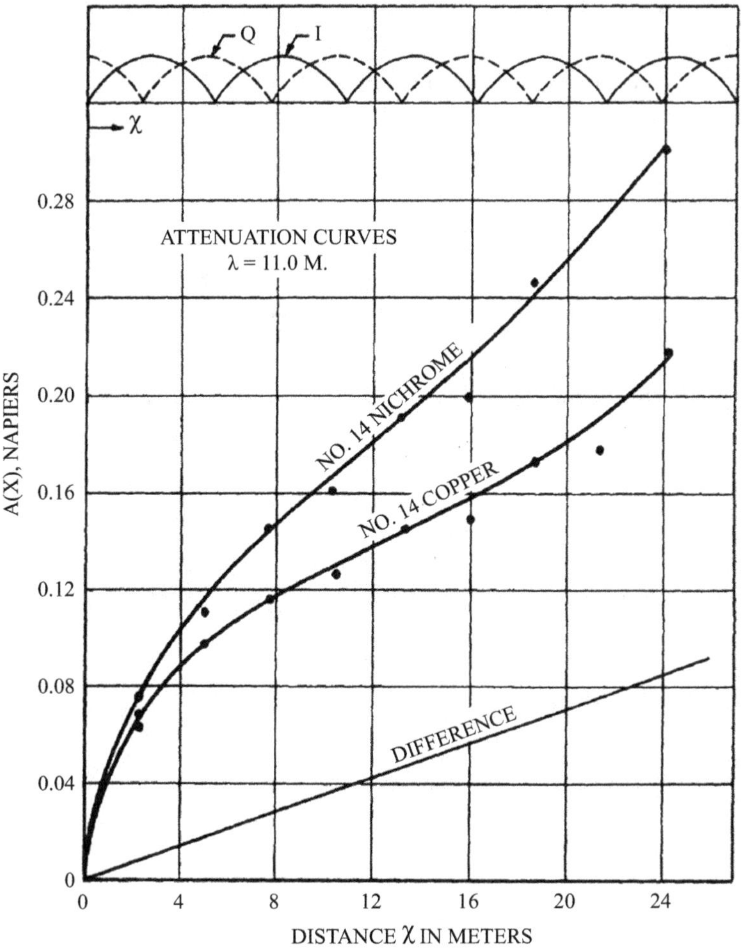

Fig. 6

III. MATHEMATICAL ANALYSIS

Consider a cylindrical wire of radius a (Fig. 3) so energized that the electric current is parallel to the axis; then, the electric intensity E_z at the surface of the wire and parallel to it is

$$E_z = -i\omega\mu\Pi = -\frac{dU}{dx}, \tag{15}$$

where the retarded electric scalar potential U and the retarded magnetic vector potential A are given [9] by (2).

Everywhere along the wire, except at the generator, the electric intensity E_z is also given by

$$E_x = ZI,\ Z = \frac{1}{2\pi a}\sqrt{\frac{i\omega\mu}{g}}, \tag{16}$$

where Z is the surface impedance of the wire and g is the conductivity. Substituting from (16) in (15), we have

$$\frac{dU}{dx} = \frac{\cos\,\beta r}{r} - i\frac{\sin\,\beta r}{r}. \tag{17}$$

We shall now carry out two successive approximations: First, we shall see what happens when the radius of the wire is approaching zero; then, starting from this approximation we shall compute certain correction terms. Consider the following factor in the integrand of (2)

$$\frac{e^{-\beta r}}{r} = \frac{\cos\,\beta r}{r} - \frac{\sin\,\beta r}{r} \tag{18}$$

If a approaches zero, the first term approaches infinity in the immediate vicinity of $x' = x$ while the second term remains everywhere finite. Hence, from (2) we obtain the following asymptotic relations

$$\varepsilon V(x) = A(x)q(x),\ \Pi(x) = A(x)I(x); \tag{19}$$

where

$$A(x) = \frac{1}{4\pi}\int_{-L}^{L}\frac{dx}{r}. \tag{20}$$

Substituting from (19) in (17), we have

$$\frac{dq}{dx} = -\left[i\omega\mu\varepsilon + \frac{\varepsilon Z}{A(x)}\right]I - \frac{A'(x)}{A(g)}q. \tag{21}$$

Assuming that Z remains finite as A(x) approaches infinity, we have

$$\frac{dq}{dx} = -i\omega\mu\varepsilon I. \tag{22}$$

Next we calculate the effect of the second term in (18) on the values of the integrals in (2), assuming that I and q satisfy (22) and that the wire is energized at $x = 0$. Thus we have [10]

$$\begin{aligned} -i\int_{-L}^{L}\frac{\sin\,\beta r}{4\pi r}I(x')dx' &= \frac{1}{4\pi\omega\mu\varepsilon}\int_{-L}^{L}\frac{\sin\,\beta r}{4\pi r}\frac{dq(x')}{dx'}dx' \\ &= \frac{1}{4\pi\omega\mu\varepsilon}\left[q(L)\frac{\sin\,\beta r_1}{r_1} - q(+0)\frac{\sin\,\beta r_0}{r_0} + q(-0)\frac{\sin\,\beta r_0}{r_0}\right. \\ &\quad \left. -q(-L)\frac{\sin\,\beta r_2}{r_2} - \int_{-L}^{L}q(x')\frac{d}{dx'}\left(\frac{\sin\,\beta_r}{r}\right)dx'\right] \end{aligned} \tag{23}$$

where

$$r_0 = \sqrt{a^2 + x^2},\ r_1 = \sqrt{a^2 + (L-x)^2},\ r_2 = \sqrt{a^2 + (L+x)^2}. \tag{24}$$

Similarly, we have

$$\int_{-L}^{L}\frac{\sin\,\beta r}{4\pi r}q(x')dx' = \frac{i}{\omega}\int_{-L}^{L}\frac{\sin\,\beta r}{4\pi r}\frac{dI(x')}{dx'}dx' = \frac{i}{\omega}I(x')\frac{\sin\,\beta r}{4\pi r}\Bigg]_{-L}^{L} \tag{25}$$

In this expression the first term vanishes if the current is equal to zero at the ends of the antenna. Neglecting the last terms (the integrals) in (23) and (25), and substituting in (17), we obtain (neglecting also $A'(x)/A\ (x)$)

$$\begin{aligned} \frac{dq}{dx} &= -i\omega\mu\varepsilon I - \frac{\varepsilon ZI}{A(x)} - \frac{1}{4\pi A(x)}\left[q(L)\frac{\sin\,\beta r_1}{r_1} - q(+0)\frac{\sin\,\beta r_0}{r_0}\right. \\ &\quad \left. + q(-0)\frac{\sin\,\beta r_0}{r_0} - q(-L)\frac{\sin\,\beta r_2}{r_2}\right] \end{aligned} \tag{26}$$

Integrating (20), we have

$$A(x) = \frac{1}{4\pi}\log\frac{\sqrt{a^2 + (L-x)^2} + (L-x)}{\sqrt{a^2 + (L+x)^2} - (L+x)} \equiv \frac{1}{4\pi}\log\frac{4(L^2 - x^2)}{a^2} \tag{27}$$

At the center of the wire and at the ends we have

$$A(0) = \frac{1}{2\pi}\log\frac{2L}{a},\ A(L) = A(-L) = \frac{1}{4\pi}\log\frac{4L}{a}. \tag{28}$$

The average value of $A\ (x)$ along the wire is

$$A(x)]_{av} = \frac{1}{2\pi}\left(\log\frac{2L}{a} - 2 + 2\log 2\right). \tag{29}$$

This average value of $A(x)$ is related to the average capacitance C between two halves of the wire; thus we have very nearly

$$A(x)]_{av} = \frac{\varepsilon}{2C}. \tag{30}$$

Replacing $A(x)$ in (26) by its average value from (30), we have

$$\frac{dq}{dx} = -i\omega\mu\varepsilon I - 2ZCI$$
$$- \frac{iC}{2\pi\varepsilon}\left[q(L)\frac{\sin\beta r_1}{r_1} - q(+0)\frac{\sin\beta r_0}{r_0} + q(-0)\frac{\sin\beta r_0}{r_0} - q(-L)\frac{\sin\beta r_2}{r_2}\right] \tag{31}$$

This approximate equation has been obtained on the assumption that the wire is energized at the center; however, the same expression holds for any position of the generator, in which case x is measured from the generator.

For a centrally disposed generator we have in the first approximation

$$\begin{aligned} I(x) &= I_o \sin\beta(L-x),\ x \geq 0 \\ I(x) &= I_o \sin\beta(L+x),\ x \leq 0 \end{aligned} \tag{32}$$

where I_o is the maximum amplitude of the current. Hence,

$$q(L) = -\frac{i\beta I_0}{\omega} = -i\sqrt{\mu\varepsilon}I_0,\ q(-L) = i\sqrt{\mu\varepsilon}I_0,\ q(+0) = -q(-0)$$
$$= -i\sqrt{\mu\varepsilon}I_0\cos\beta L. \tag{33}$$

Substituting these values in (31) we obtain

$$\frac{dq}{dx} = -i\omega\mu\varepsilon I - 2ZCI - 60CI_0\left[\frac{\sin\beta r_1}{r_1} - 2\frac{\sin\beta r_0}{r_0}\cos\beta L + \frac{\sin\beta r_2}{r_2}\right] \tag{34}$$

Since a is assumed to be small, we can write (34) as follows

$$\frac{dq}{dx} = -i\omega\mu\varepsilon I - 2ZCI$$
$$- 60CI_0\left[\frac{\sin\beta(L-x)}{L-x} - 2\frac{\sin\beta x}{x}\cos\beta L + \frac{\sin\beta(L+x)}{L+x}\right]. \tag{35}$$

It is quite obvious that the last term, representing approximately the effect of radiation on the electric charge and the electric current distribution in the antenna, differs in form from the second term (-2ZCL) representing the effect of dissipation in the wire and the effect of the internal inductance of the wire. The latter term is always proportional to the current in the particular section of the antenna while the radiation term is not, as a rule. If, however, the lengths of the wire is equal to an odd number of half wavelengths, then (35) becomes

$$\frac{dq}{dx} = -[i\omega\mu\varepsilon + (2Z + 2R_{rad})C]I,$$

$$R_{rad}\frac{60L}{L^2-x^2}. \tag{36}$$

In this case, to the extent of our approximations, radiation is represented by a term similar to the dissipation term.

The approximate formula (7) for R_{rad} was also obtained by a different method in an earlier paper [2]. There a point was made of a curious paradox, which seems to us important enough to bear a repetition. Assuming a sinusoidal current distribution ($I_o \sin\beta x$) in an antenna whose length is equal to an integral number of half wavelengths, A. A. Pistolkors [12] obtains the following expression for the electric intensity parallel to the wire at its surface

$$-E_x = 30\left(\frac{1}{x_L}+\frac{1}{2L-x_L}\right)I_0 \sin \beta\bar{x} + 30i\left(\frac{1}{x_L}-\frac{1}{2L-x_L}\right)I_0 \cos \beta\bar{x} \tag{37}$$

where $x_L = L+x$ is the distance from an end of the wire. It is evident that the real part $-E_x/I(x)$ gives (7) for the radiation resistance. This cannot be regarded as anything more than a coincidence. If we were fortunate enough (or, perhaps, unfortunate) to know the exact current distribution in the antenna, then we should have found that E_x and, hence, $-E_x/I(x)$ are equal to 0 everywhere on the surface of the wire for the simple reason that $E_x = O$ is the boundary condition from which the correct current distribution must be determined. In an antenna energized at a point there is really no relation between R_{rad} and E_x.

[1] P. O. Pedersen, "Radiation from a vertical antenna over flat perfectly conducting earth," *Ingeni&rvidenskabelige Skrifter*, ser.A, no. 38, 1935.

[2] 2 S. A. Schelkunoff, "Theory of antennas of arbitrary size and shape," *PROC. I.R.E.*, vol. 29, pp. 493–521; September, 1941.

[3] John R. Carson, "The guided and radiated energy in wire transmission," *Jour. A.I.E.E.*, pp. 906–913; October, 1924.

[4] The space with wires embedded in it is replaced by space without wires.

[5] S. A. Schelkunoff, "Some equivalence theorems of electromagnetics and their application to radiation problems," *Bell Sys. Tech. Jour.*, pp. 92–112; January, 1936.

[6] S. A. Schelkunoff, "On different fraction and radiation of electromagnetic waves," *Phys. Rev.*, pp. 308–316; August 15, 1939.

[7] S. A. Schelkunoff, "A general radiation formula," *PROC. I.R.E.*, vol. 27, pp. 660–666; October, 1939.

[8] J. A. Stratton, *Electromagnetic Theory,* McGraw-Hill Book Company, New York, N. Y. 1941, pp. 464–470.

[9] In the present case the vector potential is parallel to the wire.

[10] If the wire is energized at $x = 0$, $q(+O)$ is not equal to $q(0)$.

[11] In fact, to any number of half wavelengths if the generator is located at a current antinode.

[12] A. A. Pistolkors, "The radiation resistance of beam antennas," *PROC. I.R.E.*, vol. 17, pp. 562–579; March, 1929.

12.1 Discussion and extension

In the above discussion, Schelkunoff and Feldman (S–F) use the transmission-line theory to develop a radiation model for the SCF. Their analysis involves a possibly unusual concept, that of a position-dependent radiation resistance per unit length. They obtain experimental results for an elevated thin wire from which they deduce, after correcting for the resistive loss of the wire, an attenuation rate that is attributed to radiation loss. They derive a general expression for the distributed radiation resistance $R_{rad}(x)$ as

$$2R_{rad}^{(SF)}(x)CI = 60CI_o\left[\frac{\sin\,\beta(L-x)}{L-x} - 2\frac{\sin\,\beta x}{x}\cos\,\beta L + \frac{\sin\,\beta(L+x)}{L+x}\right] \tag{12.1a}$$

or

$$R_{rad}^{(SF)}(x) = 30\frac{I_o}{I}\left[\frac{\sin\,\beta(L-x)}{L-x} - 2\frac{\sin\,\beta x}{x}\cos\,\beta L + \frac{\sin\,\beta(L+x)}{L+x}\right] \tag{12.1b}$$

with L being the half-length of the current. Note that the $\cos\beta L$ term accounts for the current-slope discontinuity at the origin of a center-fed sinusoidal current filament. Since

$$I(x) = I_o\,\sin\,\beta(L-x), x \geq 0, \text{ and } I(x) = I_o\,\sin\,\beta(L+x), x \leq 0 \tag{12.2}$$

then

$$R_{rad}^{(SF)}(x) = 30\frac{1}{\sin\,\beta(L-x)}\left[\frac{\sin\,\beta(L-x)}{L-x} - 2\frac{\sin\,\beta x}{x}\cos\,\beta L + \frac{\sin\,\beta(L+x)}{L+x}\right].x \geq 0 \tag{12.3a}$$

and

$$R_{rad}^{(SF)}(x) = 30\frac{1}{\sin\,\beta(L+x)}\left[\frac{\sin\,\beta(L-x)}{L-x} - 2\frac{\sin\,\beta x}{x}\cos\,\beta L + \frac{\sin\,\beta(L+x)}{L+x}\right].x \leq 0. \tag{12.3b}$$

Their experimental results are found to agree to within 10% or so of their analytical expression when specialized to the special case of a current filament whose half-length L is an odd number of half wavelengths long as given by

$$R_{rad}^{(SF)}(x) = \frac{60^L}{(L^2 - x^2)}. \tag{7}$$

When evaluated for a range of values of L, plots of $R_{rad}^{(SF)}(x)$ from (12.4) are given in Figure 12.1 for SCFs of 9–10 wavelengths long. There is a dramatic

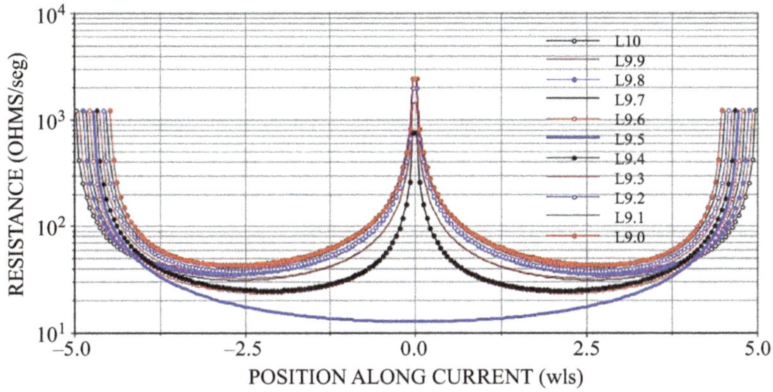

Figure 12.1 The S–F distributed radiation resistance for sinusoidal currents of variable lengths

difference between these results upon approaching the center feedpoint of the SCF due to the difference in the current-slope discontinuity there, which is zero only for the special case of SCF filaments an odd number of half-wavelengths long.

It is interesting that Schelkunoff and Feldman did not use the expression they derived for the per-unit-length radiation resistance to obtain radiated linear-power density (LPD) of the SCF. The LPD for the SCF from the S–F theory, denoted here as $p_{rad}^{(SF)}(x)$, is simply given by

$$p_{rad}^{(SF)}(x) = \frac{1}{2} I^2(x) R_{rad}^{(SF)}(x)$$
$$= 15 I(x) \left[\frac{\sin \beta(L-x)}{L-x} - 2\frac{\sin \beta x}{x} \cos \beta L + \frac{\sin \beta(L+x)}{L+x} \right] \quad (12.4)$$

with the total radiated power then obtained from

$$P_{rad}^{(SF)} = \int_{-L}^{L} p_{rad}^{(SF)}(x) dx = \sum_{m=1}^{N} p_{rad}^{(SF)}(x_m) \Delta x \quad (12.5)$$

where $p_{rad}^{(SF)}(x_m)$ is the LPD at the center of segment m. Results from the S-F model for the LPDs are compared with those from the FARS approaches for 9.3, 9.5, 9.7, and 10-wavelength SCFs in Figures 12.2–12.5. The LPDs for a 10-wavelength NEC dipole are compared with those for a SCF in Figure 12.6. The agreement between the S–F results using the distributed radiation resistance and the IEMF method are within a few percent for these examples, providing mutual validation of these various approaches.

The concept of a distributed radiation resistance developed in the S–F approach seems logically extendible for implementation with FARS, using

$$R_{rad}^{(FARS)}(x) = \frac{2P^{(FARS)}(x)}{I^2(x)}. \quad (12.7)$$

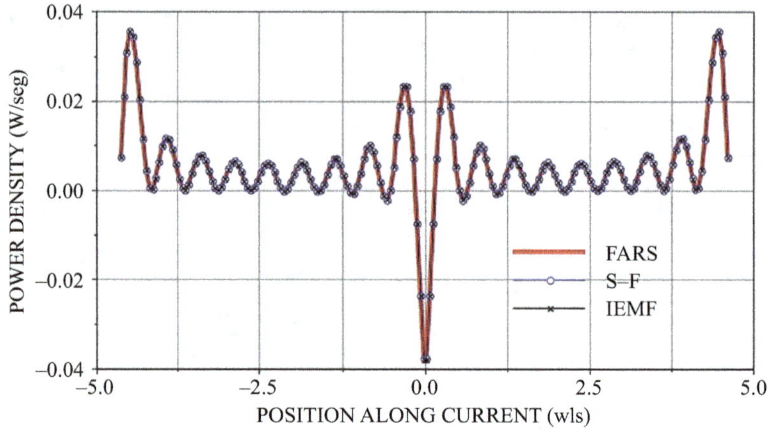

Figure 12.2 The linear power densities for a 9.3-wavelength sinusoidal current obtained using FARS, the S–F distributed radiation resistance and the IEMF method

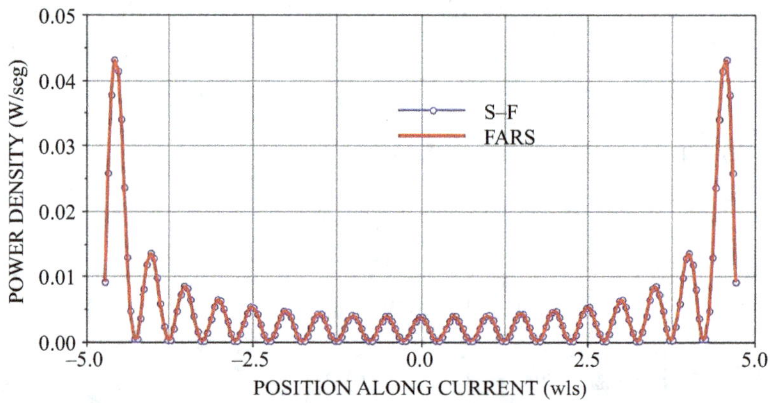

Figure 12.3 The linear power densities for a 9.5-wavelength sinusoidal current obtained using the S–F distributed radiation resistance and FARS.

A potential difficulty in using (12.7) is that both $P^{(FARS)}(x)$ and $I^2(x)$ become small together so that their ratio is likely to be numerically noisy as illustrated in Figure 12.7 for a 10-wavelength NEC dipole. An alternate way of estimating $R_{rad}^{(FARS)}(x)$ is to filter, or average, these results. Examples of using averaging windows of varying widths are shown in Figures 12.8–12.10.

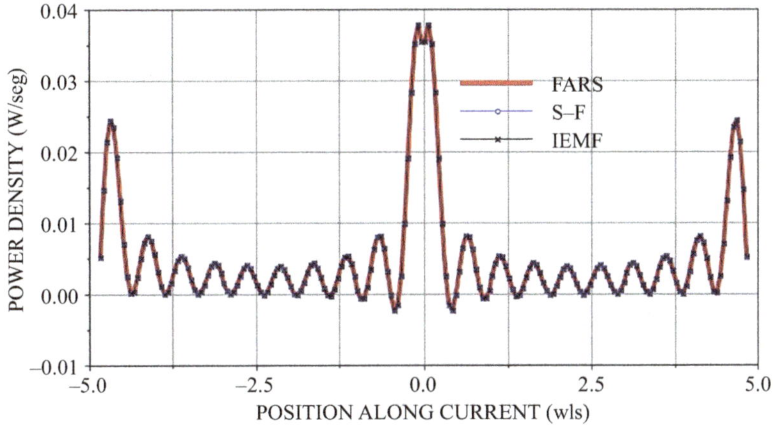

Figure 12.4 *The linear power densities for a 9.7-wavelength sinusoidal current obtained using FARS, the S–F distributed radiation resistance and the IEMF method*

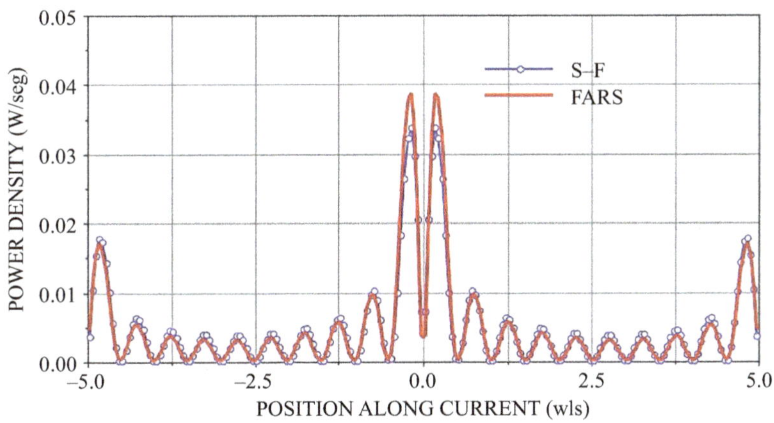

Figure 12.5 *The linear power densities for a 10-wavelength sinusoidal current obtained using the S–F distributed radiation resistance and FARS*

12.2 Summary

This chapter begins with the 1942 article by Schelkunoff and Feldman "On Radiation From Antennas" in which the concept of a distributed radiation resistance for the sinusoidal current filament (SCF) was introduced. The general formula they developed for this quantity was evaluated explicitly in their article only for a current an odd number of half-wavelengths long. Their formula was then

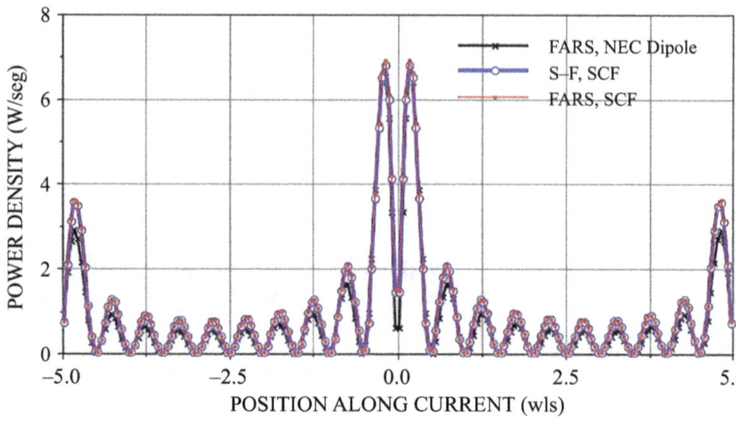

Figure 12.6 The linear power densities for a 1-A 10-wavelength NEC dipole, the S–F distributed radiation resistance and FARS for a sinusoidal current

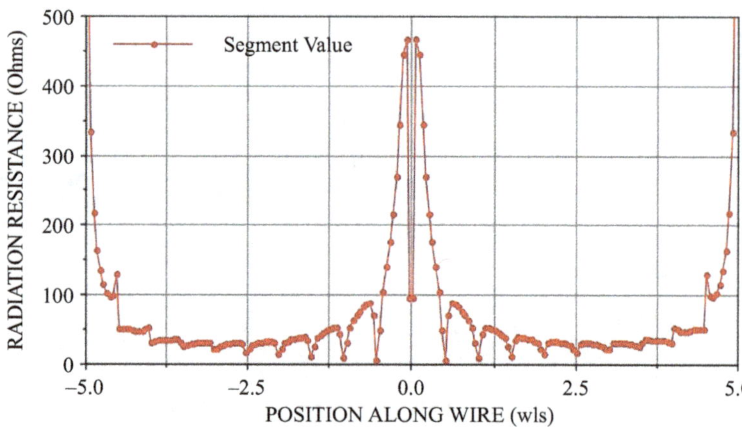

Figure 12.7 The distributed radiation resistance for a 10-wavelength NEC dipole obtained using 12.7

generalized for arbitrary lengths and upon using $\frac{1}{2}I^2(x)R_{rad}^{(SF)}(x)$ to compute the distribution of radiation from a SCF for comparison with FARS. The results agree to within a percent or so to validate both and to also validate the concept of a distributed radiation resistance. It is worth noting that the position-dependence of radiated power found between FARS and the distributed radiation resistance involves the agreement of results obtained from a far-field quantity for the former and a near-field quantity for the latter.

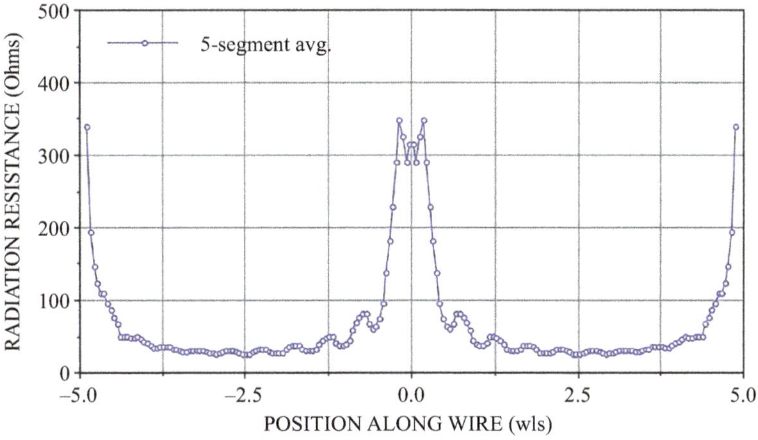

Figure 12.8 The distributed radiation resistance for a 10-wavelength NEC dipole obtained using 12.7 averaged over a 5-segment interval

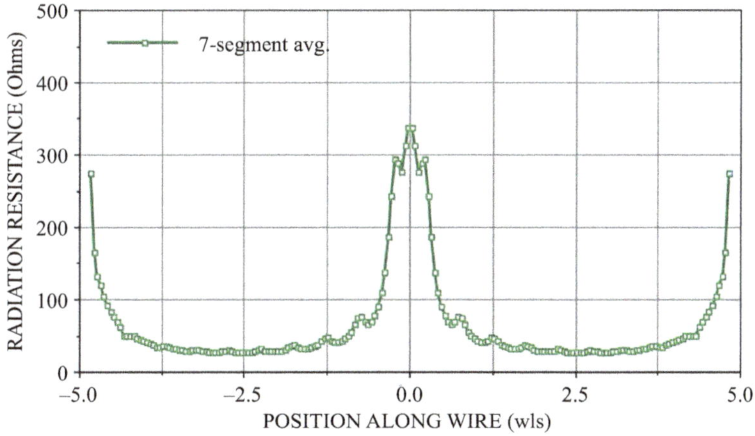

Figure 12.9 The distributed radiation resistance for a 10-wavelength NEC dipole obtained using 12.7 averaged over a 7-segment interval

Also investigated was the feasibility of numerically determining $R_{rad,SF}(x)$ from FARS results. This is analytically viable but numerically challenging because it can involve the quotient of two nearly zero quantities along the current and to consequently produce "noisy" results. Averaging or filtering of the noisy results can improve the outcome.

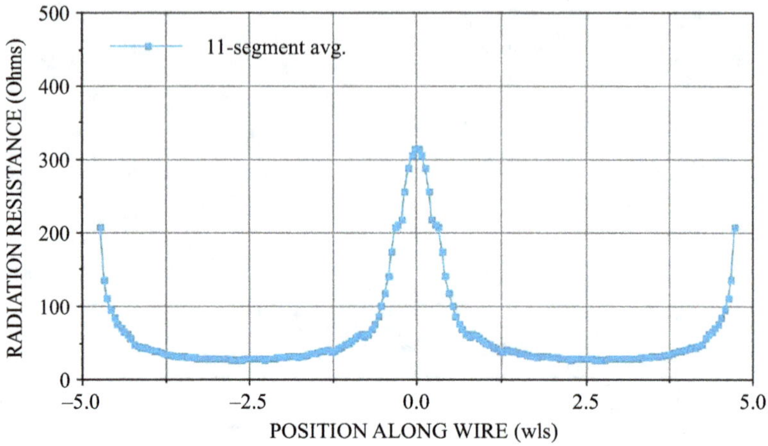

Figure 12.10 The distributed radiation resistance for a 10-wavelength NEC dipole obtained using 12.7 averaged over an 11-segment interval

Chapter 13

Radiation from a near-zero-radius dipole and a sinusoidal current filament

Abstract

The electric field of a sinusoidal current filament (SCF) is one of the relatively few problems in electromagnetics whose solution can be developed in closed form. This is why the SCF was routinely used to approximate the radiation characteristics of the thin-wire, center-fed, dipole antenna. This proved to be useful because the current on a perfect electric conductor (PEC) wire of near-zero radius (NZR) closely approximates a sinusoidal current as does its radiation patterns.

However, while the SCF and NZR dipole are very similar in some respects, they are quite different in others as will be show here. These similarities and differences are explored below and how these two sources actually produce a far radiation field is explained.

Keywords: Dipole; Distributed radiation resistance; Near-zero-radius dipole; Power density; Radiation; Sinusoidal current filament

13.1 Introduction

A comparison of the electromagnetic behavior of a PEC dipole as its radius approaches zero is compared in this chapter with the characteristics of the SCF. The Numerical Electromagnetics Code (NEC) [1] is used to obtain results for a PEC dipole of radius, a, varying from 10^{-2} to 10^{-20} wavelengths, where the latter case is referred to throughout as being of "near-zero radius" (NZR). A segment size of 0.05 wavelengths is used for all calculations presented here, with the two center segments of the dipole excited as the antenna source. The relevant quantities of the SCF [2] were evaluated using their closed-form analytical expressions. Unless otherwise indicated, all results shown are for a maximum current magnitudes of 1 A. Ways in which the SCF and NZR dipole are found to be similar include their:

- variation of the total radiated power as a function of current length;
- radiation patterns;
- linear power density (LPD) [3] or length-wise distributions of radiated power;
- the distributed radiation resistance of Schelkunoff–Feldman [4].

On the other hand, ways in which they differ include their:

- slightly, but significantly different, current distributions;
- parallel Poynting vectors along their currents;
- spatial distribution of their tangential electric fields;
- consequent distributions of induced electromotive force (IEMF) power.

These two sets of results are first discussed in Sections 13.2 and 13.3, with a more detailed examination of their different radiation processes in Section 13.4. The chapter concludes with a brief summary in Section 13.5.

13.2 Ways in which the SCF and PEC dipole are similar

13.2.1 Total radiated power as a function of current length

The total power radiated by a 10-wavelength PEC dipole having a maximum current magnitude of 1-A as a function of the dipole radius, *a/wavelength*, is shown in Figure 13.1. The total power radiated by a 10 wavelength SCF given by

$$I(z) = I_o \sin\left[k\left(\frac{L}{2} - |z|\right)\right], -\frac{L}{2} \le z \le \frac{L}{2} \tag{13.1}$$

is also shown for comparison where for all SCF results that follow $I_o = 1$-A, L is its length and $k = 2\pi/\lambda$ is the free-space wave number. A discontinuity in the current slope can exist at the effective "feedpoint" which for this particular SCF is at $z = 0$. For this discussion, the feedpoint will only be located at $z = 0$ but could be located elsewhere using different lengths for the positive and negative halves of the SCF. The far E-field and radiated power for the current of (13.1) can be expressed as [2]

$$E_\theta(\theta) = j\eta\frac{I_o e^{-jkr}}{2\pi r}\left[\frac{\cos\left(\frac{kL}{2}\cos\theta\right) - \cos\left(\frac{kL}{2}\right)}{\sin\theta}\right] \tag{13.2a}$$

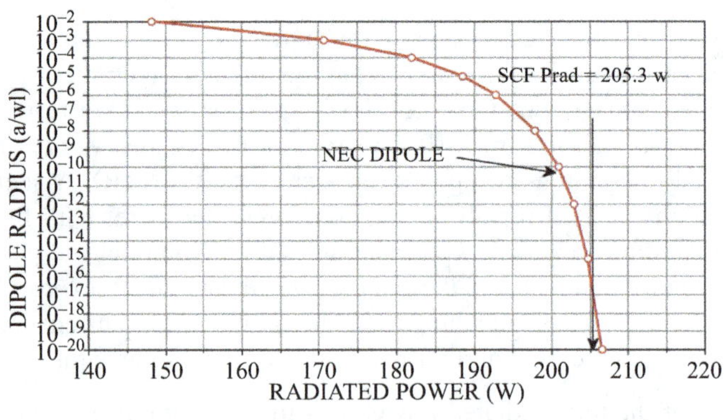

Figure 13.1 Total power radiated by a 10-wavelength PEC dipole as a function of the wire radius a/wavelength, the maximum value of which is 206.6 W, where the arrow indicates the power radiated by a 10-wavelength SCF

and

$$P_{rad} = \eta \frac{|I_o|^2}{4\pi} \left[C + \ln(kL) - C_i(kL) + \frac{1}{2} \sin(kL)[S_i(2kL) - 2S_i(kL)] \right.$$
$$\left. + \frac{1}{2} \cos(kL) \left[C + \ln\left(\frac{kL}{2}\right) + C_i(2kL) - 2C_i(kL) \right] \right] \tag{13.2b}$$

respectively, where the quantities $C_i(x)$ and $S_i(x)$ are the cosine and sine integrals and C is Euler's constant. As the dipole radius in Figure 13.1 decreases toward 0 the PEC-dipole power asymptotically approaches (from below) to within ~0.5% the power radiated by the SCF.

Plotted in Figure 13.2 are the total powers radiated by The NZR dipole and SCF as a function of length where they agree to within a few percent over a length variation of 0.2–10 wavelengths. They exhibit periodic maxima in the radiated power for $L = n(\lambda/2)$ with n even, and minima occur for n odd. Possibly more significant is the monotonic increase in the minima and maxima power levels with increasing current length for which an explanation is included in Section 13.4.

It may not be surprising that the radiated power of a dipole of vanishing radius asymptotically approaches that of the SCF. Since various authors [5–7] have previously mentioned that the current on a thin wire of vanishing radius approaches a sinusoid, it follows that the power radiated by the NZR would approach that of the SCF. It will be seen in Section 13.2, however, that the situation is not quite as simple as the results of Figures 13.1 and 13.2 might imply.

13.2.2 Radiation pattern

The comparison of the radiation patterns shown in Figure 13.3 for the NZR 10-wavelength dipole and SCF are also found to agree within a few percent. This result is consistent with the LPDs (or power density) for these same cases that follow in Figure 13.4.

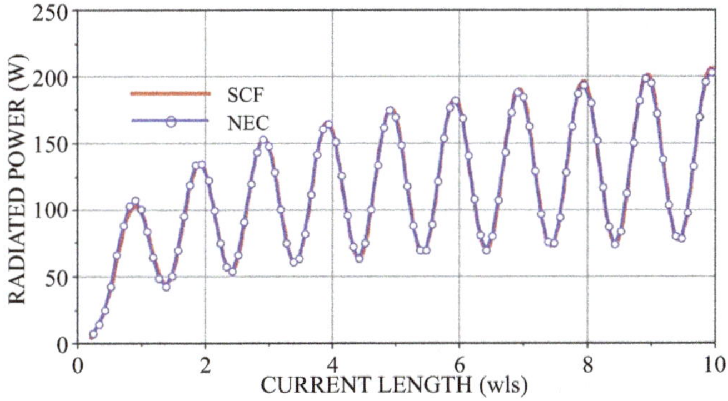

Figure 13.2 The power radiated by a NZR dipole of radius 10^{-20} wavelengths and a SCF, both having a 1-A maximum current, as a function of length

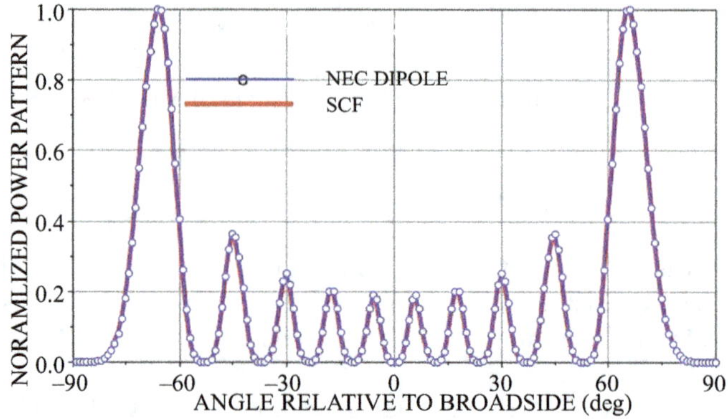

Figure 13.3 *The radiation patterns of a NZR 10-wavelength PEC dipole and a SCF*

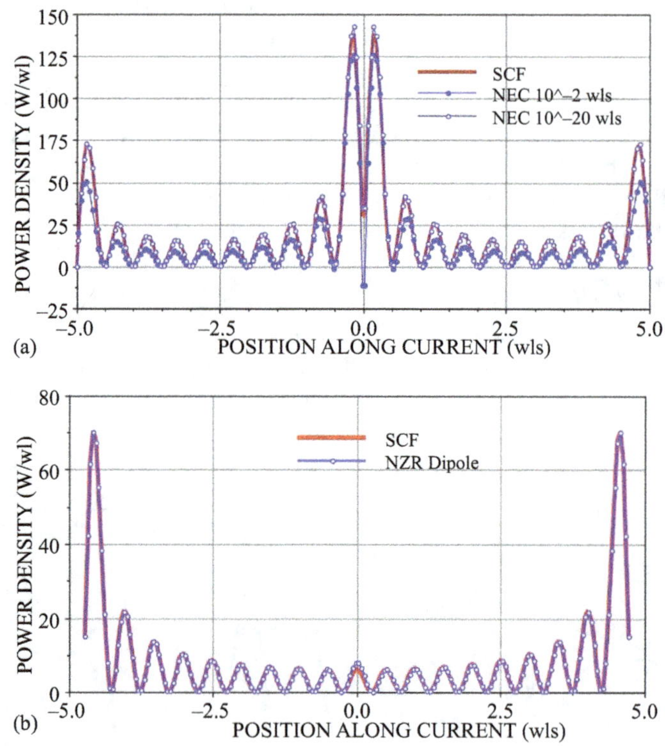

Figure 13.4 *The FARS LPD vs. position for NEC dipoles with* $a/\lambda = 10^{-2}$ *and* 10^{-20} *(the NZR dipole) and the SCF, with* $L/\lambda = 10$ *(a) and the NZR and SCF with* $L/\lambda = 9.5$ *(b), all having a maximum current magnitude of 1 A. The total radiated powers for the three cases in (a) are 148.2, 206.6, and 205.3 W, and in (b) are 81.2 and 81.7, respectively*

13.2.3 Linear power density

The approach taken in the discussion that follows is that the power or energy radiated by a PEC object originates from over its surface in a manner determined by the space and time dependence of the charge and associated current distributions on it, source quantities that are determined by how that object is excited. All radiated power is due to that excitation and the charge acceleration that it causes, whether from an antenna that is excited locally, or from a scatterer that is excited globally. A tool called far-field analysis of radiation sources (FARS) [3] determines the quantitative, per-unit-length or per-unit-area, distribution of radiated power from the object of interest. For a wire-like object such as considered here, this is referred to as the LPD. See Chapter 8 for further discussion and examples about FARS.

Illustrated in Chapter 2 was the fact that the Lienard–Wiechert (L–W) potentials [8] demonstrate analytically that $1/r$ electric and magnetic (or radiation) fields are due only to accelerated charge. The kinks in the E-field lines in the Electric-Field Kink Model comprise the radiation field caused by charge acceleration [9]. For a PEC object and where there is no space (or free) charge, the charge distribution caused by the applied excitation resides on that object's surface. This means that the radiation it emits must therefore originate from its surface, a conclusion consistent with both the L–W potentials and the E-field kink model as well as with the results predicted by FARS.

LPD results for a 10-wavelength NZR dipole and SCF are shown in Figure 13.4a. For comparison results are included for a dipole 10^{-2} wavelengths in radius. Radiation occurs near the feedpoint and ends of these current distributions, and an intermediate radiation along their lengths as a series of lobes. The LPDs of the NZR and SCF are graphically indistinguishable, while that of the 10^{-2} wavelength-radius dipole is less than them all along the dipole's length. The integrated FARS results for the SCF and NZR agree to within a percent at 205.3 and 206.6 W, respectively. Additional results are presented in Figure 13.4b for a 9.5 wavelength NZR dipole and SCF showing that they both lack the feedpoint maximum exhibited by their 10-wavelength examples. Their total radiated powers again agree to within a percent but differ from each other in that the NZR exhibits a slight increase around the feedpoint that is absent for the SCF.

13.2.4 Schelkunoff–Feldman distributed radiation resistance

The SCF was analyzed from quite a different perspective by Schelkunoff and Feldman [4] as discussed in Chapter 12. They introduced the novel idea of a spatially distributed radiation resistance $R_{rad,S\text{-}F}(s')$ for the SCF. In particular, they derived an expression for the spatial distribution of the SCF's radiation resistance for an SCF whose length $L = 2h$, is an odd number of half-wavelengths. Their derivation can be generalized to an SCF of arbitrary length to yield

$$R_{rad}^{(SF)}(s') = \frac{30}{I(s')}\left[\frac{\sin\left[k(h-s')\right]}{h-s'} - 2\frac{\sin\left(ks'\right)}{s;}\cos\left(kh\right) + \frac{\sin\left[k(h+s')\right]}{h+s'}\right]. \quad (13.3a)$$

For the special cases of an SCF an odd number (the original S–F result) or even number of half wavelengths long, this simplifies to

$$R_{rad}^{(SF)}(s') = \frac{60h}{h^2 - s'^2}, 2h = n\frac{\lambda}{2}, n = 1, 3, 5, \ldots, \tag{13.3b}$$

$$R_{rad}^{(SF)}(s') = \frac{60h^2}{(h^2 - s'^2)|s'|}, 2h = n\frac{\lambda}{2}, n = 2, 4, 6, \ldots, \tag{13.3c}$$

An independent result for the LPD of the SCF can be obtained from the Schelkunoff–Feldman $R_{rad}^{(SF)}(s')$ using (13.3) as

$$LPD^{(SF)} = I^2(s')R_{rad}^{(SF)}(s')/2. \tag{13.4a}$$

As shown in Figure 13.5, the FARS and S–F results from (13.4a) for the LPD of the 1-A SCF are graphically indistinguishable. Assuming that the S–F approach itself is correct this provides a further confirmation that FARS yields the spatial distribution of radiation from a SCF. The distributed power for the SCF as obtained from the IEMF method is also included in Figure 13.5 and displays an agreement with FARS comparable to that from the S–F distributed radiation resistance.

Using the LPD results for a NEC model as given by FARS, this process can be reversed to provide an alternative to estimate the NEC, or PEC-dipole, distributed radiation resistance from

$$R_{rad}^{(NEC)}(s') = 2LPD^{(NEC)}(s')/I^2(s'). \tag{13.4b}$$

Some $R_{rad}^{(NEC)}(s')$ results are presented in Figure 13.6 derived from (13.4b) for a NZR 10-wavelength dipole for comparison with the SCF. In order to smooth the numerical artifact that can occur near the coincident minima of the LPDs and current values, the $R_{rad}^{(NEC)}(s')$ values of Figure 13.6a are averaged over spatial intervals of various numbers of segments. For increased clarity, the S–F formulation for the SCF and the 11-segment average for the NZR dipole are repeated in

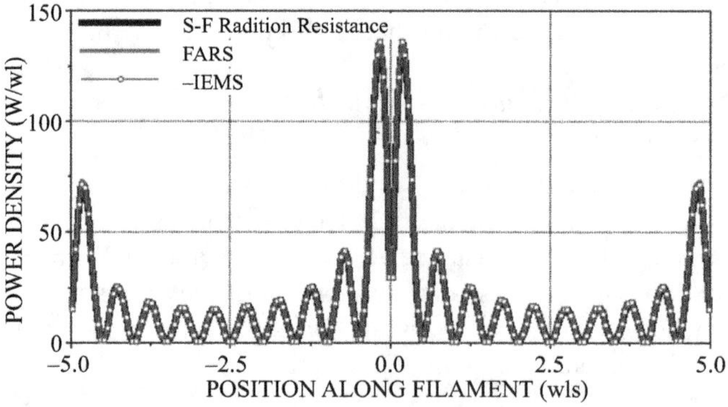

Figure 13.5 The LPDs for a SCF as obtained using FARS, the S–F distributed radiation resistance and the IEMF method

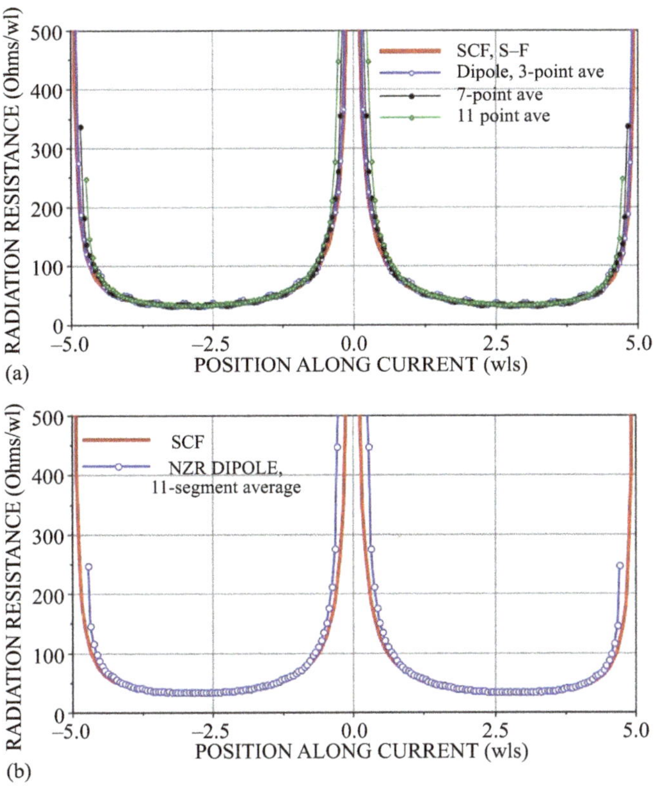

Figure 13.6 Distributed radiation resistance for the SCF from Schelkunoff and Feldman and the FARS-derived LPD values, averaged over various numbers of segments for the NZR 10-wavelength dipole (a) and for clarity the SCF results compared with only the 11-sgment average for the NZR dipole (b)

Figure 13.6b. The distributed radiation resistances of the NZR and S–F agree to within a few percent except near the ends and center where smoothing shifts the peaks. The distributed radiation resistance for a more general geometry would likely depend on its geometry and source distribution.

13.3 Ways in which the SCF and PEC dipole are different

13.3.1 Current distributions

The normalized current of the SCF is compared with that of the 10-wavelength NZR dipole in Figure 13.7. The imaginary component of dipole current is essentially identical to that of the SCF in Figure 13.7a. However, in Figure 13.7b, the NZR dipole has a small real component for which there is no SCF counterpart. The

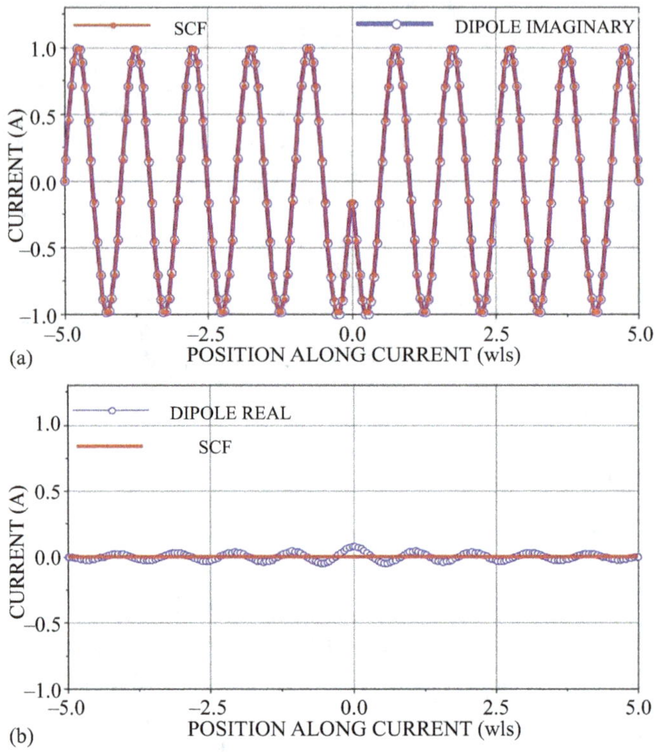

Figure 13.7 Imaginary (a) and real (b) current distributions on a SCF and NZR dipole

other properties that the SCF and NZR dipole display as summarized next are obviously due to the difference between their currents.

13.3.2 Time-average Poynting vectors parallel to the currents

The normalized value of the quantity Re[Q(s')I*(s')]/2 obtained along the NZR dipole is presented in Figure 13.8 [10]. This quantity, proportional to the time-average Poynting vector (TAPV) at the surface of the wire, is equivalent to computing on-surface Poynting vector from the near E- and H-fields. The TAPV for the SCF (nominally computed at a radius of 10^{-3} wavelengths) is identically zero, since it is a standing wave of constant maximum magnitude. The TAPV for the NZR dipole on the other hand decreases monotonically away from the feedpoint toward the dipole ends. It is positive on the right-hand side of the dipole and negative on the left since the power flows away from the center feedpoint for the NZR. This decrease with distance from the feedpoint in the dipole TAPV is caused by "shedding" of power from the dipole current and charge distributions due to radiation, a point that is discussed in more detail in Chapter 6.

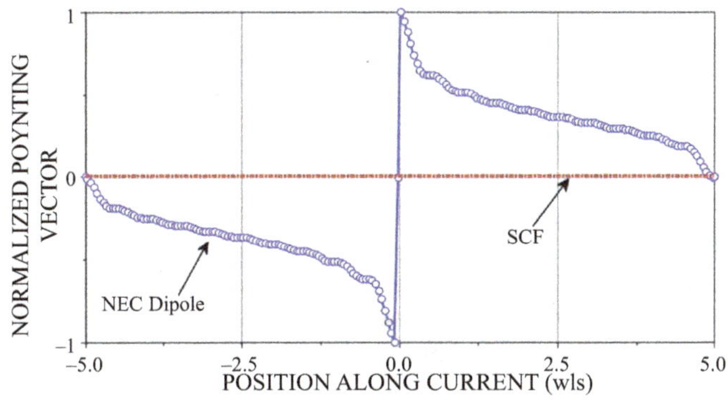

Figure 13.8 The normalized TAPV(s') at the surface of the NZR dipole and 10^{-3} wavelengths away from a SCF

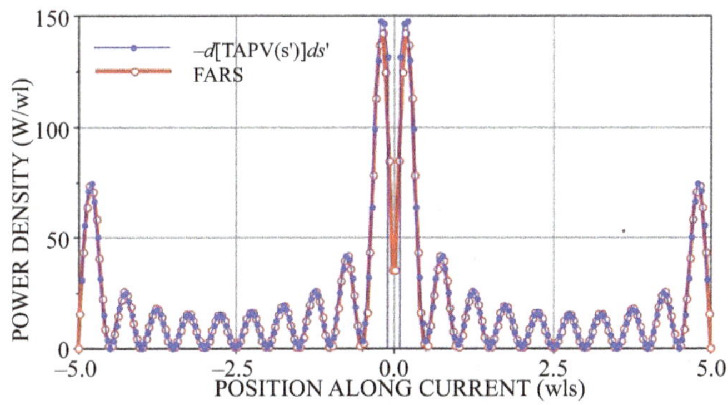

Figure 13.9 The quantity –d*[TAPV(s')]*ds' *normalized to the ends peaks of the LPD, and the LPD for the NZR dipole*

As previously demonstrated in Chapter 6, the spatial distribution of radiated power emitted by the NZR can be estimated by differentiating the result of Figure 13.8 with respect to distance. The quantity DTAPV = $-d$(TAPV)/dz, normalized to the end peaks of the LPD for the 1 A, NZR dipole, is plotted in Figure 13.9 together with the FARS result. Except in the vicinity of the feedpoint, these two results agree to within a few percent. This feedpoint difference occurs because the TAPV explicitly includes the effect of the source field while the LPD results, including only the radiated power, do not directly exhibit any source effect. This result for a straight dipole provides additional confirmation for the mutual validation of FARS and the TAPV. However, as also discussed in Chapter 6, the DTAPV and the LPD can exhibit larger differences for other wire objects, such as a circular loop. Never the less, it is intriguing to find that such a straightforward quantity as the DTAPV can apparently show where radiation originates from on a dipole.

13.3.3 Tangential electric fields

Plotted in Figure 13.10 are the tangential electric fields at the surface of the NZR dipole, and 10^{-3} wavelengths away from the SCF. The tangential electric field of the SCF is essentially independent of radius for radii smaller than about 10^{-2} wavelengths. It exhibits a series of minima as a function of position, but is otherwise non zero as it is not subject to a boundary condition. The dipole tangential electric field generally varies in a numerically noisy way and is effectively zero everywhere, except within the source region. This difference between the two tangential E-fields is expected, of course, since the dipole current has been obtained as a solution to a boundary-value problem.

13.3.4 Distributions of induced EMF power

The IEMF power flow at the surface of the NZR dipole and 10^{-3} wavelengths away from the SCF is shown in Figure 13.11. Because the tangential electric field on the

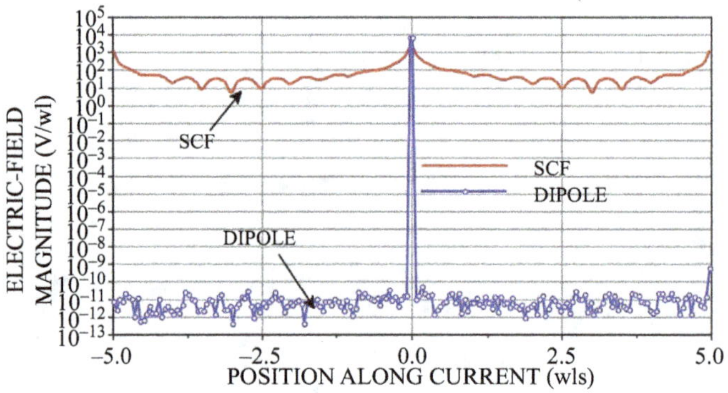

Figure 13.10 The magnitude of the tangential electric field at the surface of the NZR dipole (o) and 10^{-3} wavelengths away from a SCF (——)

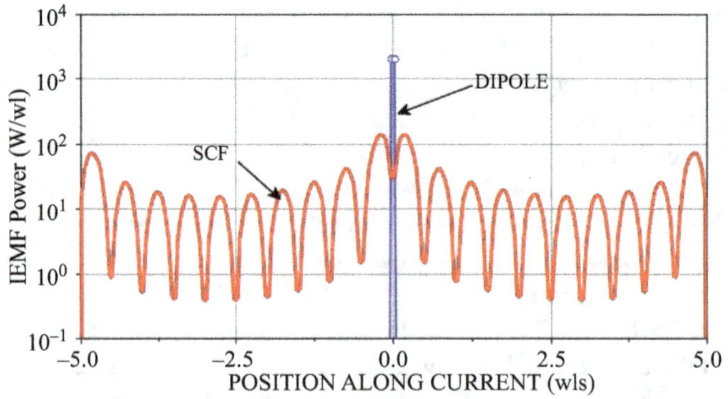

Figure 13.11 The distributions of the IEMF power for the NZR dipole (o) and 10^{-3} wavelengths away from a SCF (——)

dipole is effectively zero outside the feedpoint, the radial power on its surface is necessarily limited to the source region. The radial power for the SCF, on the other hand, is distributed all along the filament, and is essentially identical to its FARS results as shown in Figure 13.5. Observe that the standing-wave SCF has a zero time-average power flow along its length, the IEMF input power must match the radiated power. Since a radiating current cannot exist without an input power that balances the power lost via radiation, this means that IEMF power for the SCF represents the distribution of input power as well that is needed to maintain the current itself.

13.4 The different radiation mechanisms for the near-zero-radius dipole and the sinusoidal current filament

Given the similarities and differences exhibited above for the NZR dipole and SCF, it is appropriate to ask what these might imply about how these two different sources actually radiate power? Considering that for both cases, charge acceleration is the controlling factor, there must be some equivalence between them to yield such similar distributions of radiated power as revealed by FARS. Noting their differences in longitudinal power flow, IEMF power and current distribution, it is reasonable to expect that their respective charge accelerations might have some fundamentally different causes. Discussions by several authors [11–17] concerning the physics of electromagnetic radiation provide various perspectives about this subject from what follows. Concerning the SCF in particular as well as the wire dipole there is generally agreement that radiation originates from the ends and feedpoint. Whether radiation occurs from elsewhere is sometimes questioned. Results presented in this chapter and elsewhere in this book show that charge acceleration, and consequently radiation, originates from all along these objects.

13.4.1 Radiation from the sinusoidal current filament

Consider again the plot of the radiated power as a function of the SCF length in Figure 13.2. Periodic maxima and minima of slowly increasing magnitudes occur with increasing current length L. The maxima occur when $L = n\lambda$ with n an integer and the minima alternately occur when $L = (n + ½)\lambda$, i.e., when the SCF length is an even or odd number of half wavelengths, respectively. The SCF is even symmetric about the center (equivalently the feedpoint) of the SCF when $L = (n + ½)\lambda$, with zero slope discontinuity. But as L varies from $(n + ½)\lambda$ toward $(n \pm 1)\lambda$, its center-current slope varies to reach a maximum value of ± 1. This produces an alternating charge doublet of varying magnitude and sign at the feedpoint.

Referring to Figure 13.4a, the FARS radiated power near the center feedpoint is also a doublet for the even-half-wavelength case. Conversely, a power minimum occurs when the charge is zero at the feedpoint as in Figure 13.4b for an odd-half-wavelength SCF. These two effects demonstrate the charge influence on feedpoint radiation. The monotonic increase in the minima and maxima levels and the average radiated power with increasing current length in Figure 13.2 is due to the increasing number of radiation lobes exhibited by the FARS results in Figure 13.4. This result is consistent with the ln(kL) terms in (13.1c), and implies that radiation is taking place along the increasing length of the SCF. Radiation also occurs from the ends of the SCF where the outgoing propagating charge and associated current are reflected. All of these phenomena are associated with accelerated charge.

A more detailed display of the current and charge on a SCF between 0.5 and 1.0 wavelength long is displayed in Figure 13.12, where L changes from an odd to an even number of half wavelengths in extent. The current and charge are smoothly continuous for $L = 0.5$ but exhibit increasing discontinuities at the feedpoint with increasing L, an effect that maximizes at $L = 1.0$. Thus, the varying charge doublet at the feedpoint causes a corresponding behavior in FARS and is the cause of the oscillation in the total radiated power with increasing L. Some additional aspects of the FARS and current length dependencies can be found in Chapter 10.

The induced tangential electric field along the SCF plotted in Figure 13.10 provides the accelerating force to the SCF charge. This tangential electric field maintains the assumed unit-amplitude sinusoidal current distribution. It also

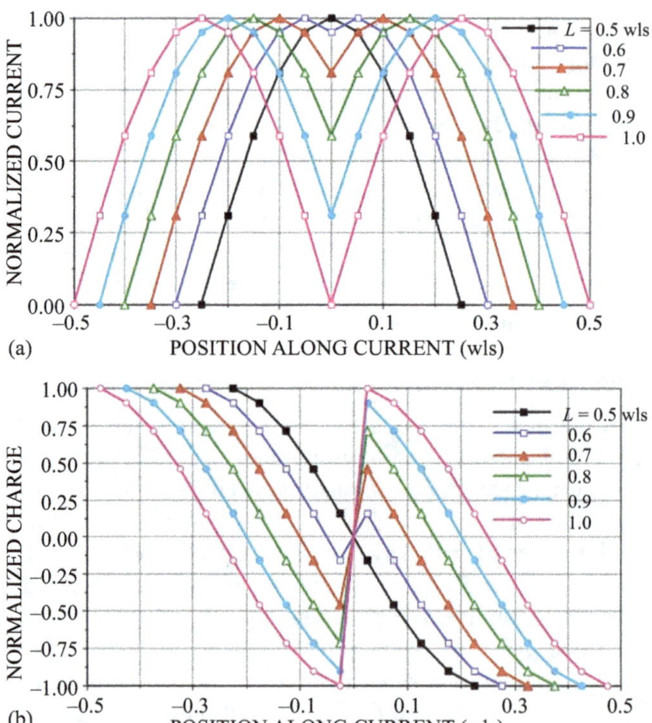

Figure 13.12 Normalized sinusoidal current filaments varying from 0.5 to 1.0 wavelengths long (a) and the corresponding normalized charge distributions (b)

provides the input power to balance the loss caused by radiation as demonstrated by the IEMF results in Figure 13.11. Although the tangential electric field of the SCF is entirely different from the zero total field of the PEC NZR dipole, their spatial distributions of radiation are very similar as demonstrated in Figure 13.4.

13.4.2 Radiation from the PEC dipole

As previously mentioned, there is zero tangential electric field outside the source region due to the NZR dipole current in contrast with the SCF. But since their FARS results are essentially identical the dipole charge must experience a spatial acceleration similar to that implied by the induced tangential electric field of the SCF. It would seem that the two effects of feedpoint and end-reflection acceleration on the NZR charge are similar to those to which the SCF is subject.

However, the FARS radiation results along the length of the dipole must have a different explanation. It is perhaps more revealing first to refer to time-domain results obtained using TWTD in seeking the cause of the length-wise dipole charge acceleration for time-harmonic excitation, as is being considered here. A center-fed, impulsively excited (Gaussian in time) dipole antenna exhibits two outward-propagating charge pulses of opposite sign [16] as demonstrated by Miller and Burke [18] and elsewhere. These opposite-signed charge pulses propagating in opposite directions constitute same-sign currents. When these charge-density, or Q, pulses are multiplied by light speed c, they closely replicate the current, or I pulses, as shown in Figure 13.13. The I and Qc pulses differ most in the "undershoot" or trailing parts behind the peaks as discussed in [18]. In other words, the effective charge speed along the wire is effectively the speed of light. Observe also that the amount of charge in the outward-propagating pulses decreases with distance from the feedpoint since the charge peaks and their areas reduce monotonically.

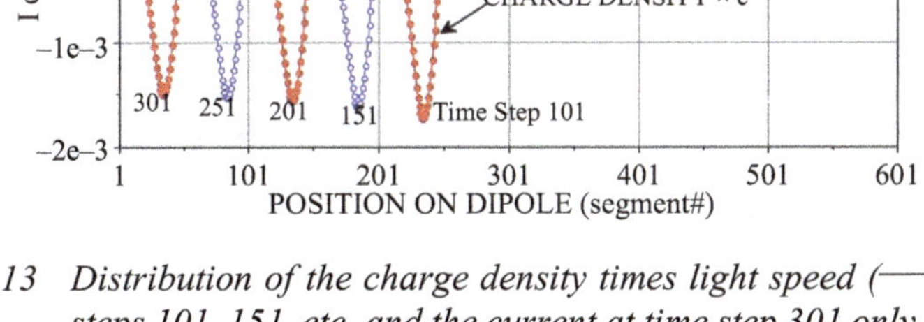

*Figure 13.13 Distribution of the charge density times light speed (——) at time steps 101, 151, etc. and the current at time step 301 only, for clarity, for an impulsively excited dipole 599 time steps long (*dx = cdt*) and having a radius of* dx/3

Observe that the charge/current (Q/I) pulse initially launched at the center feedpoint does not propagate along the entire length of a finitely conducting wire. Instead, the electrons that primarily constitute the current and charge flow move in a fashion somewhat analogous to beads distributed along a string. Whereas a negative charge pulse is a region of excess electrons, a positive pulse is a region of electron depletion. The negative pulse of excess electrons forms, in its direction of movement, a moving region of negative charge. The positive pulse of election depletion attracts electrons to form a moving region of positive charge. The propagation of a sound wave in air or a vibration wave along a string is other examples of similar phenomena.

There are only a few ways for the charge to decrease with distance away from the feedpoint as demonstrated in Figure 13.13. One, a phenomenon not included in the TWTD model, is for some charge to leak off the wire as the pulse propagates. A second possibility, a phenomenon not permitted by the PEC boundary conditions in the TWTD model, would be for some charge to slow down or stop. A third

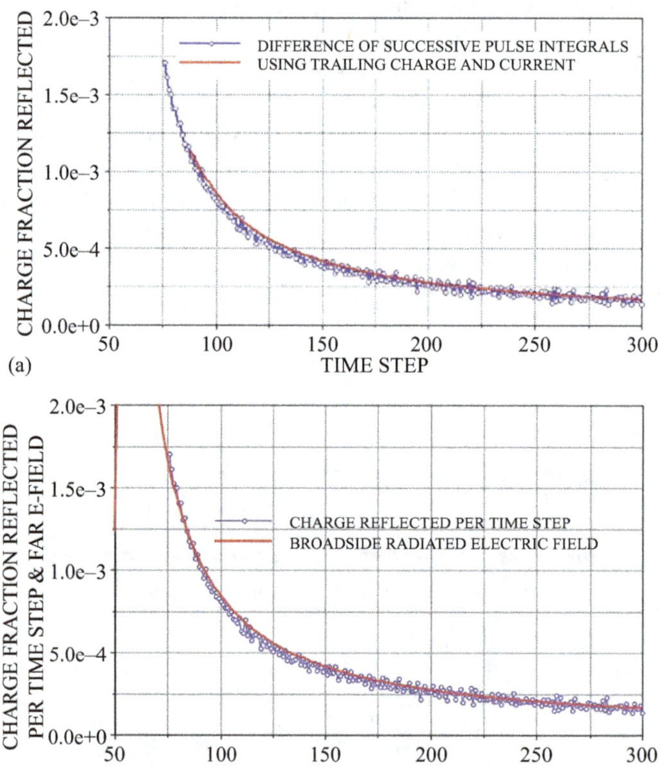

Figure 13.14 The proportion of charge reflected per time step as a charge pulse propagates outward on a center-fed, PEC dipole (a) and compared with the broadside-radiated electric field (b)

possibility, the actual cause of the charge decrease in the outward-propagating pulse, is for some charge to be reflected from the outward-propagating pulses, This reflected charge flows back towards the feed point at the same speed as the outward-flowing charge. The reflection occurs because the wave impedance presented by a constant-radius wire to a propagating pulse varies with distance along the wire, as shown by frequency-domain analysis, e.g., that due to Anderson [19]. Thus, the wave experiences a partial, continuous, impedance-mismatch reflection as it propagates.

An illustration of this reflection phenomenon in Figure 13.14a shows the fractional amount of reflected charge as a function of time (in time steps) as obtained in two different ways, The more "noisy" of the two reflected-charge curves is obtained from subtracting the charge in two successive time steps to determine the charge reflected per time step as a function of position. The difference of two nearly equal numbers accounts for its numerically noisy appearance. The other involves a computation that uses the opposite-sign trailing current just behind the charge pulse [18]. These different results agree to within a few percent and clearly demonstrate that charge is removed from the outward-propagating charge pulses. As explained further below, the reflected charge accelerates through a speed change of 2*c*, resulting in a frequency-domain, intermediate radiation exhibited in Figure 13.4 and in the time-domain radiation of Figure 13.14b. In the latter case, the reflected charge and its broadside-radiated electric field are closely correlated.

First, observe that the counter-propagating waves on a thin-wire antenna produce standing waves of current and charge whose peaks are displaced by one-quarter wavelength as shown in Figure 13.15 for one-half of the NZR dipole. Second, note that charge peaks are found where same-signed and opposite-propagating charge pulses meet while current peaks occur where opposite-signed and opposite-propagating charge pulses meet. Third, it follows that charge-reflection acceleration fields are subtractive in the former case and additive in the latter as illustrated conceptually in Figure 13.16. The peaks of the current magnitude and the LPD as a result are coincident as shown in Figure 13.17 while further

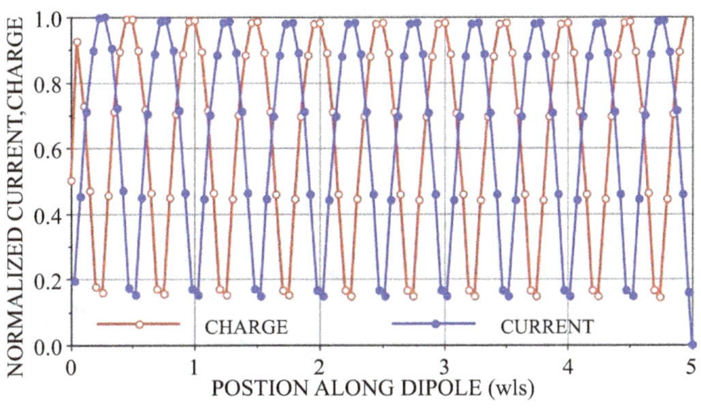

Figure 13.15 The normalized current and charge on the right side of a half-wave, center-fed dipole 10-wavelengths long

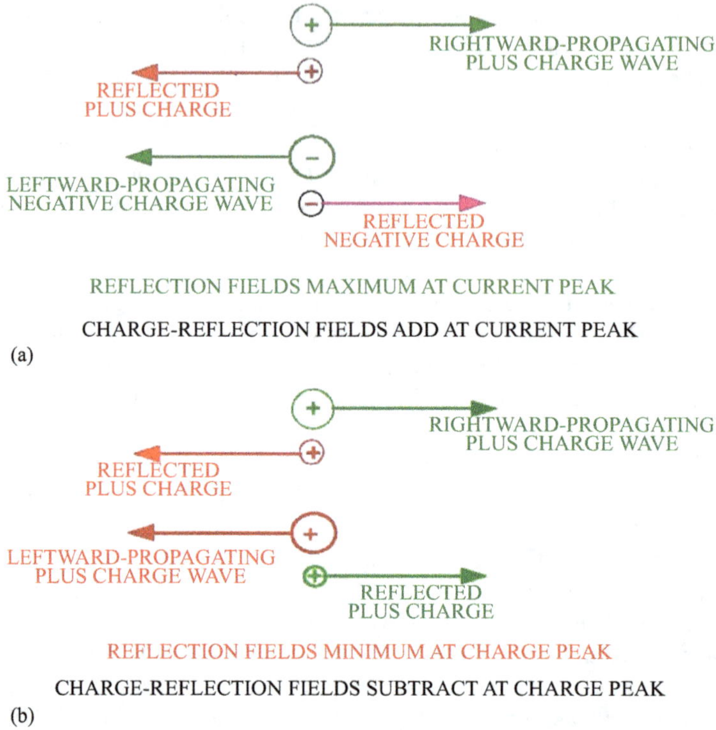

Figure 13.16 Conceptual demonstration of how oppositely propagating, opposite-signed charge pulses produce a current peak and additive reflection-acceleration fields (a) and how same-signed charge pulses produce a charge peak and reflection–acceleration fields that subtract (b)

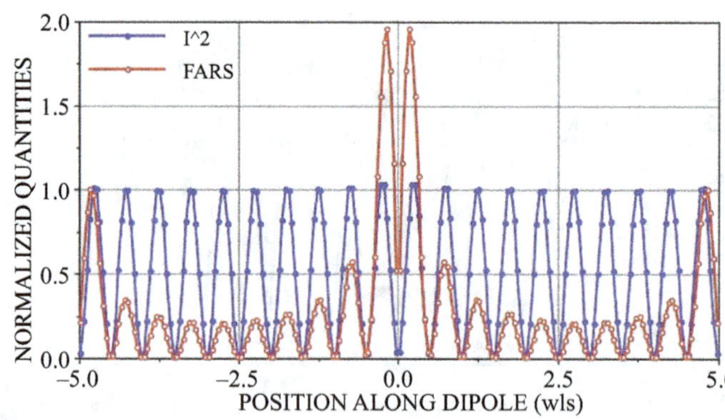

Figure 13.17 The squared current and LPDs on the dipole of Figure 13.14 normalized to their end peaks

providing a qualitative validation of FARS. Thus, the somewhat surprising result is that even though the SCF and NZR dipole experience quite different charge–acceleration mechanisms for their intermediate radiation, the former being due to a tangential electric field while the latter is due to charge reflection, their distributed radiation is nearly identical.

13.5 Summary

Discussed in this chapter is a comparison of the radiation properties of a sinusoidal current filament (SCF) and a PEC dipole of near-zero radius (NZR). Being very different radiators as the SCF has an assumed current distribution while the current on the NZR dipole satisfies the PEC boundary condition they might reasonably be expected to have quite-different radiation properties. Nevertheless, the SCF and NZR dipole are found to be nearly identical with respect to their:

- total radiated power as a function of current length;
- radiation patterns;
- length-wise distributions of radiated power, or LPD [3];
- the Schelkunoff–Feldman [4] distributed radiation resistance.

On the other hand, ways in which they differ include their:

- current distributions, which differ slightly, but significantly;
- Poynting vectors parallel to their currents;
- tangential electric fields along the current;
- distributions of IEMF power.

Specific explanations have been discussed with respect to these observations in terms of their charge–accelerations and consequent radiation mechanisms. Charge acceleration for the SCF is due to the tangential E-fields needed to maintain the (assumed) sinusoidal spatial variation of the current along the filament as well as the feedpoint current-slope discontinuity and at the filament ends. Charge acceleration for the PEC dipole, on the other hand, is caused by reflection of propagating charge along the wire due to its varying wave impedance, as well in the vicinity of the feedpoint and at the dipole ends.

In conclusion, the following more general observations can be made for a problem involving only electric charge and a PEC object:

1. As derived from the Lienard–Wiechert potentials a radiation, or $1/r$, field is seen to be caused only by accelerated charge.
2. For a PEC, charge resides only on that object's surface.
3. Charge separation and acceleration occur when the PEC object is subjected to an exciting field.
4. Acceleration of the electric surface charge on the PEC object produces a transverse "kink" in its E-field.
5. This E-field kink propagates outward at light speed from the accelerated charge.
6. Therefore, it follows that radiation from a PEC object originates from its surface.

References

[1] G. J. Burke, *Numerical Electromagnetics Code – NEC-4 Method of Moments, Part II: Program Description – Theory*, Lawrence Livermore National Laboratory, Report USRL-MA-109338, January 1992.

[2] C. Balanis, *Antenna Theory, Analysis and Design*, Harper & Row, New York, NY, 1982.

[3] E. K. Miller, Some Further Results from FARS: far-field analysis of radiation sources, in: *Proceedings of 16th Annual Review of Progress in Applied Computational Electromagnetics*, Naval Postgraduate School, Monterey, CA, pp. 278–285, 2000; E. K. Miller, Further investigation using far-field analysis of radiation sources (FARS), in: *IEEE International Symposium on Antennas and Propagation*, Salt Lake City, UT, Hilton Hotel, July 16–21, 2000, pp. 2058-61.

[4] S. A. Schelkunoff and C. B. Feldman, On radiation from antennas, *Proceedings of the IRE*, 30, pp. 511–516, 1942.

[5] H. C. Pocklington, Electrical oscillations in wires, *Proceedings of the Cambridge Philosophical Society*, 9, pp. 324–332, 1897.

[6] C. Manneback, *Radiation from Transmission Lines*, *AIEE Journal*, 42, pp. 95–105, 1923.

[7] S. A. Schelkunoff, *Advanced Antenna Theory*, John Wiley & Sons, Inc., New York, NY, 1952.

[8] W. K. H. Panofsky and M. Phillips, *Classical Electricity and Magnetism*, Addison-Wesley Publishing Co., Inc., Cambridge, MA, 1955.

[9] G. S. Smith, *An Introduction to Classical Electromagnetic Radiation,* Cambridge University Press, New York, NY, 1997.

[10] E. K. Miller, Differentiating the surface Poynting vector to determine the rate of power loss on a perfect conductor, submitted to the *AP-S Magazine*.

[11] G. S. Smith, On the interpretation for radiation from simple current distributions, *IEEE Antennas Propagation Magazine*, 40, pp. 9–14, 1998.

[12] G. S. Smith, On the interpretation for radiation from simple current distributions, *IEEE Antennas Propagation Magazine*, 40, pp. 39–44, 1998.

[13] R. G. Martin, A. R. Bretones, and S. G. Garcia, Some thoughts about transient radiation by straight thin wires, *IEEE Antennas Propagation Magazine*, 41, pp. 24–33, 1999.

[14] Y. Beers, The role of the Larmor radiation formula in the classical theory of electromagnetic radiation, *IEEE Antennas Propagation Magazine*, 41, pp. 51–56, 1999.

[15] H. G. Schantz, Electromagnetic energy around Hertzian dipoles, in: *1999 IEEE Int. Antennas Propagat. Symp. Dig.*, Vol. 2, pp. 1320–1323, June 1999.

[16] C. C. Bantin, Radiation from a pulse-excited thin wire monopole, *IEEE Antennas Propagation Magazine*, 43, pp. 64–69, 2001.

[17] G. S. Smith and T. W. Hertel, On the transient radiation of energy from simple current distributions and linear antennas, *IEEE Antennas Propagation Magazine*, 43, pp. 49–63, 2001.
[18] E. K. Miller and G. J. Burke, A multi-perspective examination of the physics of electromagnetic radiation, *Applied Computational Electromagnetics Society Journal*, 16(3), pp. 190–201, 2001.
[19] J. B. Anderson, "Admittance of infinite and finite cylindrical metallic antenna," *Radio Science*, 3(6), pp. 607–621, 1968.

Chapter 14

The pattern rank and spatial radiation distribution of radiation emitted by a sinusoidal current filament*

Abstract

Two propositions are explored in this chapter that relate to the issue of where radiation originates from a specified current source. One is that if some incremental portion of a source distribution increases (changes) the total far field at any observation angle, then by definition radiation originates from that portion of the source distribution. The other is that the greater the number of degrees of freedom in a radiation pattern the greater will be the proportion of the source distribution that actually radiates. These issues are explored here in terms of the radiation from a sinusoidal current filament.

Keywords: Degrees of freedom; FARS; Pattern rank; Radiation; Sinusoidal current filament; Singular values; Spatial distribution of radiation

14.1 Introduction

The most important phenomenon in electromagnetic theory is radiation. In spite of that, how and where radiation occurs from over a specified source such as the sinusoidal current filament (SCF) or a perfect-electric conductor (PEC) object continues to be uncertain. While charge acceleration is known to be the fundamental cause of electromagnetic radiation, uncertainty persists about these questions. The places where radiation originates seem to be debatable for even such simple objects as PEC straight wires or simple variations thereof. Does radiation from a linear current filament or PEC wire come only from the ends and feedpoint as some claim [1–4] or does the radiation instead come from along the length of the current [5–10], the latter being the one that the author holds [11,12].

*Excerpted and rewritten from E. K. Miller, "The Incremental Far Field and Degrees of Freedom of the Sinusoidal Current Filament," *IEEE Antennas and Propagation Society Magazine*, 49 (4), pp. 13–21, August, 2007.

Some examples of the former include [28]:

- SCF: "Radiates from only its center and ends" [1].
- Hertzian dipole: "The aim of this present paper is to show how to determine where static and reactive energy are converted into radiation in the fields around an antenna" [2].
- Pulse-excited straight wire: "In this case, after the positive pulse due to the initial charge acceleration, the current propagates along the antenna without radiating until it reaches the end of the wire, where the charges are decelerated and the radiation of the negative pulse takes place" [3].
- Pulse-excited straight wire: "Radiation is produced when the spatially integrated value of the product of the charge density and its acceleration is non-zero. This is supported by the results presented in this article. Between the feed point and tip (or subsequently between the tip and its image), the charge-density pulse travels along the monopole at the speed of light, with no dispersion. During these times, the integrated value of charge density times acceleration is zero," i.e., there is no radiation along the monopole [4].

Examples of the latter viewpoint, which have the common theme of current decay as an indicator of radiation for PEC wires, include [28]:

- Finite dipole antenna: "If damping due to the emission of radiation is neglected . . . the current along the antenna can be taken as sinusoidal . . . " [5].
- Finite dipole antenna: "An outgoing traveling wave . . . decays slowly . . . as a result of radiation" [6].
- Infinite cylindrical antenna: "The current wave on the infinite rod decreases slowly with distance and it is damped by radiation losses" [7].
- Conical spiral antenna: "Most of the radiation takes place in the exponential decay (of the current) region" [8].
- Infinite cylindrical antenna: "The current wave decays slowly with distance along the rod . . . gradually radiating energy" [9].
- Pulse-excited traveling-wave and a standing-wave (dipole) current elements: "The results presented above show that there is a net amount of energy leaving every unit length of both filamentary current distributions: the traveling-wave element and the standing-wave dipole. . . . For the traveling-wave element, some of this energy is radiated away from the element, and the rest is stored in the field of the point charges that remain at the two ends of the element. For the standing-wave dipole, this energy is all radiated away" [10].

That there are various opinions about where radiation takes place for even simple geometries is demonstrated by the above quotations. While having a definitive answer to where radiation originates is obviously not necessary to solving real-world problems, radiation is a crucial phenomenon of electromagnetics the better understanding of which would add to our knowledge base. That is the motivation of the discussion that follows.

Radiation by the SCF from two different perspectives is to be explored in this chapter. One is to examine the fundamental mathematical operation of determining a far field from integrating over a source distribution. It will be explicitly demonstrated that radiation indeed does occur from the interior of the SCF away from its ends and feedpoint using a simple and straightforward approach. The other is to explore the number of degrees of freedom (DoF), or rank, of the SCF's radiation pattern. This is motivated from a synthesis viewpoint, where the goal is to determine the minimum number of discrete sources needed to radiate a given pattern to some prescribed accuracy [13]. If the SCF really were to only radiate from three isotropic point sources, then its radiation-pattern rank should be expected to be no greater than three. But if radiation from along its length results in needing more than three such sources to produce its pattern, the rank of its radiated field would be greater than three. Observe that although what follows is restricted to the frequency domain, its implications apply to time-domain results as well since they are related via a Fourier transform.

14.2 The far field of the SCF

First note that an infinitesimal (or Hertzian), z-directed current moment $I_z dz$ located at the origin on the z-axis produces a differential, θ-directed, total electric field given by

$$dE_\theta(\theta, r) = j\eta \frac{kI_z dz \sin(\theta)}{4\pi} \left[\frac{1}{r} + \frac{1}{jkr^2} - \frac{1}{(kr)^3} \right] e^{-jkr} \tag{14.1}$$

where

$$r = \sqrt{x^2 + y^2 + z^3}$$

is the field-observation point. Observe also that the directivity of the current is $\sin(\theta)$ with the observation angle θ measured from the positive z-axis. The radiation E-field is the $1/r$, or first term in the brackets of (14.1). It might be implied from (1) that such a current distribution would radiate from wherever the current is non-zero since a current filament of arbitrary spatial variation can be approximated by a piecewise sequence of Hertzian current elements. Indeed, given that each non-zero Hertzian element produces a $1/r$ E-field, a relevant question to ask is "How could it not radiate"? This is the premise and question examined below for the specific problem of the SCF.

The far field of a z-directed SCF of length L

$$I(z') = I_o \sin\left[k\left(\frac{L}{2} - |z'| \right) \right], -\frac{L}{2} \le z' \le \frac{L}{2} \tag{14.2}$$

can then be obtained by integrating over the field of the infinitesimal current element as [14]

$$E_\theta(\theta) = \int_{-\frac{L}{2}}^{\frac{L}{2}} dE_\theta(\theta, z') = j\eta \frac{ke^{-jkr}}{4\pi r} \sin(\theta) \int_{-\frac{L}{2}}^{\frac{L}{2}} I_z(z') e^{jkz' \cos(\theta)} dz'. \tag{14.3}$$

In the above expression,

$$R = \sqrt{x^2 + y^2 + z^3} - z' \cos(\theta) = r - z' \cos(\theta)$$

has been used for phase terms and $R = r$ for amplitude terms. This integral can be evaluated analytically to give [14]

$$E_\theta(\theta) = j\eta \frac{I_o e^{-jkr}}{2\pi r} \left[\frac{\cos\left(\frac{kL}{2} \cos \theta\right) - \cos\left(\frac{kL}{2}\right)}{\sin \theta} \right] \quad (14.4a)$$

where it can be seen that the far E-field is purely in phase quadrature with the current.

An alternate way of writing (14.4a) is in which the interior, square-bracketed term accounts for the far field of three isotropic point sources, two of unit amplitude at $\pm kL/2$ and another at the origin of amplitude $2\cos(kL/2)$.

$$E_\theta(\theta) = j\eta \frac{I_o e^{-jkr}}{\pi r} \left(\left[e^{j(kL/2)\cos(\theta)} + e^{-(jkL/2)\cos\theta} - 2\cos(kL/2) \right] \frac{1}{\sin\theta} \right) \quad (14.4b)$$

This will be referred to as case (a) in the following. This expression is used to distinguish a point-source pattern from results obtained when that pattern is multiplied by $\sin(\theta)$ [case (b)] and divided by $\sin(\theta)$ [case (c)], respectively. Obviously, case (c) pertains specifically to the far E-field pattern of the SCF. Omitting the $1/\sin(\theta)$ term as case (a) does has erroneously led to the conclusion that the SCF far field is that from three-point sources. However, the effect of the $\sin(\theta)$ term in the denominator invalidates this conclusion as is demonstrated in the following. To summarize we have

Case a:

$$E_\theta(\theta) = j\eta \frac{I_o e^{-jkr}}{\pi r} \left[e^{j(kL/2)\cos(\theta)} + e^{-(jkL/2)\cos\theta} - 2\cos(kL/2) \right] \quad (14.5a)$$

Case b:

$$E_\theta(\theta) = j\eta \frac{I_o e^{-jkr}}{\pi r} \left[\left(e^{j(kL/2)\cos(\theta)} + e^{-(jkL/2)\cos\theta} - 2\cos(kL/2) \right) \sin\theta \right] \quad (14.5b)$$

Case c:

$$E_\theta(\theta) = j\eta \frac{I_o e^{-jkr}}{\pi r} \left[\left(e^{j(kL/2)\cos(\theta)} + e^{-(jkL/2)\cos\theta} - 2\cos(kL/2) \right) \frac{1}{\sin\theta} \right] \quad (14.5c)$$

14.3 The implications of numerically evaluating the far field

It can be instructive to examine the integral of the far field of the SCF as given in (14.3) from a numerical perspective. Thus, (14.3) can be written approximately as

$$E_\theta(\theta) \approx \int_{-\frac{L}{2}}^{\frac{L}{2}} dE_\theta(\theta, z') = j\eta\frac{ke^{-jkr}}{4\pi r}\sin(\theta)\sum_{m=1}^{N} e^{jkz_n\cos(\theta)}I_z(z_m)dz$$

$$= \sum_{m=1}^{N} e_{m,\theta}(\theta) \tag{14.6a}$$

with

$$z_m = -L/2 + (m - 1/2)dz \tag{14.6b}$$

as a sum of incremental terms of the far E-field from each incremental current element extending from $z_m - dz/2$ to $z_m + dz/2$ and where $dz << 1$. By varying z_m from $-L/2 + dz/2$ to $L/2 - dz/2$, the field of the incremental current is scanned over the entire SCF.

Each incremental non-zero current element $I_z(z_m)$dz in (14.6) contributes an amount $e_{\theta,,m}(\theta)$ to the total far field. At some observation angles, an individual $e_{\theta,,n}(\theta)$ may increase, and at others, it may decrease, the total far E-field. The only exceptions to this would be that all the $e_{\theta,,m}(\theta)$ are identically zero away from the ends and center, or that their sum cancels identically. Either situation would imply that the far field from the interior part of the SCF would be everywhere zero over all observation angles from which it would follow that only the ends and center actually radiate. That neither of these conditions is met will be demonstrated below. Whether it increases or decreases the total field, the conclusion inevitably follows that if the field of a given current element alters the total far field it is because that current element radiates.

14.4 Examination of the far-field analytical expression

Given in Figure 14.1 for L/wavelength = 10 are three plots of the squared far field obtained from the square-bracketed terms of (14.5). All three patterns overlap graphically within ±20 degrees or so of broadside, but beyond that region the differences grow significantly. The three-point sources produce a pattern identifiable by its equal-amplitude maxima. The pattern away from broadside is tapered by the effect of the sin(θ) multiplier. The actual SCF pattern exhibits the greatest difference from the three-point-source pattern. The most substantial difference from that of three-point sources due to the 1/sin(θ) term in the SCF radiation pattern is the generation of large, end-fire, or traveling-wave, lobes.

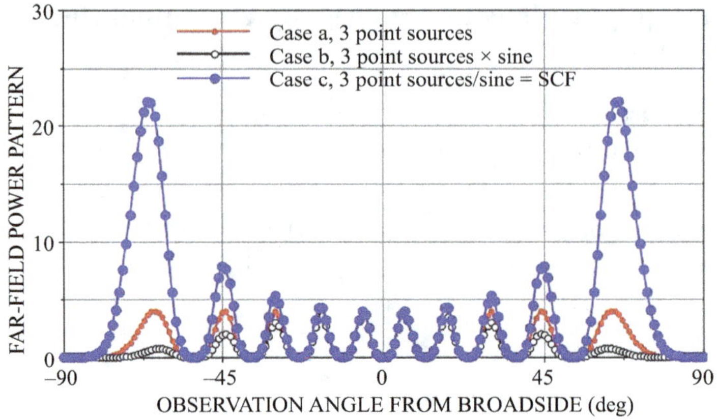

Figure 14.1 Square of the SCF far E-field from the square-bracketed term of (14.5) for L/*wavelength* = *10*

14.5 Examination of numerically evaluating the far field

First observe that each incremental non-zero current element $I_z(z_m)dz$ in (14.6) contributes an amount $e_{\theta,m}(\theta)$ to the total far field. At some observation angles, an individual $e_{\theta,,m}(\theta)$ may decrease, and at others it may increase, the summed far E-field. Only if all the $e_{\theta,,m}(\theta)$ are identically zero away from the ends and center, or if their sums canceled identically, would this not be the case. Either situation would imply that the far field from the interior part of the SCF would be everywhere zero over all observation angles from which it would follow that only the ends and center actually radiate. It will be demonstrated below that neither of these conditions is met. The conclusion inevitably follows that if the field of a given current element alters the total far field it is because that current element radiates whether it increases or decreases the total field.

Plotted in Figure 14.2 for the 10-wavelength SCF at an observation angle at the peak of the traveling-wave lobe (24.7 deg) are the individual segment contributions, $e_{m,\theta}$, to the total far field from (14.6) for dz/wavelength = 0.05. Positive segment fields increase the positive traveling-wave lobe and the far-field radiated power that it carries. Demonstrated in Figure 14.2 is that radiation occurs not only from the ends and center but also from various places along the length of the SCF. An integral of the segment fields along the SCF as shown in Figure 14.3 where its value is seen to be nearly monotonically increasing. This is also found to be the case for other observation angles within the traveling-wave lobe and elsewhere as in the overall pattern. This is consistent with a traveling wave being due to the fields from along the current filament being primarily additive.

These results are consistent with the conclusion that SCF radiation occurs from not only the ends and center of the current filament but also from along its length. Concerning the conflicting interpretations that have arisen over radiation from the SCF one additional observation can be made.

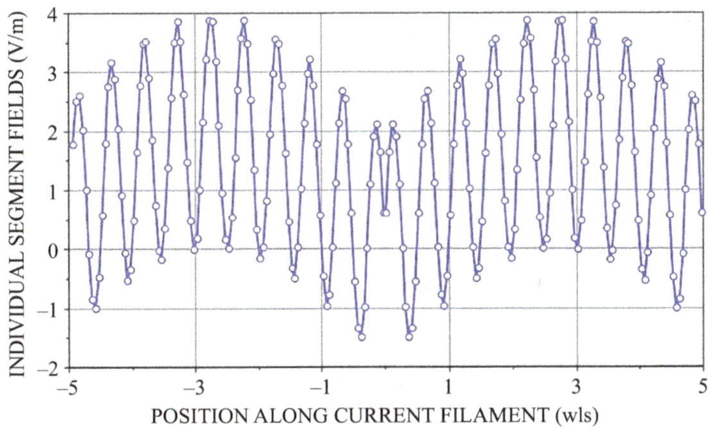

Figure 14.2 The $e_{m,\theta}$ far E-fields from incremental current elements as a function of position along the L/wavelength = 10 SCF at the center of the traveling-wave lobe (24.7 deg)

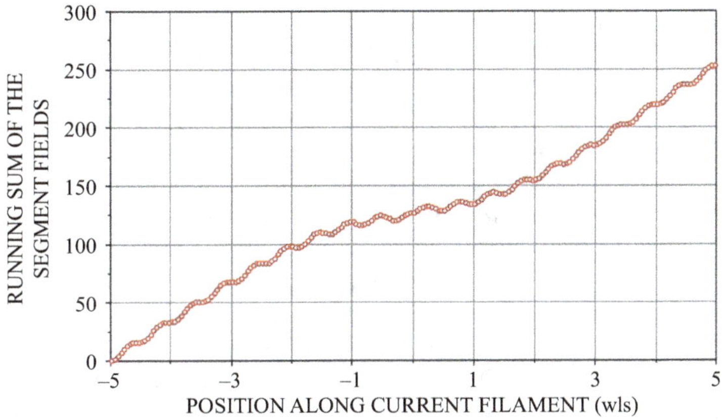

Figure 14.3 The cumulative sum of the incremental $e_{m,\theta}$ far fields in Figure 14.2 as a function of position along the 10-wavelength SCF of Figure 14.2

Recall that the basic proposition of the "ends-and-center" origin of SCF radiation is the expressions for the far field as given by (14.4), a result made possible by the fact that the integral for the far field in (14.3) can be evaluated in closed form. This suggests considering the question "What is implied by the existence of a closed-form result for a definite (as here) or indefinite integral?" Observe that a closed-form result involves only the ends of the integration range, i.e., no abscissa values from within that range. Should this lead to a conclusion that only the end points of the integration range contribute to the value of the integral? This is not true, of course, as demonstrated by simply approximating the integral by a Riemann sum as was done above with the entire integration range clearly contributing to the final answer.

This point can be further clarified by a simple example. Consider the integral of $y = f(x) = A$ from x_1 to x_2 where $x_2 - x_1 = B$. We thus have

$$\int_{x_1}^{x_2} A dx = A(x_2 - x_1) = AB \tag{14.7}$$

which is the "formula" for the area of a rectangle of sides A and B. No one would argue that only these end points contribute to a final closed-form result involving only the end points of the integration range. The area of the rectangle involves its height A over x_1 to x_2 and not just the end points. The situation is similar here in evaluating the integral in (14.3) to get (14.4).

14.6 The degrees of freedom of various patterns

One measure of the complexity of a given radiation pattern is its rank, the number of degrees of freedom (DoF) it has, or alternatively, the number of discrete sources or modes that a current source would require to replicate it. This concept is not only useful to gain some insight about how the source distribution under consideration radiates but is also relevant to pattern synthesis. As will be shown in the following, one way to "measure" the pattern rank is to determine the number of nonzero singular values that it has. As a general rule, the SV distribution is determined by the maximum number of point sources or modes in an array or line source of L wavelengths will be on the order of it $2L + 1$. This is because a good estimate of the far-field spatial resolvability of a source from its pattern is generally about ½ wavelength. On the other hand, a pattern produced by N discrete sources would require N or fewer sources to produce it, fewer being needed if the average spatial distance between the sources is less than a half wavelength.

Numerous authors [13,15–20] have investigated various aspects of the pattern-rank concept in the context of special characteristic modes. The use of spatially discrete sources, as is done here, is somewhat different but similarly motivated. Of particular relevance to the present discussion are results presented in [19,20] where entire-domain, "generalized characteristic" modes are described. With increasing mode number, or number of oscillations along the object these orthogonal modes radiate a decreasing amount of power. The power radiated per mode is presented in [20] for a wire dipole of length $L = 1.91$ wavelengths. These results imply that it radiates from along its length since each mode extends over the entire wire.

An alternate way to determine a radiation-pattern rank is to compute its singular values from a matrix of the pattern samples. With its angle dependence being proportional to $e^{ikd\cos(\theta)}$, the radiation pattern is sampled in uniform steps of $\cos(\theta)$ in the same way as a time series is sampled in equal steps of the time variable to apply Prony's method. Similarly, each successive row of the pattern-sample matrix is constructed by shifting the window of pattern samples in θ by that same $\cos(\theta)$

step. As obtained from Prony's method [21] for $L = 10$ wavelengths the singular-value (SV) spectra of E-field-pattern samples from the three cases defined above in connection with (14.5) are plotted in Figure 14.4. A P-term exponential-series (or fitting) model using 2P samples of the actual pattern to compute its parameters (poles, s_α, and residues, R_α) was used as given by

$$E_\theta(\theta) = \sum_{\alpha=1}^{P} R_\alpha e^{s_\alpha \cos\theta} \tag{14.8}$$

The computations were done in 24-digit precision based on 42 samples uniformly spaced in cos(θ) over a window extending from −0.99 to +0.99 in cos(θ) to obtain the fitting-model parameters. The real part of the pole in (14.8) determines its non-isotropy in cos(θ) while the imaginary part determines the distance of the corresponding point source from the origin. Note that the 3-point-source pattern [case (a)] has only three significant singular values and then falls off abruptly to the noise floor of about 10^{-23}. The other two SV spectra for the point-source pattern multiplied or divided by sin(θ) [cases (b) and (c)] are substantially above the noise floor out to the 21st singular value. They are also quite similar in their variation, showing that their effective pattern ranks are substantially greater than 3. The SV spectrum for case (c) shows that there are equivalently more radiation sources along the interior part of the SCF, i.e., away from the ends and center.

Note that sampling finely enough in cos(θ) to avoid aliasing is required for this computation while also using a large-enough value for P to cover the anticipated SVD spectral range. Consequently, when sampling over the entire observation range in cos(θ) as is done here, it should be expected that $P >= 2L$/wavelength + 1 would be needed. Alternatively, a succession of smaller windows in cos(θ) might

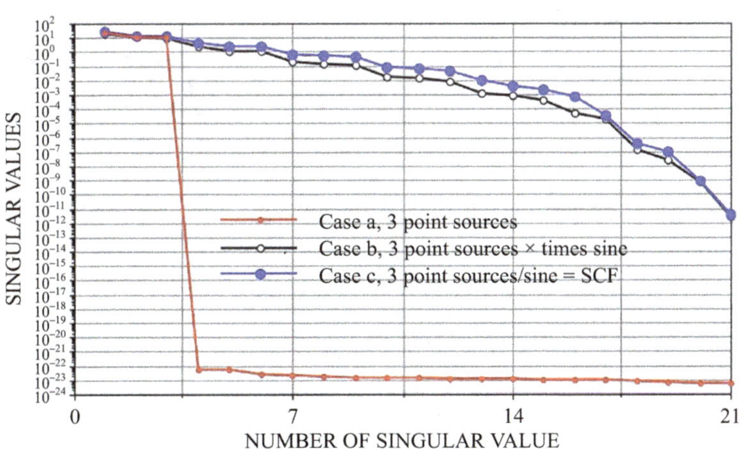

Figure 14.4 The singular-value spectra for the three patterns computed from (14.4b) with small-dot plot being that of three isotropic point sources

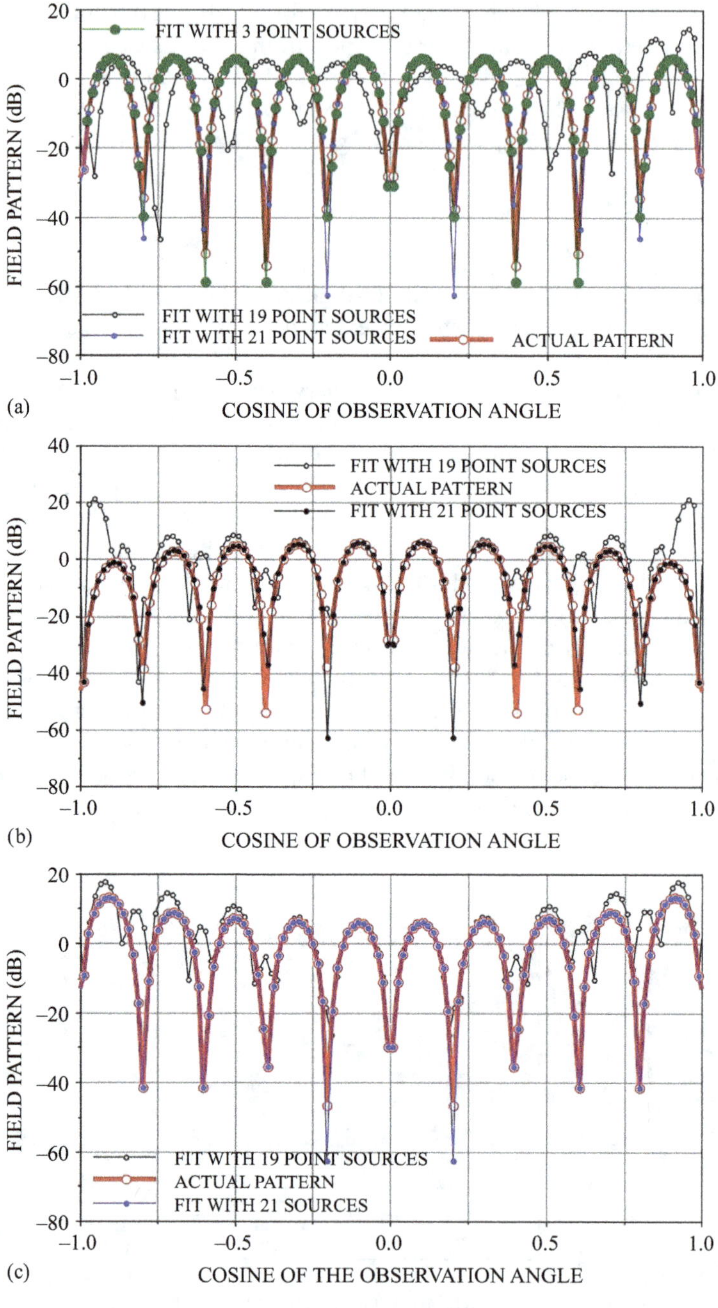

Figure 14.5 Power patterns of the successive bracketed terms in (14.5) for cases (a) through (c) respectively. For the pattern fit obtained using 19 and 21 point sources, the sampling interval is from -0.99 to $+0.99$ in $\cos(\theta)$. For case (a) and 3 point sources, the sampling interval is -0.1 to $+0.1$. The curves of the actual pattern and using 21 point sources are essentially graphically indistinguishable as is that for case (a) with three-point sources

be used instead to then estimate the total pattern rank from the sum of the ranks of these individual windows [22].

Results for synthesizing the three patterns of (14.4b) from point sources using Prony's method are presented in Figure 14.5 [23]. A sampling interval of -0.99 to $+0.99$ was used in $\cos(\theta)$ for all three cases with 19 and 21 point-source exponential fitting models while a sampling interval of -0.1 to $+0.1$ and 3 point sources was used for case (a). Note that the pattern for case (a) can be synthesized using only three-point sources, as expected. On the other hand and as expected due to the SV spectra shown in Figure 14.4, those for cases (b) and (c) required 21 point sources to be accurately reproduced over a range of -0.99 to $+0.99$ in $\cos(\theta)$. Also observe that when the observation interval is -0.99 to $+0.99$, even case (a) requires 21 samples to avoid aliasing although only the three-point sources contribute significantly to the synthesized pattern.

The synthesized patterns shown in Figure 14.5 are graphically indistinguishable from the actual patterns when using 21 point sources, with the RMS difference between them being on the order of 10^{-11}.

14.7 Current and charge relationship to radiation

As shown by the Lienard–Wiechert potentials [24], it is well known that source radiation is due to accelerated charge. However, attention here has so far been directed only to the current to identify where radiation originates. It is relevant to ask "Are these two apparently different concepts mutually consistent?"

Discussed elsewhere [25] is the observation that for standing-wave currents such as the SCF, current maxima occur where the maxima of oppositely propagating charge waves of opposite sign meet. These current maxima are also locations where the local tangential electric field is a maximum. Associated with these two phenomena is a maximum in the induced EMF (IEMF or input) power there through the acceleration this field imposes on the charge. Thus, consistent with the discussion above concerning the current contribution to the far field, a current maximum should be expected to produce a spatial maximum in the distributed radiation along the standing-wave current filament. These results are demonstrated by farfield analysis of radiation sources (FARS) [28] in Chapters 8 and 9, and are consistent with various other observations. However, as shown by FARS the distribution of radiated power along the SCF, being largest at the ends and at the center feedpoint when the filament length is an even number of half wavelengths, is not equally proportional to the current magnitude everywhere, as shown in Chapter 10.

A somewhat similar situation holds for the current on a PEC wire. However, in that case the local, or IEMF input power, is zero away from the feedpoint. Instead, a partial reflection of the counter-propagating charge waves causes the charge acceleration. As for the SCF, radiation maxima occur at current maxima where opposite-signed maxima in the charge waves meet. Their reflection in opposite directions produces an additive radiated E-field of the same sign.

Thus, the current distribution that the charge comprises provides a simple and direct connection to the radiated field while charge acceleration is the fundamental physical cause of far-field radiation. This appears consistent with the far field coming from an integral over the current. So, a qualitative estimate of where radiation comes from does not require explicit consideration of accelerated charge. Instead, it can be identified from the current distribution. Thus, the simplest, although a somewhat superficial answer from a physics' viewpoint to the question "Where does radiation originate from a source distribution?" is after all, "It's the current," the answer most-often given by undergraduate EM students.

Finally, note that radiation in the time domain can come from time-domain solutions or can be evaluated by Fourier-transforming frequency-domain results. It follows from frequency-domain results like those presented above that time-domain radiation also occurs from along the length of a PEC wire or a specified current. This conclusion is consistent with time-domain solutions [10,26], the former for an assumed Gaussian current/charge pulse and the latter for a PEC wire.

14.8 Summary

The preceding results and observations presented are consistent with various others in supporting the conclusion that radiation originates from along the length of a linear current distribution. This is true whether that distribution satisfies the boundary condition on a PEC wire or one that has no boundary condition, as is the case of the SCF. To summarize:

1. For the most part, the current all along the SCF contributed to the far field at the maximum of the traveling-wave lobe.
2. The singular-value spectrum for the SCF extends well beyond the three SVs for the spectrum of three-point sources.
3. The total power radiated by a SCF grows in proportion to Log(kL) showing a length dependence inconsistent with radiation from three isotropic point sources located at the ends and feedpoint. The power from three-point sources as in (14.5a) instead varies between two fixed limits with increasing length [25].
4. A lobed, distributed radiated power along the entire length of the SCF is exhibited by FARS results [25].
5. Because the time-average power flow along the standing wave SCF is zero, the IEMF distributed input power is essentially identical to FARS, in accord with the fact that the input power and radiated power must be equal as a function of position as shown in Chapter 10 and [25].
6. The Schelkunoff–Feldman (S–F) distributed radiation resistance when multiplied by $[I_z(z_n)]^2/2$ yields results essentially identical to FARS and the IEMF distributed power in Chapter 10 and [25].
7. The radial power flow a few thousands of a wavelength from the SCF is within a few per cent of the FARS, IEMF, and S–F powers except at the center feedpoint whose effect is included explicitly in the radial power flow whereas it is not in the others as discussed in Chapter 6 and [25].

8. In the time domain, Gaussian-pulse-excited current/charge pulses propagating on a straight PEC wire decrease monotonically in amplitude with increasing distance from the feedpoint, as do their associated energies measured by spatial integrals of I^2 and $(Qc)^2$ [27].

References

[1] R. C. Hansen, Duality in Antennas, North America Radio Science Meeting, The Queen Elizabeth Hotel, Montreal, Canada, July 1997.

[2] H. G. Schantz, Electromagnetic energy around Hertzian dipoles, in: *IEEE Int. Antennas Propagat. Symp.*, Renaissance Orlando Resort, Orlando, FL, June 21–26, June 1999.

[3] Rafael Gómez Martin, A. R. Bretones, and S. G. Garcia, Some thoughts about transient radiation by straight thin wires, *IEEE Antennas Propagat. Mag.*, 41, pp. 24–33, 1999.

[4] C. C. Bantin, Radiation from a pulse-excited thin wire monopole, *IEEE Antennas Propagat. Mag.*, 43, pp. 64–69, 2001.

[5] J. D. Jackson, *Classical Electrodynamics*, John Wiley & Sons, New York, NY, 1962, p. 401.

[6] L.-C. Shen, T. T. Wu, and R. W. P. King, A simple formula of current in dipole antennas," *IEEE Trans. Antennas Propagat.*, 16, pp. 542–547, 1968.

[7] J. Bach Anderson, Admittance of infinite and finite cylindrical metallic antenna, *Radio Sci.*, 3(6), pp. 607–621, 1968.

[8] Y. S. Yeh and K. K. Mei, Theory of conical equiangular-spiral antennas. Part II – current distributions and input impedances, *IEEE Trans. AP-S*, AP-16 (1), pp. 14–21, 1968.

[9] D. S. Jones, *Methods in Electromagnetic Wave Propagation*, Clarendon Press, Oxford, 1994, p. 295.

[10] G. S. Smith, On the interpretation for radiation from simple current distributions, *IEEE Antennas Propagat. Mag.*, 40, pp. 39–44, 1998.

[11] E. K. Miller and G. J. Burke, Time-domain far-field analysis of radiation sources, in: *IEEE Antennas and Propagation Society International Symposium Digest*, Salt Lake City, UT, pp. 2058–2061, 2000.

[12] E. K. Miller and G. J. Burke, A multi-perspective examination of the physics of electromagnetic radiation, *Appl. Computat. Electromagn. Soc. J.*, 16(3), pp. 190–201, 2001.

[13] E. B. Ojeba and C. H. Walter, On the cylindrical and spherical wave spectral content of radiated electromagnetic fields, *IEEE Trans. Antennas Propagat.*, 27, pp. 634–639, 1979.

[14] C. Balanis, *Antenna Theory, Analysis and Design*, Harper & Row, New York, NY, 1982.

[15] R. J. Garbacz and R. H. Turpin, A generalized expansion for radiated and scattered fields, *IEEE Trans. Antennas Propagat.*, 19, pp. 348–358, 1971.

[16] R. F. Harrington and J. R. Mautz, Theory of characteristic modes for conducting bodies, *IEEE Trans. Antennas Propagat.*, 19, pp. 622–628, 1971.

[17] R. F. Harrington and J. R. Mautz, Computation of characteristic modes for conducting bodies, *IEEE Trans. Antennas Propagat.*, 19, pp. 629–639, 1971.

[18] N. Inagaki and R. J. Garbacz, Eigenfunctions of composite Hermitian operators with application to discrete and continuous radiating systems, *IEEE Trans. Antennas Propagat.*, 30, pp. 571–575, 1982.

[19] D. M. Pozar, Antenna synthesis and optimization using weighted Inagaki modes, *IEEE Trans. Antennas Propagat.*, 32, pp. 159–165, 1984.

[20] D. Liu, R. J. Garbacz, and D. M. Pozar, Antenna synthesis and optimization using generalized characteristic modes, *IEEE Trans. Antennas Propagat.*, 38, pp. 862–868, 1990.

[21] E. K. Miller, Using Prony's method to synthesize discrete arrays for prescribed source distributions and exponentiated patterns, *IEEE Antennas Propagat. Soc. Mag.*, 57 (1), pp. 147–163, 2015.

[22] E. K. Miller, Using windowed, adaptive sampling to minimize the number of field values needed to estimate radiation and scattering patterns, in: *Proceedings of 14th Annual Review of Progress in Applied Computational Electromagnetics*, Naval Postgraduate School, Monterey, CA, 1998, pp. 958–963.

[23] E. K. Miller, Model-based parameter estimation in electromagnetics: Part I. background and theoretical development, *IEEE Antennas Propagat. Soc. Mag.*, 40(1), pp. 42–52; Part II. Applications to EM observables, *IEEE Antennas Propagat. Soc. Mag.*, 40(2), pp. 51–64; Part III. Applications to EM integral equations, *IEEE Antennas Propagat. Soc. Mag.*, 40(3), pp. 49–66, 1998.

[24] W. K. H. Panofsky and M. Phillips, *Classical Electricity and Magnetism*, Addison-Wesley Publishing Co., Inc., Cambridge, MA, 1955.

[25] E. K. Miller, Comparison of the radiation properties of a sinusoidal current filament and a PEC dipole of near-zero radius, *IEEE Antennas Propagat. Soc. Mag.*, 48(4), pp. 37–47, 2006.

[26] E. K. Miller, Some further results from FARS: far-field analysis of radiation sources, in: *Proceedings of 16th Annual Review of Progress in Applied Computational Electromagnetics*, Naval Postgraduate School, Monterey, CA, pp. 278–285, 2000; E. K. Miller, further investigation using far-field analysis of radiation sources (FARS), in: *IEEE International Symposium on Antennas and Propagation*, Salt Lake City, UT, Hilton Hotel, July 16–21, 2000, pp. 2058–2061.

[27] E. K. Miller, Exploring electromagnetic physics using thin-wire time-domain (TWTD) modeling, in: *Proceedings of 14th Annual Review of Progress in Applied Computational Electromagnetics*, Naval Postgraduate School, Monterey, CA, 1998, pp. 583–588.

[28] E. K. Miller, The incremental far field and degrees of freedom of the sinusoidal current filament, *IEEE Antennas Propagat. Soc. Mag.*, 49(4), pp. 13–21, 2007.

Appendix A

The thin-wire time-domain (TWTD) computer code

Abstract

A brief description of thin-wire time-domain (TWTD) is presented together with a sampling of some of the results a time-domain model can generate. The code is based on a time-domain version of the electric-field integral equation specialized to thin wires. It originated in the late 1960s inspired by time-domain results reported at the 1968 IEEE AP-S conference. Its initial use was for simple wire antennas and scatterers and subsequently for ElectroMagnetic Pulse (EMP) applications. For the latter problems, a circuit capability was added to evaluate the damage that an EMP might do to electronic systems. This later feature was exploited in later research to design passive electronic tags for radio-frequency identification (RFID).

Keywords: Electric-field integral equation; EMP threat analysis; RFID tags; Time-domain modeling; TWTD

A.1 The origin of TWTD

Thin-wire time-domain (TWTD) was developed at MB Associates, San Ramon, CA in 1968 by Jerry Burke. It was motivated on hearing presentations by Drs Edward Sayre and C. Leonard Bennett at the 1968 IEEE AP-S conference where they described two of the first time-domain electromagnetic modeling developments. On returning from the conference, a request was made to MBA management for some seed funding to pursue this evolving technology. Jerry was given 2 or 3 weeks to evaluate the possibility of developing a time-domain computer model as a complement to a frequency-domain, thin-wire code called BRACT that was being developed for modeling chaff and re-entry decoys. Working alone, Jerry developed a preliminary program in a couple of weeks that over time was eventually given the name TWTD by some of the MBA staff who had moved to Lawrence Livermore National Laboratory. Preliminary reports of this work were initially presented at various conferences [1–5]. Two articles [6,7] and a book chapter [8] followed to describe the approach accompanied with numerous example applications [9–19].

A.2 A brief description of TWTD and sample results

TWTD is based on a time-domain version of the electric-field integral equation specialized to a thin-wire geometry. It is somewhat similar to Numerical Electromagnetics Code (NEC) in its numerical treatment, but lacks two major features of NEC. It offers no surface-modeling capability aside from approximating a surface with a wire mesh. It can only model a perfectly conducting half space in contrast to the sophisticated capability in NEC for handling penetrable half spaces with objects on either side of, or penetrating, the interface. It is similar to NEC in using point matching to test the tangential electric field of a wire that can be a perfect-electric conductor or have discrete or continuous impedance loading. Also, as does NEC, TWTD uses a tangential electric field as an antenna source model, or as an alternative a two-wire transmission line. Whereas NEC uses a three-term basis function to represent the wire current, TWTD uses 9-point, second order in space–time, Lagrangian polynomial interpolation for the current. Also, while NEC provides solutions at discrete frequencies but for arbitrary excitations, TWTD yields solutions over a range of frequencies for a specific excitation.

The time-domain, electric-field integral equation used in TWTD specialized to a thin wire is

$$\mathbf{s}\cdot\mathbf{E}_{\mathbf{exc}}(\mathbf{r},\mathbf{t}) = \frac{\mu_o}{4\pi}\int_{C(\mathbf{r})}\left[\frac{\mathbf{s}\cdot\mathbf{s}'}{R}\frac{\partial}{\partial t'}\mathbf{I}(s',t') + c\frac{\mathbf{s}\cdot\mathbf{R}}{R^2}\frac{\partial}{\partial s'}\mathbf{I}(s',t') - c^2\frac{\mathbf{s}\cdot\mathbf{R}}{R^3}q(s',t')\right]ds' \tag{A.1}$$

where

$$\mathbf{r}\in C(\mathbf{r})+a(\mathbf{r}),$$

C(***r***) is the spatial contour of the wire being modeled and ***s*** is a unit vector. The charge q is expressed in terms of the current I as

$$q(s',t') = \int_{-\infty}^{t'}\frac{\partial}{\partial s'}I(s',\tau)d\tau \tag{A.2}$$

and *a*(***r***) is the wire radius at point ***r***.

The numerical solution is developed by marching on in time, or time stepping, usually employing a Gaussian-pulse excitation voltage for an antenna or a plane-wave equivalent for scattering. This provides a result with a spatial resolution determined by the time duration of the excitation, and a corresponding frequency bandwidth. An especially important feature of TWTD compared with NEC is that nonlinear or time-varying impedances can be modeled. This capability can include electronic circuits as part of the model, to compute the effect, for example, of an ElectroMagnetic Pulse (EMP) on some electronic system exposed to such a threat. An example of modeling the field scattered from a center-loaded wire having a time-varying load whose resistance $R(t)$ varies in time as described by (A.3) is shown in Figure A1. The incident-wave frequency is 24 MHz and the load frequency variation is 4 MHz. Upper and lower sidebands are generated at 4 MHz intervals about the frequency of the excitation. This effect makes if feasible to modulate a field scattered from an otherwise passive object to enable two-way communication having only one active node as discussed in [20]. The same effect can modify scattering from a radar target.

$$R(t) = 500[1 + \sin(\omega t)]\text{ohms}. \tag{A.3}$$

The TWTD code played a role in the development of the radio frequency identification (RFID) tag in a research program at Los Alamos National Laboratory. A project led by Landt [21] was tasked with developing an electronic tag for cattle. This tag was to transmit back to a transceiver a signal with an ID code and the animal's temperature. This passive tag had no battery but instead used collected energy from the transceiver to power a modulating circuit to encode the scattered field with that information. The system was intended to work over a distance of a few meters, enough to read the tag as cattle passed through a gate between fields. The collected energy is used to power a response circuit that encodes information on the scattered field. The modified TWTD code was used in optimizing the Los Alamos National Laboratory (LANL) electronic tag [now generally referred to as the RFID tag]. By adding a small battery to the tag as

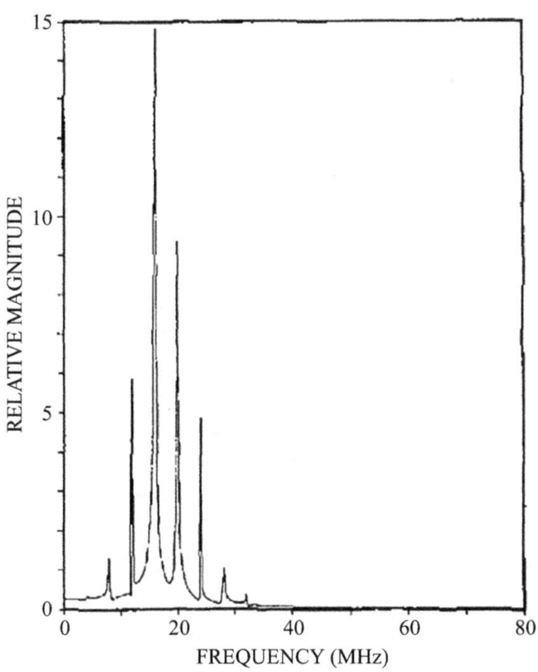

Figure A1 The spectrum of a 16-MHz incident electric field scattered from a wire having a center load varying at a frequency of 4 MHz [12]

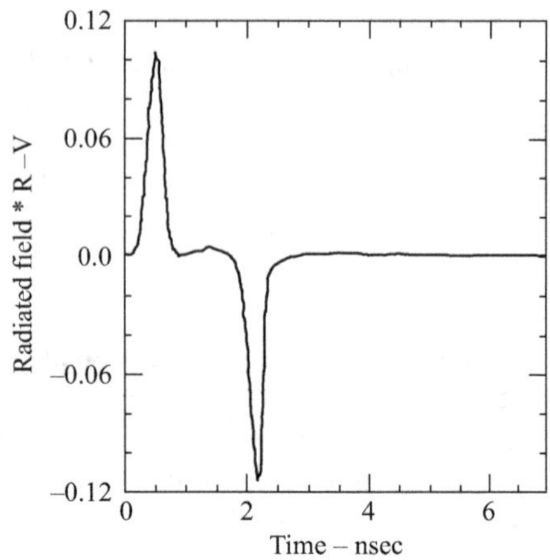

Figure A2 The far radiated field from an impulsively excited dipole loaded with ideal diodes [12]

implemented by the system described in [20], the range can be extended to several kilometers.

Another application enabled by TWTD modeling is determining the effect of nonlinear loads on an antenna or scatterer, an example of which is presented in Figure A2. A 1-m center-excited dipole is modeled using 60 segments loaded with a series of ideal diodes, one per segment [12]. The broadside far field is shown as a function of time in Figure A2. Two radiated pulses are produced, one from the feedpoint as the exciting pulse is turned on and the other when the outward-propagating charge is stopped because the diodes block the normal reflection that would otherwise occur. Being decelerated to zero velocity by the diodes, the charge radiates a pulse opposite in sign from that due to the exciting source. This result demonstrates the potential of using diode loads for pulse shaping. Note that the two radiated pulses in Figure A2 are of nearly the same magnitude as compared with Figure 3.4 where the magnitude of the pulse reflected from the end of a wire is about twice that of the feedpoint pulse. This is because the diodes stop the outgoing pulse so that its velocity changes from $+c$ to 0 rather than $+c$ to $-c$.

A.3 Summary

A brief description of the TWTD program is included with some representative numerical results to illustrate the special features provided by a time-domain electromagnetics computer model. A user's manual is also available [22] as is a Fortran listing of the program [23].

References

[1] A. J. Poggio, E. K. Miller, and G. J. Burke (1971), "Transient Scattering by Arbitrary Wire Structures", In: *International Symposium on Antennas and Propagation*, Tokyo, Japan, September 1–3.

[2] A. J. Poggio, E. K. Miller, and G. J. Burke, (1971), "Time-Domain Analysis of Wire Antennas", In: International Symposium on Electromagnetic Wave Theory, Tblisi, USSR.

[3] A. J. Poggio, E. K. Miller, and G. J. Burke (1971), "Scattering from Thin-Wire Structures in the Time Domain", *1971 USNC-URSI Spring Meeting*, Washington, DC, April 8–10.

[4] E. K. Miller, G. J. Burke, and A. J. Poggio (1971), "Time-Domain Analysis of Wire Antennas", *1971 USNC-URSI Fall Meeting*, University of California, Los Angeles, CA, September 21–23.

[5] C. L. Bennett and E. K. Miller (1972), "Some Computational Aspects of Transient Electromagnetics", *1972 USNC-URSI Spring Meeting*, Washington, DC, April 13–15.

[6] E. K. Miller, A. J. Poggio, and G. J. Burke (1973), "An Integro-Differential Equation Technique for the Time-Domain Analysis of Thin-Wire Structures. Part I: The Numerical Method", *Journal of Computational Physics*, 12, pp. 24–48.

[7] A. J. Poggio, E. K. Miller, and G. J. Burke (1973), "An Integro-Differential Equation Technique for the Time-Domain Analysis of Thin-Wire Structures. Part II: Numerical Results", *Journal of Computational Physics*, 12, pp. 210–233.

[8] A. J. Poggio and E. K. Miller (1973), "Integral-Equation Solutions of Three-Dimensional Scattering Problems", *In: Computer Techniques for Electromagnetics*, Pergamon Press, Oxford, pp. 159–264.

[9] J. A. Landt and E. K. Miller (1974), "Short-Pulse Response of a Circular Loop", *IEEE Transactions on Antennas and Propagation,* AP-22, pp. 114–116.

[10] J. A. Landt and E. K. Miller (1976), "Transient Response of the Infinite Cylindrical Antenna and Scatterer", *IEEE Transactions on Antennas and Propagation,* AP-24, pp. 246–250.

[11] E. K. Miller and J. A. Landt (1977), "Short-Pulse Characteristics of the Conical Spiral Antenna", *IEEE Transactions on Antennas and Propagation, AP*-25, pp. 621–626.

[12] E. K. Miller and J. A. Landt (1980), "Direct Time-Domain Techniques for Transient Radiation and Scattering from Wires", Invited Paper in *Proceedings of the IEEE*, 68, pp. 1396–1423.

[13] J. A. Landt, E. K. Miller, and F. J. Deadrick (1983), "Time-Domain Modeling of Nonlinear Loads", *IEEE Transactions on Antennas and Propagation, AP*-31, pp. 121–126.

[14] E. K. Miller (1987), "An Overview of Time-Domain Integral-Equation Models in Electromagnetics", *Journal of Electromagnetic Waves and Applications, VNU Science Press*, 1(3), 269–293.

[15] E. K. Miller (1994), "Time-Domain Modeling in Electromagnetics," *Journal of Electromagnetic Waves and Applications*, 8(9/10), pp. 1125–1172.

[16] E. K. Miller (1998), "Exploring Electromagnetic Physics Using Thin-Wire Time-Domain (TWTD) Modeling," In: *Proceedings of 14th Annual Review of Progress in Applied Computational Electromagnetics*, Naval Postgraduate School, Monterey, CA, pp. 583–588.

[17] E. K. Miller (1999), "Introduction" to *Time Domain Electromagnetics*, S. M. Rao, Ed. (eds.) , Academic Press, San Diego, CA, pp. 1–48.

[18] E. K. Miller, G. J. Burke, and T. K. Sarkar (2001), "TDFARS: A Procedure to Quantitatively Identify Time-Domain Radiation Sources", International Scientific Radio Union, 2001 Electromagnetic Theory Symposium, May 13–17, Victoria, Canada.

[19] E. K. Miller (2011), "Time-Domain Far-Field Analysis of Radiation Sources," *IEEE Antennas and Propagation Society Magazine*, 53 (5), pp. 81–97.

[20] R. Bracht, E. K. Miller and T. Kuckertz (1999), "An Impedance-Modulated-Reflector System," *IEEE Potentials*, 18(4), pp. 29–33.

[21] J. A. Landt (1980), "The Transient Response of Thin-Wire Antennas and Scatterers Loaded with Circuits and Nonlinear Elements," Los Alamos National *Laboratory, LA-8242*-MS.

[22] J. A. Landt, E. K. Miller, and M. Van Blaricum (1974), "WT-MBA/LLL1B: A Computer Program for the Time-Domain Electromagnetic Response of Thin-Wire Structures," Lawrence Livermore Laboratory, AFWL Interaction Notes, No. 210.

[23] FORTRAN version in PDF available as :TWTD FARS: @edmundmiller.academia.edu.

Appendix B
The Numerical Electromagnetics Code (NEC)

Abstract

The Numerical Electromagnetics Code is one of the longest-lived computational electromagnetics code, having its origin in the 1960s. As NEC exists now it has little resemblance to its beginnings, benefiting from a changing team of contributors over time. Its primary manager-developer over most of those years was Jerry Burke. Various enhancements and features have been added from the original NEC to NEC-4. Some problems that can arise when validating computer models are demonstrated as part of the process of validating NEC that has continued through various applications as illustrated by some additional examples.

Keywords: Computational electromagnetics; Electric-field integral equation; Interface effects; NEC; Validating CEM results; Wire modeling

B.1 The origin and development of NEC

Numerical Electromagnetics Code (NEC) originated in the late 1960s at MB Associates in San Ramon, CA, initially being called BRACT. A brief history of the evolution of BRACT into NEC is available [1]. Although including a capability for modeling surfaces using the magnetic-field integral equation, NEC is primarily known for modeling objects comprised of wires. In its early years, the code was used to model chaff and decoys for re-entry vehicles on a project for Space And Missile Systems Organization (SAMSO) in Riverside, CA. In 1968, a request for proposal (RFP) was issued from what is now called the US Army Communications-Electronics Command at Fort Monmouth, NJ for a modeling program for aircraft antennas (called in the RFP Mathematical Modeling of Aircraft Antennas and Supporting Structures). This was given the acronym MAS Squared at MBA after the company surprisingly won the proposal. It was then that BRACT became ANTBRCT (for ANTenna BRACT), differing from BRACT only in how the exciting field is incorporated in the solution process.

An extension of ANTBRACT that soon followed was to the modeling of antennas near an interface. AM broadcast stations operating in the low MHz range and a variety of other communication systems involve antennas located on or near the earth's surface. A computationally affordable approach for modeling antenna–ground interaction was developed for ANTBRACT called the reflection–coefficient approximation (RCA) [2]. The rigorous Sommerfeld integrals were also implemented as a check on the RCA, though their numerical evaluation was too expensive to be routinely used at the time.

Further development of the ANTBRACT code continued with US Coast Guard support at Lawrence Radiation Laboratory in the early 1970s for modeling of LORAN (LOng RAnge Navigation) antennas operating at 100 kHz. This was done using an extended version of ANTBRACT renamed WAMP (Wire Antenna Modeling Program) [2]. As described in [1], three of the MBA staff rejoined as part of a growing computational electromagnetics activity at LRL With support from various military organizations, and going through various name changes, the code acquired its final name as NEC with sequential numbered versions ending eventually at NEC-4, presently considered military critical technology. The features offered by these various versions of NEC improved the code's performance and capabilities being developed and implemented by Jerry Burke. Complementing these improvement was the accompanying documentation [3]. An essential feature than made NEC so widely used, even with the distribution restrictions, was this documentation and related support provided by Burke.

B.2 A brief description and some representative results

As is its time-domain counterpart, the wire treatment in NEC is based on the thin-wire approximation to the electric-field integral equation, given in the frequency domain by

$$-s \cdot E_{inc}(r) = \frac{j\eta}{4\pi k} \int_{c(r')} I(s') \left(k^2 s \cdot s' - \frac{\partial^2}{\partial s \partial s'} \right) g(r, r') ds' \tag{B.1}$$

where

$$g(\mathbf{r}, \mathbf{r}') = \frac{e^{-(jk|\mathbf{r}-\mathbf{r}'|)}}{|\mathbf{r} - \mathbf{r}'|} \tag{B.2}$$

and

$$k = \sqrt{\mu_o \varepsilon_o} \text{ and } \eta = \sqrt{\mu_o / \varepsilon_o}.$$

The basis function for the current is

$$I(z) = A_m + B_m \sin(z - z_m) + C_m \cos(z - z_m) \tag{B.3}$$

where z_m is the z-coordinate of the center of segment m. Two of the three unknowns in (B.3) are determined by equating the currents on adjacent segments at their connecting ends. A similar approach is used for junctions of multiple wires.

There are six electric field components due to the current (B.3), a radial and longitudinal field for each of the three terms. Five of these fields have closed-form, analytical expressions. Only the integral for the longitudinal field of the constant term needs numerical integration. This is done in NEC using an adaptive method based on Romberg quadrature.

NEC has various features that make its application more efficient such as:

- Rotational and plane symmetries in the objects being modeled can be exploited to shorten the matrix fill and solution times.
- NEC provides a numerical Green's function that enables an object matrix to be solved and stored when evaluating the effects of varied antenna locations and types interacting with it.
- NEC uses model-based parameter estimation in connection with the interface problem to make the rigorous Sommerfeld treatment much less time consuming.
- It has been used for modeling lossy media such as a plasma as well as objects being located on either side of or penetrating a lossy interface.
- NEC has also been extended for modeling objects located in a chiral media and over a chiral half-space.
- A useful pattern-synthesis feature in NEC is that of exciting arrays where the element excitations can be specified as currents rather than voltages.
- There is a version of NEC in which FARS has been implemented.

A 194-segment wire-grid model was developed for the US Army OH-6A observation helicopter, as shown in Figure B1a [4]. A scaled physical replica of the numerical wire-grid model was constructed for making scattering measurements, one of which is shown in Figure B1b. Radiation-pattern measurements were also made of the actual helicopter for comparison with the wire-grid model as in Figure B1c [4]. Results from this latter experiment initially were not in good agreement. This was found to be due to the numerical model having the flying configuration of the helicopter while the actual helicopter was mounted on a pylon. This resulted in its landing struts being compressed. Correcting the numerical model for this configuration produced the better results in Figure B1c.

When BRACT was being developed for RCS applications, there was little experimental data available at the time for even such a fundamental problem as scattering from a long, straight wire. Our SAMSO sponsors had an Aerospace consultant monitoring our work, a person who was skeptical about BRACT's accuracy. About that time we were able to obtain the measured backscatter pattern for an 11-wavelength wire. But the first model computation we generated as shown in Figure B2a exhibits a significant difference from the measurement [5]. Further discussion with the measurement provider revealed the interesting fact that the target in this case was so thin that it was not self-supporting but had to be taped to a Styrofoam rod to do the experiment. Since the relative permittivity of the road was 1.03, we repeated the model computation using a wire length increased by the square root of 1.03 to compensate for this with the new result presented in Figure B2b. The previous skepticism expressed about the BRACT code was essentially resolved.

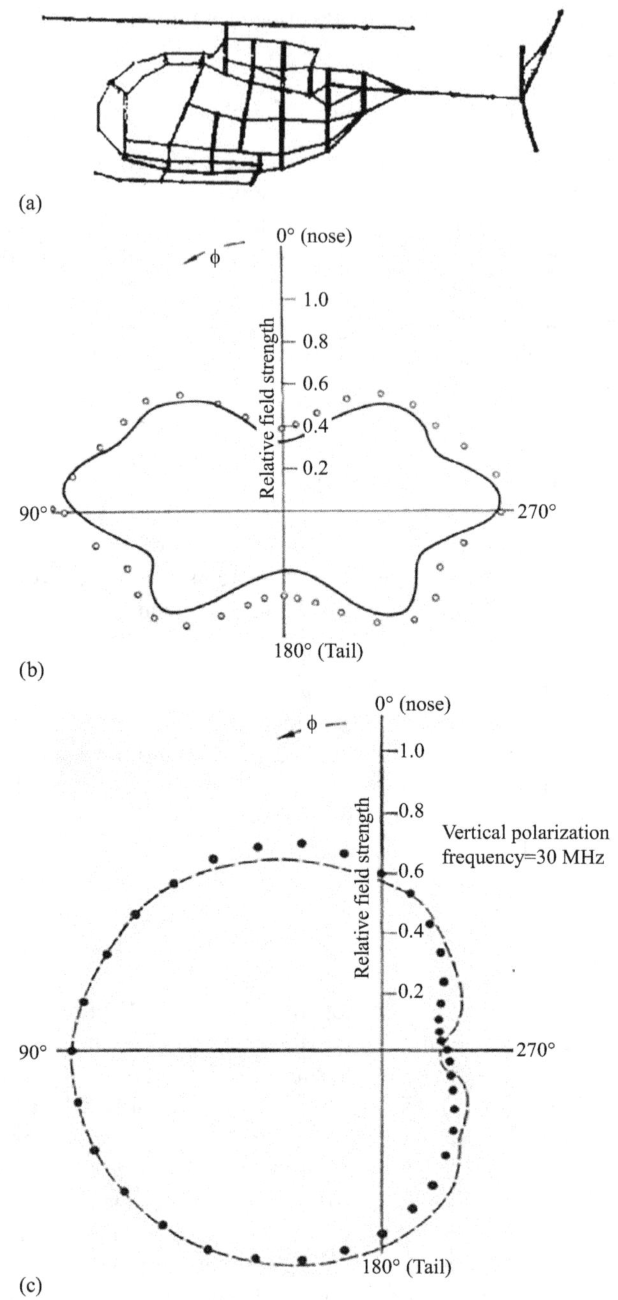

Figure B1 *The scaled wire-grid model of the OH-6A helicopter (a), measured and computed scattering patterns of the scale model (b), and measured radiation pattern of actual helicopter and computed radiation patterns (c) [4]*

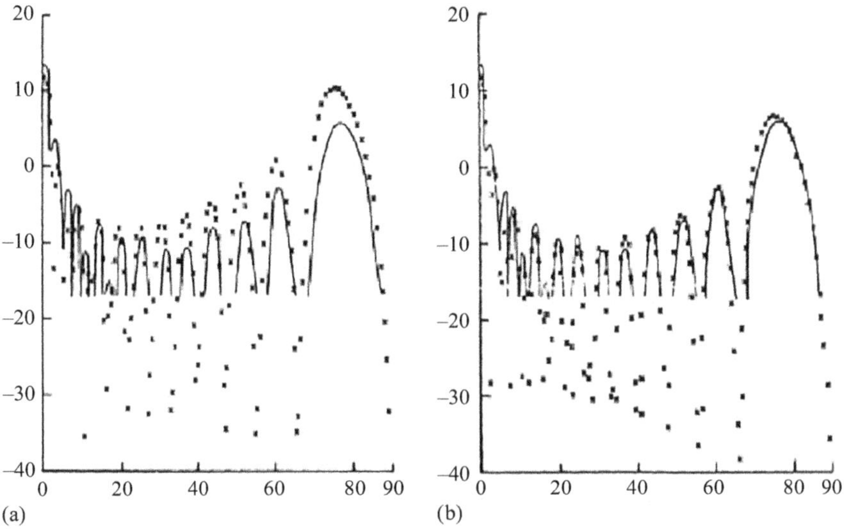

Figure B2 The RCS (square wavelengths in dB) vs. angle of Incidence (degrees) for an 11-wavelength straight wire of radius 1.47 × 10^{-2} wavelengths, (a) without correction for the Styrofoam support and (b) with correction for Styrofoam support [5]

The TWTD and NEC codes were made available to Lawrence Radiation Laboratory when former MBA staffers transferred there. The US Coast Guard soon provided funding to LRL to model the Sectionalized LORAN transmitting (SLT) antenna. At the time, it was not computationally feasible to use ANTBRACT to model such a complex structure using the Sommerfeld treatment. This was especially the case because of the ground screen that was essential for acceptable performance of the LORAN antennas. But ANTBRACT had been shown to produce acceptable results for the interface problem by modeling the antenna-ground interaction using a reflection coefficient approximation (RCA). The code required only a slight modification to include the ground screen in the RCA approach.

The 237-segment computer model of the SLT antenna is shown in Figure B3a with the computer-model results for the antenna input impedance compared with Coast Guard measurements from a scale-model compared in Figure B3b. The ground parameters used for the computer model are a relative permittivity of 15 and a conductance of 2 × 10^{-2} ohms/m together with a 120 radial-wire ground system.

The current on a quarter-wavelength monopole antenna with a wire radius of 0.0025 wavelengths and driven by a 1-V source against ground stakes of various lengths is presented in Figure B4 [7]. The relative complex permittivity of the ground is 16–j16. Being able to model this kind of problem can be useful in evaluating the tradeoff between using a ground stake or buried ground screen.

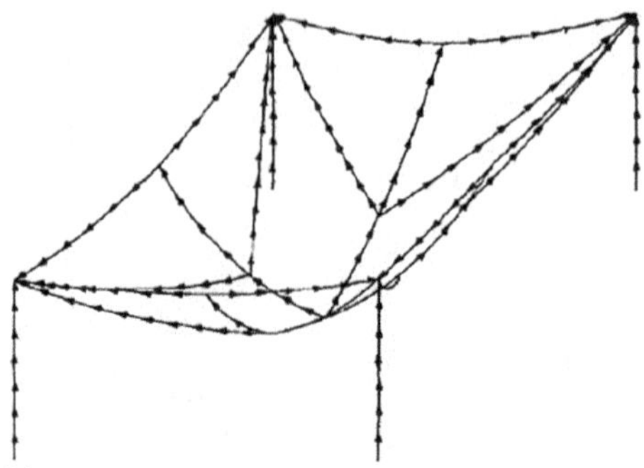

(a)

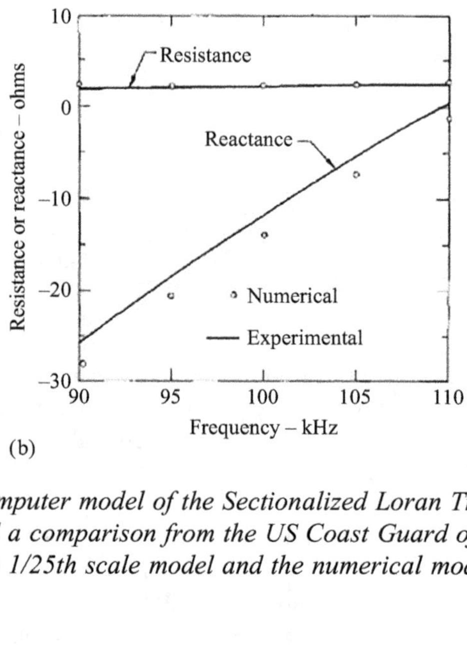

(b)

Figure B3 The computer model of the Sectionalized Loran Transmitting antenna (a) and a comparison from the US Coast Guard of impedance made using a 1/25th scale model and the numerical model (b) [6]

The magnitude of the current on a wire 30-wavelengths long with the parameter *a* the wire radius in a chiral medium having a chirality γ of 0.01 obtained by NEC is displayed in Figure B5 [8]. The faster-decay slope region beginning at the source exhibits a marked increase with increasing wire radius that transitions to a slower, rather uniform decay region with wire length for larger wire radii.

B.3 Summary

A brief history of the development of NEC is given [1] together with some representative numerical results to demonstrate various application of this widely used model. A user's manual and Fortran listing of an earlier version of the program are available [3].

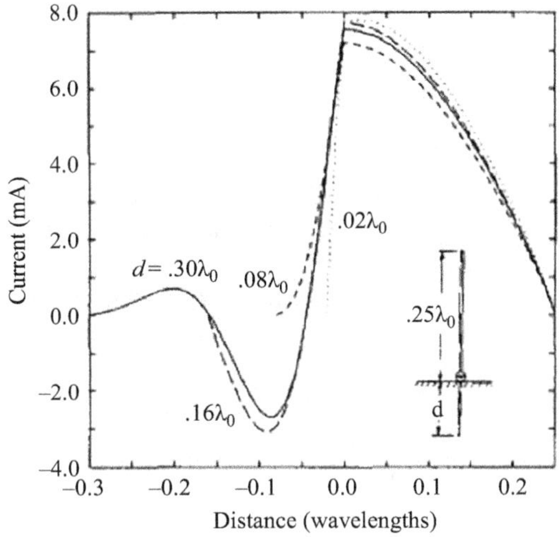

Figure B4 The real part of the current on a quarter-wave monopole driven against a ground stake of variable length over a ground of relative permittivity 16–j16 [7]

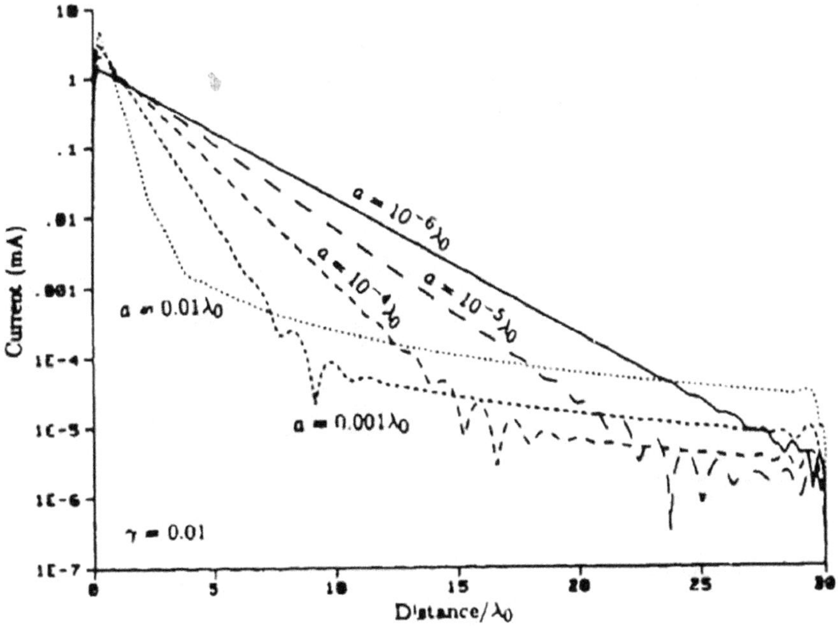

Figure B5 The current magnitude of a 30-wavelength wire with a *the wire radius and for a chirality of* $\gamma = 0.01$ *[8]*

References

[1] G. J. Burke, E. K. Miller, and A. J. Poggio, "The Numerical Electromagnetics Code (NEC) – a brief history," IEEE Antennas and Propagation, Monterey, CA, 2004, pp. 2871–2874.

[2] F. J. Deadrick and E. K. Miller (1971), "WAMP: a users manual for the wire antenna modeling program," Lawrence Livermore Laboratory, AFWL Interaction Note 207.

[3] G. J. Burke and A. J. Poggio, Numerical Electromagnetics Code-(NEC) – Method of Moments: Part I, Program Description – Theory; Part II, Program Description-Code; Part III, User's Guide, January 1981.

[4] E. K. Miller and B. J. Maxum, "Mathematical modeling of aircraft antennas and supporting structures," Final Report, ECOM Contract ADDB 07-68-C-0456, Report No. ECOM-0456-1.

[5] E. K. Miller, G. J. Burke, B. J. Maxum, G. M. Pjerrou, and A. R. Neureuther (1969), "Radar cross section of a long wire", *IEEE Transactions on Antennas and Propagation, AP*-17, pp. 381–384.

[6] E. K. Miller, F. J. Deadrick, and W. O. Henry (1973), "Computer evaluation of large, low-frequency antennas", *IEEE Transactions on Antennas and Propagation, AP*-21, pp. 386–389.

[7] G. J. Burke and E. K. Miller (1984), "Modeling antennas near to and penetrating a lossy interface", *IEEE Transactions on Antennas and Propagation, AP*-32 (10), pp. 1040–1049.

[8] J. Burke, E. K. Miller, and A. K. Bhattacharyya (1994), "Current on a long, thin wire in a chiral medium," *IEEE Transactions Antennas and Propagation, AP*-42 (6), pp. 827–832.

Appendix C

Notation

Note that any superscripts that denote a type of quantity such as radiation (RAD), FDFARS (FDF), TDFARS (TDF), the Induced EMF (IEMF), etc. are enclosed in parentheses. In some instances, a subscript such as "rad" may be used when there is no possibility of confusion. Only superscripts that serve as exponents are shown without parentheses. Subscripts that denote physical quantities such as coordinates, field components, etc. are all without parentheses.

AF	acceleration factor
c	speed of light in free space
ε_o	permittivity of free space
DPV	$DPV_{\|\|}(a,z)$ on-surface differentiated Poynting vector
Δ, dx	length of wire segment
Δ_m	the length of segment m
dt	time step
DTAPV	$-d$(TAPV)/dz
ELR	energy loss rate
$E_{//}$	broadside far field parallel to a wire segment
$E_z^{(S)}$, $E_\rho^{(S)}$	the z and ρ components of the SCF electric field
$E(z)_{inc}^{(\tan)}$	incident, tangential electric field at z
FARS	farfield analysis of radiation sources
FDFARS	frequency-domain FARS
η_o	wave impedance of free space
$I(x,t)$	space- and time-dependent current
I_o	steady-state current
$I(r)$	bicone antenna current
$I(z)$	wire current at space coordinate z
$I_\infty(z)$	infinite antenna current
IEMF	induced electromotive force
IIEMF	incremental IEMF

IFARS	incremental FARS
IPV	integrated Poynting vector
$IPV_{\perp}$, $IPV_{\parallel}$	integrated Poynting vector normal and parallel to wire surface
$IPV_{\perp}\ (R, z_{m\pm})$	integrated Poynting vector flowing radially out of pillbox at segment m
$IPV_{\parallel}(z)$	integrated Poynting vector flowing longitudinally out of annular segment at z
L	loop inductance
LED	linear energy density (time domain)
LPD	linear power density (frequency domain)
μ_o	permeability of free space
NEC	numerical electromagnetics code
NZR	near-zero radius
PEC	perfect electric conductor
$P^{(FDF)}$, $P^{(FARS)}$	FDFARS total power
$p(z)^{(IEMF)}_{inc}$	the incident IEMF power at z
$p^{(FDF)}_m(\vartheta, \varphi)$	incremental FDFARS power from segment m at location ϑ, φ
$PV_{\perp}(R,z) = PV_{\perp}$	the PV normal to current axis at the pillbox surface R and z
$PV_{\parallel}(r,z) = PV_{\parallel}$	the PV parallel to current axis at radial coordinate r and z
$P^{(FDF)}_m$	integrated total FDFARS power from segment m
$P^{(FDF)}$	FDFARS total power
$p^{(IEMF)}(z)$	the IEMF power density at z
$p^{(IEMF)}(z_m)$	the IEMF power density at z_m
$P^{(IEMF)}$	the total IEMF power
$p^{(TDF)}_m$	TDFARS power from segment m
$P^{(RAD)}(t)$	radiated power at time t
$P^{(RAD)}$	total radiated power
P_{coll}	power collected by wire scatterer
$P_{rad}(s)$	distributed radiated power density from Schelkunoff–Feldman
PV	Poynting vector
$PV_{QI}(z)$	on-surface PV computed from the current and charge
$PV_{\parallel}(a,z)$	the parallel on-surface PV
$\boldsymbol{PV}(\boldsymbol{r})$	Poynting vector at space point $\boldsymbol{r}$
$Q(x,t)$	space- and time-dependent charge density
$R_{rad}(s)$, $R^{(SF)}_{rad}(s)$	Schelkunoff–Feldman distributed radiation resistance for a SCF
$R^{(NEC)}_{rad}(s')$	radiation resistance derived from NEC
SCF	sinusoidal current filament
TAPV	time-average Poynting vector

TDEM	time-domain energy measure
TDFARS	time-domain FARS
TWTD	thin-wire time domain
$V_{ex}(t)$	excitation voltage
$W_D(t)$	time-dependent dipole energy
$W_E(t)$, $W_{Qc}(t)$	time-dependent electric-field energy
$W_H(t)$, $W_I(t)$	time-dependent magnetic-field energy
$W_L(t)$	time-dependent loop energy
$W^{(RAD)}(t)$	radiated energy up to time t
$W_T(t)$	total time-dependent energy

Index

www.ingramcontent.com/pod-product-compliance
Lightning Source LLC
LaVergne TN
LVHW020055220926
848484LV00002BB/406
* 9 7 8 1 8 3 9 5 3 8 1 3 1 *